EW 103

Tactical Battlefield Communications Electronic Warfare

DISCLAIMER OF WARRANTY

The technical descriptions, procedures, and computer programs in this book have been developed with the greatest of care and they have been useful to the author in a broad range of applications; however, they are provided as is, without warranty of any kind. Artech House, Inc. and the author and editors of the book titled *EW 103: Tactical Battlefield Communications Electronic Warfare* make no warranties, expressed or implied, that the equations, programs, and procedures in this book or its associated software are free of error, or are consistent with any particular standard of merchantability, or will meet your requirements for any particular application. They should not be relied upon for solving a problem whose incorrect solution could result in injury to a person or loss of property. Any use of the programs or procedures in such a manner is at the user's own risk. The editors, author, and publisher disclaim all liability for direct, incidental, or consequent damages resulting from use of the programs or procedures in this book or the associated software.

For a complete listing of the
Artech House Electronic Radar Series,
turn to the back of this book.

EW 103

Tactical Battlefield Communications Electronic Warfare

David L. Adamy

ARTECH HOUSE

BOSTON | LONDON
artechhouse.com

Library of Congress Cataloging-in-Publication Data
A catalog record for this book is available from the U.S. Library of Congress.

British Library Cataloguing in Publication Data
A catalogue record for this book is available from the British Library.

ISBN-13: 978-1-59693-387-3

Cover design by Yekaterina Ratner

© 2009 ARTECH HOUSE
685 Canton Street
Norwood, MA 02062

All rights reserved. Printed and bound in the United States of America. No part of this book may be reproduced or utilized in any form or by any means, electronic or mechanical, including photocopying, recording, or by any information storage and retrieval system, without permission in writing from the publisher.

All terms mentioned in this book that are known to be trademarks or service marks have been appropriately capitalized. Artech House cannot attest to the accuracy of this information. Use of a term in this book should not be regarded as affecting the validity of any trademark or service mark.

10 9 8 7 6 5 4 3 2

Contents

	Preface	***xiii***
1	**Introduction**	**1**
1.1	Nature of Communications	1
1.2	Frequency Ranges	2
1.3	Roadmap to the Book	3
1.4	dB Math	4
1.4.1	dB Values	5
1.4.2	Conversion to and from dB Form	5
1.4.3	Converting to dB Using a Slide Rule	7
1.4.4	Absolute Values in dB Form	8
1.4.5	dB Equations	9
2	**Communications Signals**	**13**
2.1	Analog Modulations	13
2.2	Digital Modulations	16
2.2.1	Transmission of Analog or Digital Information	16
2.2.2	Digitization	17
2.2.3	Digital RF Modulations	19

2.2.4	Bandwidth of Digital Signals	23
2.2.5	Digital Signal Structure	26
2.3	Noise	27
2.4	LPI Signals	30
2.4.1	Pseudo-Random Codes	33
2.4.2	Frequency Hopping Signals	36
2.4.3	Chirp Signals	41
2.4.4	Direct Sequence Spread Spectrum Signals	43
2.4.5	Combined Technique LPI Signals	47
2.4.6	Cell Phone Signals	49
2.5	Error-Correction Codes	51
3	**Communication Antennas**	**55**
3.1	Antenna Parameters	55
3.1.1	Types of Antennas	56
3.1.2	General Characteristics of Various Types of Antennas	57
3.2	Important Types of Communication Antennas	59
3.3	The Antenna Beam	59
3.4	More About Antenna Gain	62
3.5	Polarization	62
3.6	Phased Arrays	63
3.6.1	Phased Array Beamwidth and Gain	65
3.7	Parabolic Dish Antennas	66
4	**Communications Receivers**	**75**
4.1	Types of Receivers	75
4.1.1	Pulse Receivers	76
4.1.2	Superheterodyne Receiver	79
4.1.3	Tuned Radio Frequency Receiver	81
4.1.4	Fixed-Tuned Receiver	82
4.1.5	Channelized Receiver	82

4.1.6	Bragg Cell Receiver	84
4.1.7	Compressive Receiver	85
4.1.8	The Digital Receiver	86
4.2	Digitization	88
4.2.1	Sampling Rates	89
4.2.2	Digital Waveforms	89
4.2.3	Digitizing Techniques	90
4.2.4	I & Q Digitization	91
4.3	Digitized Signal Quality Issues	91
4.3.1	Chip Detection	92
4.3.2	Catching a Frequency Hopping Signal	93
4.4	Receiver System Sensitivity	95
4.4.1	kTB	97
4.4.2	Noise Figure	98
4.4.3	Required Predetection Signal-to-Noise Ratio	99
4.5	Receiver System Dynamic Range	105
4.5.1	Analog Versus Digital Dynamic Range	106
4.5.2	Analog Receiver Dynamic Range	107
4.5.3	Digital Dynamic Range	110
4.6	Typical Receiver System Configurations	111
4.6.1	Multiple Receiver Reconnaissance and Electronic Support Systems	112
4.6.2	Multiple Receiver Systems	113
4.6.3	Remote Receiving Systems	116
5	**Communications Propagation**	**119**
5.1	One-Way Link	119
5.2	The One-Way Link Equation	119
5.3	Propagation Losses	123
5.4	Line-of-Sight Propagation	124
5.5	Two-Ray Propagation	129

5.6	Fresnel Zone	134
5.7	Knife-Edge Diffraction	137
5.8	Atmospheric and Rain Losses	140
5.8.1	Atmospheric Loss	140
5.9	HF Propagation	143
5.10	Satellite Links	147
6	**Search for Communication Emitters**	**153**
6.1	Probability of Intercept (POI)	154
6.2	Search Strategies	154
6.2.1	General Search	154
6.2.2	Directed Search	154
6.2.3	Sequentially Qualified Search	155
6.2.4	A Useful Search Tool	155
6.2.5	Practical Considerations Affecting Search	156
6.3	System Configurations	158
6.3.1	Types of Receivers Used for Search	160
6.3.2	Digitally Tuned Receiver	161
6.3.3	Digital Receivers	163
6.3.4	Frequency Measuring Receivers	163
6.3.5	Energy Detection Receivers	164
6.4	The Signal Environment	167
6.4.1	Angular Coverage	168
6.4.2	Channel Occupancy	169
6.4.3	Sensitivity	169
6.5	Radio Horizon	170
6.6	Search for Low Probability of Intercept Signals	174
6.6.1	LPI Search Strategies	175
6.6.2	Frequency Hoppers	176
6.6.3	Chirp Signals	177
6.6.4	Direct Sequence Spread Spectrum Signal	177

6.7	Look Through	178
6.8	Fratricide	181
6.9	Search Strategy Examples	181
6.9.1	Narrowband Search	181
6.9.2	Hand-Off from Wideband Receiver	185
6.9.3	Search with a Digital Receiver	185
7	**Location of Communications Emitters**	**187**
7.1	Emitter Location Approaches	187
7.1.1	Triangulation	188
7.1.2	Single-Site Location	189
7.1.3	Azimuth and Elevation	191
7.1.4	Other Location Approaches	191
7.2	Accuracy Definitions	193
7.2.1	RMS Error	193
7.2.2	Circular Error Probable and Elliptical Error Probable	195
7.2.3	Calibration	198
7.3	Site Location and North Reference	198
7.4	Moderate Accuracy Techniques	203
7.4.1	Watson-Watt Direction Finding Technique	204
7.4.2	Doppler Direction-Finding Technique	204
7.5	High Accuracy Techniques	207
7.5.1	Single Baseline Interferometer	207
7.5.2	Multiple Baseline Precision Interferometer	211
7.5.3	Correlative Interferometer	212
7.6	Precision Emitter Location	212
7.6.1	Time Difference of Arrival Method	214
7.6.2	Precision Emitter Location by Frequency Difference of Arrival	218
7.6.3	Combined FDOA and TDOA	221
7.7	Emitter Location—Error Budget	222

7.7.1	Combination of Error Elements	223
7.8	Locating Spread Spectrum Emitters	225
7.8.1	Locating Frequency Hoppers	225
7.8.2	Chirp Emitters	231
7.8.3	Direct Sequence Spread Spectrum Emitters	232
7.8.4	Precision Emitter Location Techniques Against LPI Emitters	232

8 Intercept of Communications Signals — 235

8.1	The Intercept Link	236
8.1.1	Intercept of a Directional Transmission	237
8.1.2	Intercept of a Nondirectional Transmission	238
8.1.3	Airborne Intercept System	241
8.1.4	Nonline-of-Sight Intercept	242
8.2	Intercept of Weak Signal in Strong Signal Environment	244
8.3	Intercept of LPI Signals	245
8.3.1	Intercept of Frequency Hoppers	245
8.3.2	Intercept of Chirped Signals	246
8.3.3	Intercept of Direct Sequence Spread Spectrum Signals	247

9 Communications Jamming — 251

9.1	Jammer-to-Signal Ratio	252
9.1.1	Other Losses	254
9.1.2	Stand-In Jamming	254
9.2	Digital Versus Analog Signals	255
9.2.1	Pulse Jamming	258
9.3	Jamming Spread Spectrum Signals	258
9.3.1	Partial Band Jamming	259
9.3.2	Jamming Frequency Hop Signals	262
9.3.3	Jamming Chirp Signals	267
9.3.4	Jamming Direct Sequence Spread Spectrum Signals	267

9.3.5	Jamming of Combined Mode Spread Spectrum Signals	269
9.4	Impact of Error-Correction Coding on Jamming	270
9.4.1	Jamming Cell Phones	270
9.4.2	Jam the Uplink	271
9.4.3	Jam the Downlink	272

Appendix A Problems with Solutions	**275**
Appendix B Bibliography	**307**
Appendix C Using the Included CD	**311**
About the Author	**317**
Index	**319**

Preface

This is the third book in the *EW 100* series. Like the two earlier books, it is based on the *EW 101* series of tutorial articles in the *Journal of Electronic Defense*. However, this book focuses on the practical aspects of electronic warfare against enemy communications signals

The target audiences for this book, like the earlier two, are new EW professionals, specialists in some part of EW, specialists in technical areas peripheral to EW, and managers who are responsible for the efforts of EW engineers and technicians.

As this book goes to press, the *EW 101* series continues, and some of the material in this book will be in future columns. The material from past or future columns is all organized into chapters with added introductory and connective material. Like the *EW 102* book, this one also has an appendix containing problems with solutions (not just answers).

Another new feature is a slide rule for the quick calculation of antenna and propagation answers. There are many similar slide rules floating around, but this one has some newly designed scales not found on any of the others.

Finally, there is a CD with built-in formulas for the calculation of propagation losses, received signal strength, effective range, jamming-to-signal ratio, and similar important values. The formulas are in spreadsheet format because most technical folks have that program. MATLAB would be more elegant, but the program is very expensive. You are welcome to convert the spreadsheet formulas (which are provided) to MATLAB or any other program if you like.

1
Introduction

This book is intended to be an easy read. Explanations of hardware and techniques are given in physical, rather than mathematical, terms. The math is largely simple dB formulas which are easy to remember and to use.

Like the two earlier *EW 100* series books, the technical material is intended to be accurate as opposed to precise. In most cases, the formulas and examples are set up to calculate values to 1-dB accuracy. However, constants are provided to higher accuracy for the convenience of those who will be using the formulas in higher precision applications.

The focus of the book is on communications electronic warfare (EW), so there is no coverage of radar threats, search, intercept, jamming, or decoys. For those subjects, check the *EW 101* and *EW 102* textbooks.

1.1 Nature of Communications

Communications electronic warfare (EW) is above all about communication. Therefore, we will be discussing the nature of communication signals, propagation, and hardware in reasonable detail. The main emphasis will be on tactical battlefield communication in the VHF, UHF, and low-microwave frequency ranges. However, there is some coverage of lower frequency band propagation, digital command and data links, and satellite communication.

The purpose of communication is to take information from one point to some distant point, thus (unlike radar) communication is inherently one way. Although there is "burst" communication which involves very short signals, most communication is more or less continuous for periods from a few seconds to continuous.

Communication signals are typically rather narrowband, although there are some modulations which artificially spread the signal far beyond the bandwidth required to carry the information. This is done to prevent detection or to diminish the effects of unintentional or intentional interference.

Communications signals can be either analog or digital, with digital signals becoming more common as time goes by. There are significant differences in the way EW systems deal with these two classes of signals. For digital communication signals, there are many ways in which an enemy can make the EW tasks more difficult by use of sophisticated techniques for the preservation of signal integrity.

1.2 Frequency Ranges

Table 1.1 shows the different frequency ranges used for communication along with the typical applications, propagation modes, and propagation issues.

Table 1.1
Frequency Ranges

Frequency Range	Abbreviation	Type of Signal and Characteristics
Very low, low, and medium frequency: 3 kHz to 3 MHz	VLF LF MF	Very long-range communication (ships at sea, and so forth), commercial AM radio. Ground waves circle the Earth.
High frequency: 3 to 30 MHz	HF	Over-the-horizon communication, Signals reflect from ionosphere.
Very high frequency: 30 to 300 MHz	VHF	Mobile communication, TV, and commercial FM radio. Line-of-sight required.
Ultra high frequency: 300 MHz to 1 GHz	UHF	Mobile communication and TV. Line-of-sight required.
Microwave: 1 to 30 GHz	Mw	TV and telephone links, satellite links. Line-of-sight required.
Millimeter wave: 30 to 100 GHz	MMW	Very short-range communication. Requires line-of-sight, high absorption in rain and fog.

Note that propagation at the lower frequencies is characterized by reduced dependence on line of sight. Ground wave and ionospheric skip allow communication over very extended ranges. However, the lower frequencies are also characterized by narrow bandwidths. High percentage bandwidths create difficulties with antenna and amplifier performance. In general 10% bandwidth is fairly well behaved, and performance trade-offs are required for bandwidths greater than 10%.

VLF and LF links usually carry low-rate digital signals or Morse code, while MF links are wide enough to carry voice signals. Commercial AM radio is broadcast in the upper end of the MF frequency range. Above approximately 30 MHz, radio transmissions pass through the ionosphere, so higher frequency signals cannot propagate through ionospheric hops. They are dependent on line-of-sight or near line-of-sight propagation paths.

VHF and UHF transmissions can support enough bandwidth to carry not only voice and data, but also video signals—including commercial television broadcasts. Microwave frequencies are used to carry high information content signals in wide bandwidths. Wideband microwave point-to-point links carry large blocks of telephone signals, television signals, and wideband digital data. Communication satellite links are also at microwave, as are command and data links between unmanned aerial vehicles (UAVs) and their control stations.

1.3 Roadmap to the Book

- In this introductory chapter, we will discuss dB values and formulas.

- Chapter 2 deals with communication signals including analog modulations, digital modulations, and low probability-of-intercept modulations. A large part of this chapter deals with digital signals and the related issues of error correction codes.

- Chapter 3 covers the types of antennas used in communication and communication EW. It covers the applications and typical performance parameters of all common types of antennas used in the communication frequency ranges.

- Chapter 4 describes the types of receivers used for communication and for the intercept of communication signals. It also covers the calculation of sensitivity and dynamic range in communication receivers.

- Chapter 5 is about radio propagation. Its major emphasis is on VHF, UHF, and low-microwave communication, but it also covers HF and lower band propagation and communication satellite propagation.
- Chapter 6 is about search techniques. This includes both search for fixed frequency and low probability of intercept communication signals
- Chapter 7 is about the location of hostile communication transmitters. The common approaches and techniques are described. The applications, expected accuracy, and other performance and implementation issues are described for each.
- Chapter 8 covers the intercept of communication signals—both normal and low probability of intercept modulations. The related issues of search and probability of intercept are discussed for the various types of signals.
- Chapter 9 is about communications jamming. It includes the jamming of conventional signals, and then goes on to deal with techniques for jamming all common types of low probability-of-intercept signals.
- Appendix A is a series of problems covering all of the subjects discussed in the book. Each problem is solved, with all solution steps shown.
- Appendix B, the Bibliography, includes reference information for more study. This is a list of more detailed textbooks for further study. The publisher and ISBN are given for each text referenced, and the focus of each is described in a few sentences.
- Appendix C contains instructions on how to use the CD that is included with the book.
- Pockets inside the book covers contain an antenna and propagation slide rule and a CD with communications formulas.

1.4 dB Math

This section covers the basic math that underlies the electronic warfare concepts covered in the other chapters. It includes a discussion of the dB forms of numbers and dB formulas.

1.4.1 dB Values

In any professional activity which includes consideration of radio propagation, signal strength, gains and losses are often stated in dB form. This allows the use of dB forms of equations which are typically easier to use than the original forms. One of the great charms of dB numbers is that they are logarithmic, and thus conveniently allow the comparison of very large and very small values. Since the difference between the signal strength of transmitted and received signals can be as much as 20 orders of magnitude, this is an important issue.

Any number expressed in dB is logarithmic. For convenience, we will call number in non-dB form "linear" to differentiate them from the logarithmic dB numbers. Numbers in dB form also have the great charm of being easy to manipulate:

- To multiply linear numbers, you add their logarithms.
- To divide linear numbers, you subtract their logarithms.
- To raise a linear number to the nth power, you multiply its logarithm by n.
- To take the nth root of a linear number, you divide its logarithm by n.

To take maximum advantage of this convenience, we put numbers in dB form as early in the process as possible, and convert them back to linear forms as late as possible (if at all). In many cases, the most commonly used forms of answers remain in dB.

It is important to understand that any value expressed in dB units must be a ratio (which has been converted to logarithmic form). Common examples of ratios in communication are amplifier or antenna gain and losses in circuits or in radio propagation.

1.4.2 Conversion to and from dB Form

A linear form number (N) is converted to dB form by the formula:

$$N(\text{dB}) = 10 \log_{10}(N)$$

For most of the equations in this book, we just say 10 log (N), with the logarithm to the base 10 understood. To do this operation using a scientific calculator, enter the linear form number, then touch the "log" key, then multiply by 10.

dB values are converted to linear form with the equation:

$$N = 10^{N(\text{dB})/10}$$

Using a scientific calculator, enter the dB form number, divide it by 10, then press the = (equal) key, then touch the second function key, then the log key. This process is also described as taking the "antilog" of the dB value divided by 10. This can be stated as:

$$N = \text{Antilog}[N(\text{dB})/10]$$

For example, if an amplifier has a gain factor of 100, we can say it has 20-dB gain, because:

$$10\log(100) = 10 \times 2 = 20 \text{ dB}$$

Reversing the process to find the linear form gain of a 20-dB amplifier:

$$\text{Antilog}[20/10] = 100$$

Table 1.2 shows some important ratios and their dB equivalents. Note that a ratio of 2 converts to 3 dB and a ratio of ½ converts to −3 dB. In Chapter 3, we will talk about the half-power points of antenna beams as the "3-dB points." Another interesting point is that a ratio of unity (1) converts to 0 dB. Thus when two values are equal, we say they have a ratio of 0 dB.

Note that 1 dB corresponds to a ratio of 1.25. This means that when we make calculations to 1 dB, which we will do quite often, we are really

Table 1.2
Linear Form Ration and dB Equivalents

Ratio	dB Equivalent	Ratio	dB Equivalent
1/10	−10	1.25	1
1/4	−6	2	3
1/2	−3	4	6
1	0	10	10

operating at a precision of 25%. Although this seems quite crude, it is often adequate for radio propagation calculations, in which signal strengths can vary over several orders of magnitude.

Since the ratio 10 converts to 10 dB, an order of magnitude change in a value that has been converted to dB form requires only the addition or subtraction of 10 dB.

An exception to the above conversion rule is that voltage ratios are converted to dB by the formula $20\log_{10}$ (voltage ratio).

1.4.3 Converting to dB Using a Slide Rule

The slide rule included inside the book cover allows you to make many different calculations, each of which will be discussed in the appropriate chapter of this book. The first of these is the conversion of linear form numbers to dB form, and from dB form to linear form.

Figure 1.1 shows both sides of the slide rule. Note that on the left hand end of the rule, there is a number 1 on one side and number 2 at the same location on the other side. We will refer to the two sides of the slide rule by these numbers. Also, please note that the slide must be properly inserted into the body of the rule—otherwise, the scales do not align with the windows.

Figure 1.2 shows the window at the bottom of side 2 of the rule. This window allows conversion from ratios to dB values and back. This feature of the rule only works for ratio values from 0.01 to 100 and for dB values between −20 dB and +20 dB. Move the slide so that 2 is at the arrow at the top of the window as shown in this figure. You will see that +3 is next to the bottom arrow. This indicates that the ratio 2 is equal to +3 dB. Now move the slide so that 0.5 is at the top arrow. Note that the quantity at the bottom arrow is −3 dB.

This example relates to a widely used dB application. The point at which a half value occurs is often called the "3-dB point." For example, when we talk about antenna gain patterns (in Chapter 3), the boresight of the antenna is typically the direction in which it has maximum antenna gain. If the antenna is rotated to an angle at which the gain is reduced to half, we say we are looking at the half-power point, or the 3-dB point. The angle between the two half power (or 3 dB) points (on either side of the boresight) is called the 3-dB beamwidth of the antenna. When considering the frequency coverage of a receiver or a bandpass filter, we talk about its 3-dB bandwidth. This is the difference between the upper and lower frequencies at which the output of the filter or the sensitivity of the receiver is half of the maximum value.

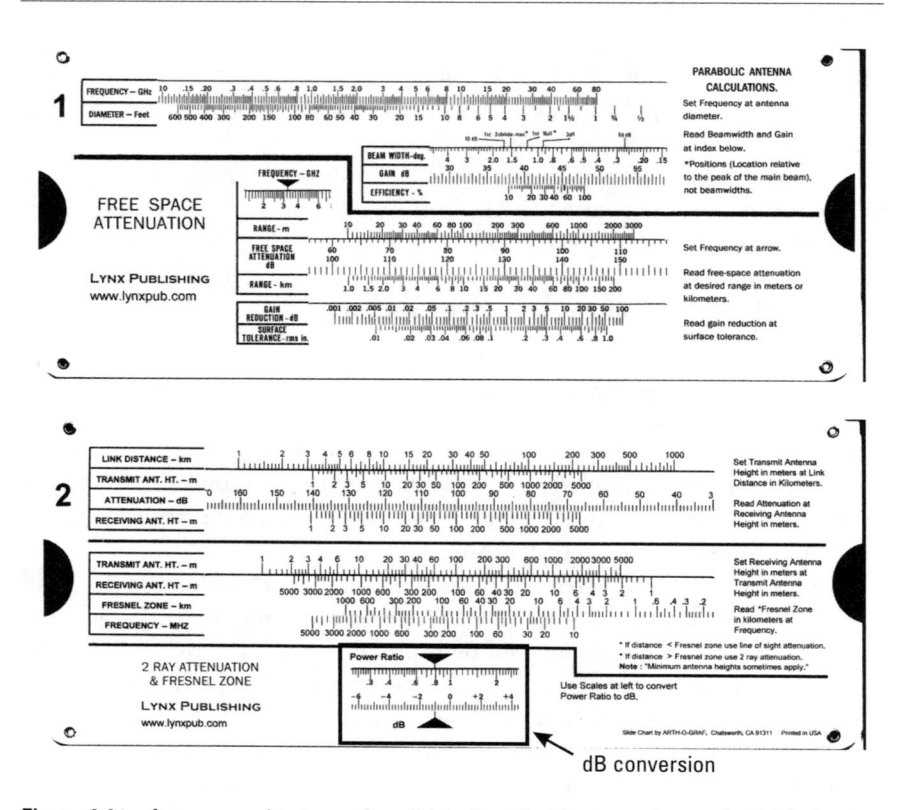

Figure 1.1 Antenna and propagation slide rule with dB conversion scale highlighted.

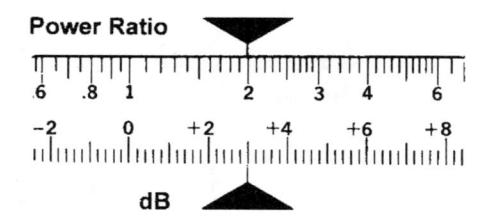

Figure 1.2 Conversion between ratio and dB values on the slide rule.

1.4.4 Absolute Values in dB Form

To express absolute values as dB numbers, we first convert the value to a ratio with some understood constant value. The most common example is signal strength expressed in dBm. To convert a power level to dBm, we divide it by 1 milliwatt and then convert to dB form. For example, 4 watts equals 4,000 milliwatts. Then convert 4,000 to dB form to become 36 dBm. The small "m" indicates that this is a ratio to a milliwatt.

$$10 \log(4,000) = 10 \times 3.6 = 36 \text{ dBm}$$

Then, to covert back to watts:

$$\text{Antilog}(36/10) = 4,000 \text{ milliwatts} = 4 \text{ watts}$$

Since signal strength in dBm is widely used in EW, the following table of signal strength in common units and in dBm is provided for your convenience (see Table 1.3). Picking values from this table: 1 milliwatt is 0 dBm, 1 watt is 30 dBm, and a kilowatt is 60 dBm. This chart is most useful when dealing with transmitted signal strengths. Since received signal strengths are typically much less than one microwatt, it is common practice to leave them in dBm form. For example, a received signal level might be −100 dBm.

Other examples of dB forms of absolute values are shown in Table 1.4.

1.4.5 dB Equations

In this book, we use a number of dB forms of equations for convenience. These equations have one of the following forms, but can have any number of terms:

$$A(\text{dBm}) \pm B(\text{dB}) = C(\text{dBm})$$
$$A(\text{dBm}) - B(\text{dBm}) = C(\text{dB})$$
$$A(\text{dB}) = B(\text{dB}) \pm N \log(\text{number not it dB})$$

where N is a multiple of 10.

This last equation form is used when the square (or higher order) of a number is to be multiplied. An important example of this last formula type is the equation for the spreading loss in radio propagation (which will be covered in detail in Chapter 5):

$$L = 32 + 20 \log(d) + 20 \log(f)$$

Table 1.3
Signal Strength Levels in dBm

Signal Strength	dBm
1 μwatt	−30
10 μwatt	−20
100 μwatt	−10
1,000 μwatt	0
1 mwatt	0
10 mwatt	10
100 mwatt	20
1,000 mwatt	30
1 watt	30
10 watts	40
100 watts	50
1,000 watts	60
1 kwatt	60
10 kwatts	70
100 kwatts	80
1,000 kwatts	90
1 Mwatt	90
10 Mwatts	100
100 Mwatts	110
1,000 Mwatts	120

Table 1.4
Common dB Definitions

dBm	= dB value of power/1 milliwatt	Used to describe signal strength.
dBW	= dB value of power/1 watt	Used to describe signal strength.
dBsm	= dB value of area/1 square meter	Used to describe antenna area or radar cross-section.
dBi	= dB value of antenna gain relative to the gain of an isotropic antenna	0 dBi is, by definition, the gain of an omni-directional (isotropic) antenna.

where
    L = the spreading loss (in dB)
    d = link distance in km
    f = transmission frequency in MHz

The factor 32 is a conversion factor that is added to make the answer come out in the desired units from the most convenient input units. It is actually 4π squared, divided by the speed of light squared, multiplied and divided by some unit conversion factors—and the whole thing converted to dB form and rounded to a whole number. The important thing to understand about this conversion factor (and the equation which contains it) is that it is correct only if exactly the correct units are used. The distance *must* be in km and the frequency *must* be in MHz—otherwise, the loss value will not be correct.

2

Communications Signals

Communication signals carry information from one place to another. The information can be either analog or digital, and is carried as modulation. This chapter discusses the various types of communication modulations and their impact on intercept, emitter location, and jamming. Both conventional and low probability of intercept (LPI) modulations are covered. For completeness, the chapter also includes discussion of signal-to-noise ratio, digitization, and error-correction codes.

2.1 Analog Modulations

Figure 2.1 shows an amplitude modulated (AM) signal in the time domain. This is the way it would appear on an oscilloscope. The carrier frequency is the frequency at which the signal is transmitted. The information is carried as changes in the amplitude of the carrier signal. In this figure, the modulation is a sign wave at much lower frequency than the carrier. Note that the ratio between the carrier frequency and the modulation is typically much greater than shown in this figure. The percentage modulation is the ratio between the amplitude of the modulation pattern and the amplitude of the signal. A 50% modulation would have a ratio between the highest and lowest amplitudes of 1.67. The maximum and minimum amplitudes would be 1.25 and 0.75 times the amplitude of the carrier without modulation respectively.

Figure 2.2 shows the AM signal in the frequency domain. This is how the signal looks on a spectrum analyzer. The two sidebands are identical and

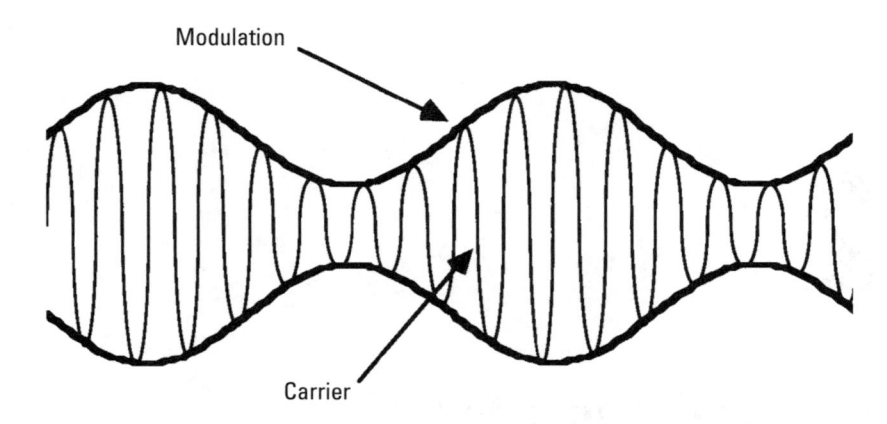

Figure 2.1 AM signal in time domain.

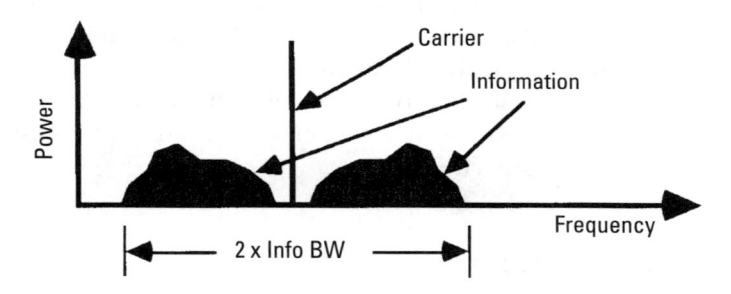

Figure 2.2 AM signal in frequency domain.

symmetrical to the carrier. The frequency bandwidth of the modulated signal is twice the highest frequency present in the modulating signal. For example, if the modulation is a voice signal, the modulation signal might be 4 kHz wide. If the 4-kHz audio signal is amplitude modulated onto the carrier, the AM signal (the carrier and both sidebands) will be 8 kHz wide. The power of the signal is divided between the carrier and the sidebands. If you were watching an audio signal on a spectrum analyzer, you would see that the sidebands go away between spoken syllables.

If the carrier and one of the sidebands are removed from an AM signal by filtering, as shown in Figure 2.3, the resulting signal is call single sideband (SSB). It can either be upper sideband (USB) or lower sideband (LSB) depending on which sideband is retained in the signal. In SSB, only one sideband is transmitted, so less bandwidth is required. The sideband signal changes its frequency and amplitude with the modulation, which creates

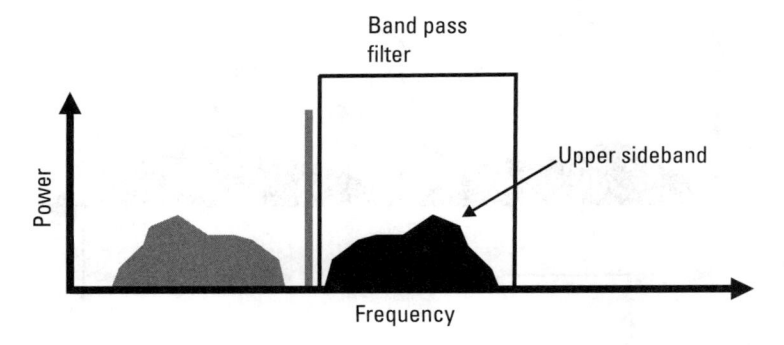

Figure 2.3 Single sideband signal.

challenges for some kinds of EW operations. We will discuss these issues later.

Frequency modulation (FM) carries the information in a signal by changes in the transmitted frequency. Figure 2.4 shows the FM signal in the time domain. The amplitude of the transmitted signal does not change with the modulation. Note that this figure, like Figure 2.1, is unrealistic in that the transmitted frequency is actually much higher than the modulating frequency. Figure 2.5 shows the FM signal in the frequency domain. Like the AM signal, it has a carrier and sidebands. The frequency modulation of the FM signal is typically much wider than would be required for an AM signal. The FM signal changes its frequency as a function of the amplitude of the modulating waveform, and the transmitted frequency change versus modulation amplitude is a design value. The ratio of the maximum frequency deviation from the carrier to the maxim frequency of the modulating signal is called the modulation index and denoted by the Greek letter β. As an example, consider commercial FM broadcast signals. The maximum deviation

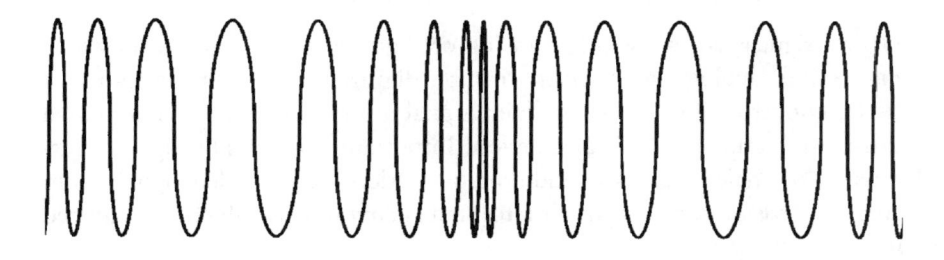

Figure 2.4 FM signal in time domain.

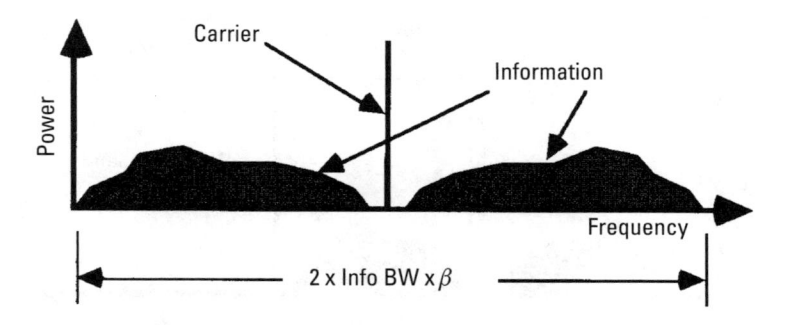

Figure 2.5 FM signal in frequency domain.

from the carrier frequency is 75 kHz and the maximum modulating frequency is 15 kHz

$$\beta = 75 \text{ kHz}/15 \text{ kHz} = 5$$

One of the advantages of FM transmission is the reduction of the effects of interference. In general, the larger the modulation index, the more tolerance to interference the received signal will have.

2.2 Digital Modulations

Digital data comprises a series of binary "ones" and "zeros," which cannot be directly transmitted. First, those binary values must be modulated onto a radio frequency carrier using one of several techniques. In this section, we will discuss the important digital signal modulations, the structure of digital signals, and the bandwidth required for digital transmission.

2.2.1 Transmission of Analog or Digital Information

Digital signals can carry either analog or digital information. As shown in Figure 2.6, analog signals must first be digitized by an analog-to-digital (A/D) converter. The resulting digital signal must then be modulated onto a carrier and amplified for transmission. Important examples of digital transmission of analog signals include digital tactical radios, video signals from analog television cameras, and intercepted enemy signals, which are digitized in the receiving system.

There are, of course, many signals that have never been analog. Important examples are computer-to-computer communication, command links

Communications Signals 17

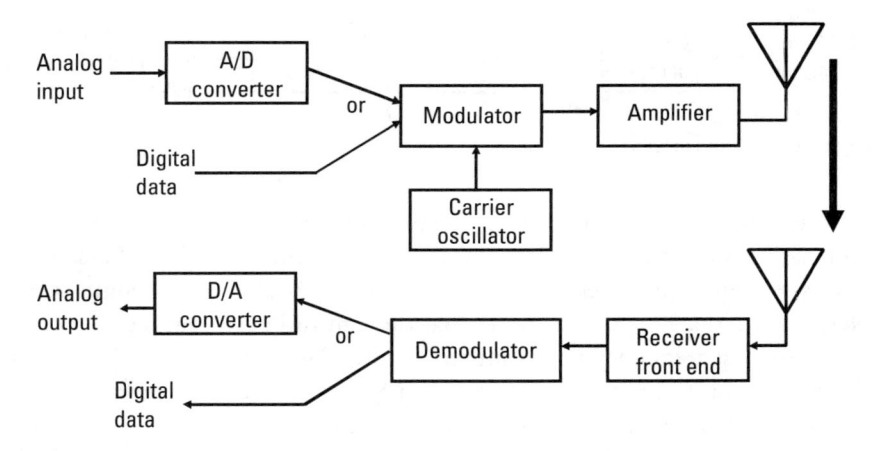

Figure 2.6 Analog or digital information can be transmitted digitally.

from control panels for remotely controlled platforms and their payloads, and data links from UAVs and other remote assets.

In either case, digital transmission provides several advantages over analog transmission:

- Compatibility with high security encryption;
- Error-correction coding;
- LPI techniques;
- The ability to pass through various transmission paths while retaining high signal quality.

As you will see later, there are also disadvantages in terms of vulnerability to jamming.

2.2.2 Digitization

Digitization is performed in an analog-to-digital converter (ADC). Figure 2.7 shows the parts of a simple ADC. First, a sample of the analog signal is stored, then a digital representation of the amplitude of the sample is generated, then the digital signal is formatted for output. It can be either parallel or series. A parallel output generates several bits simultaneously and outputs them on separate lines. A series output generates a series bits one at a time and outputs them on a single line. More digitization techniques will be discussed in Chapter 4 during the coverage of digital receivers.

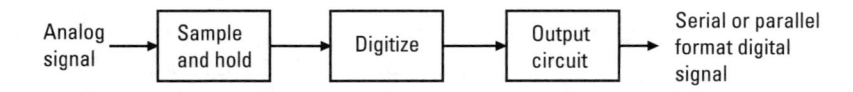

Figure 2.7 Digitizer block diagram.

Figure 2.8 shows an arbitrary analog waveform which is input to the digitization process. The horizontal scale shows the timing of the sample points. A commonly accepted value for the timing of the sample points is the Nyquist criteria. This criteria states that there must be two samples per cycle of the highest frequency you want to capture. Another way to think about this is that you want to sample often enough to capture the character of the analog signal that you care about. The amplitude of each sample is characterized by a binary word. In this case, there are 16 threshold levels in the ADC, so each digital word has four bits: 0000 is zero, 0001 is one, 0010 is two, 0011 is three, and so forth. An important concept about digitization (by any technique) is that once the signal is digitized, the analog input signal is gone.

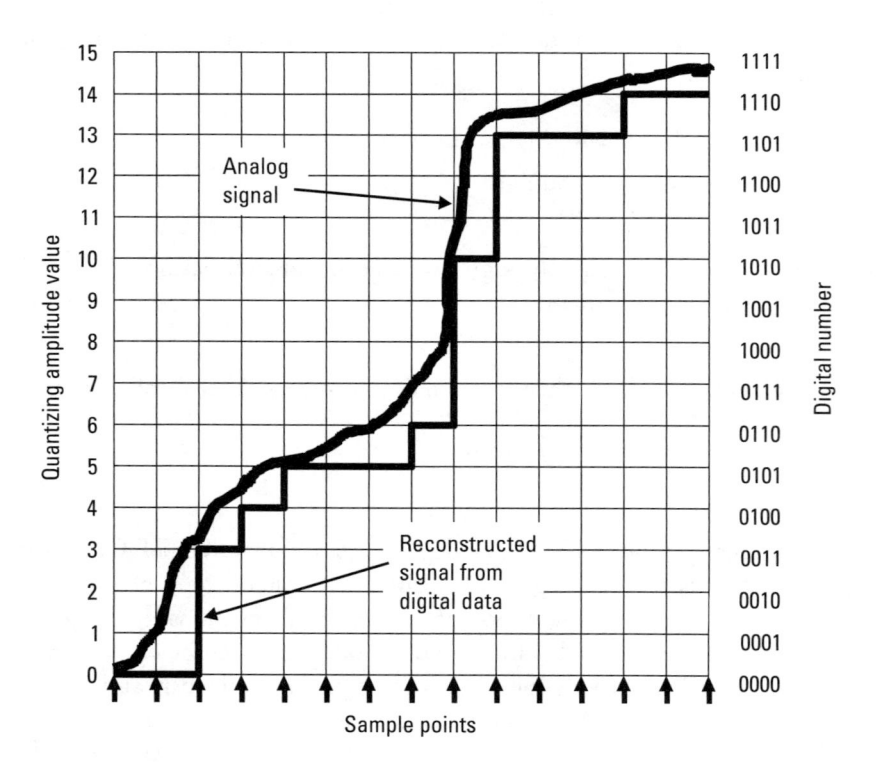

Figure 2.8 Digitization of an analog waveform.

When the digitized signal is recovered in the receiver, it is input to a digital-to-analog converter (DAC) which generates the stair step line shown in the figure. The bad news is that once the signal is digitized, it will never get any better than this quantized curve. You can run it though a filter to knock the corners off and make it sound better, but there will always be residual distortion. The good news is that if you are reasonably careful with the handling and transmission of the digital signal it will never get any worse than this digital representation.

The higher the sampling rate and the greater the number of bits in the digital words, the higher the quality of the digitized signal.

Figure 2.9 shows the raster scan that is typically used to capture a video signal for transmission. The raster covers a whole frame, which is a single picture. The raster is generated in a video camera and is reproduced to paint the captured picture on a video screen. Each point on the screen is a pixel. To digitize the video signal, each pixel is digitized. If the picture is monochromatic, only the brightness (the luminance) of the signal is digitized. If it is a color picture, the components of the color (e.g., red, blue, and green) are each digitized to capture the color (chrominance).

2.2.3 Digital RF Modulations

Digital information is a series of "ones" and "zeros," which cannot be efficiently broadcast unless they are modulated onto an RF carrier. There are

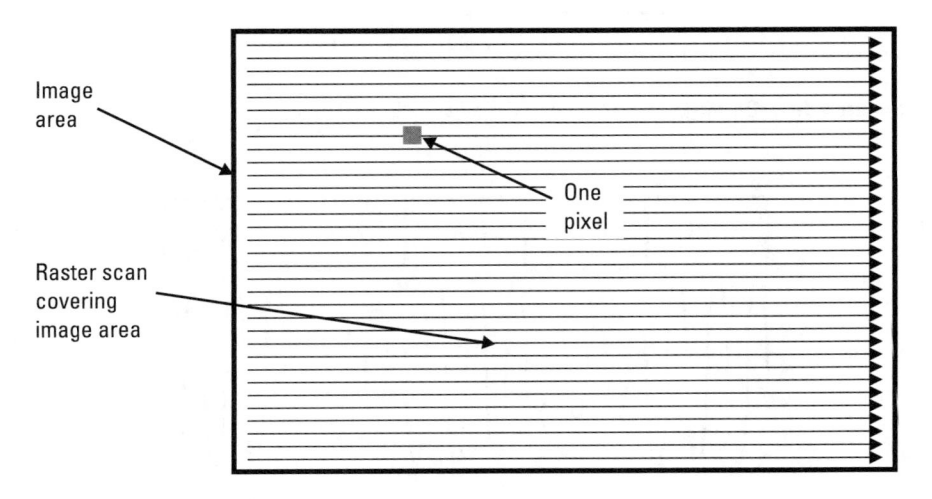

Figure 2.9 Raster scan to capture image for transmission.

many appropriate modulation approaches; the following discussion covers only a few important examples.

Each of the bits of digital information is transmitted during a fixed time period called a baud or symbol. These bits are called "chips" when the information carrying digital signal has been modified by a secondary digital modulation to spread its spectrum. The rate at which bauds are transmitted is called the "clock rate." As we will see, each baud can carry one or more bits of digital information.

The following discussion covers typical examples of the many digital modulations that are possible.

Figure 2.10 shows on-off keying (OOK) in which a signal is transmitted only during bauds which carry "ones." The transmission could as well have been during a "zero." This can be considered a special case of pulse amplitude modulation (PAM) in which there are two RF signal amplitudes as shown in Figure 2.11, one for a "one" and the other for a "zero." Figure 2.12 shows frequency shift keying (FSK) in the time domain. In FSK, one frequency in a baud transmits a "one" and another frequency transmits a "zero." Figure 2.13 shows the FSK signal in the frequency domain. The two signals can be produced by two different oscillators, causing the signal to be noncoherent. They can also be produced by a single synthesizer to produce a

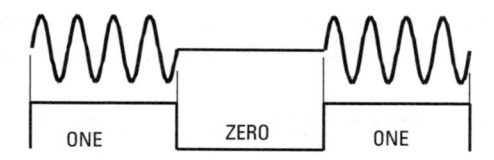

Figure 2.10 On-off-keyed modulated digital signal.

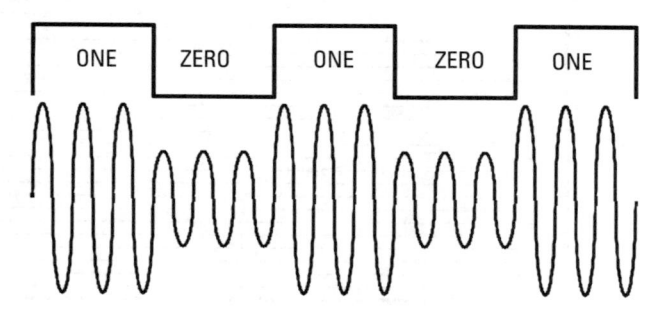

Figure 2.11 Amplitude modulated digital signal.

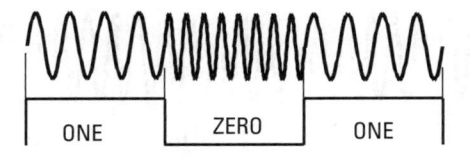

Figure 2.12 Frequency shift keyed signal in time domain.

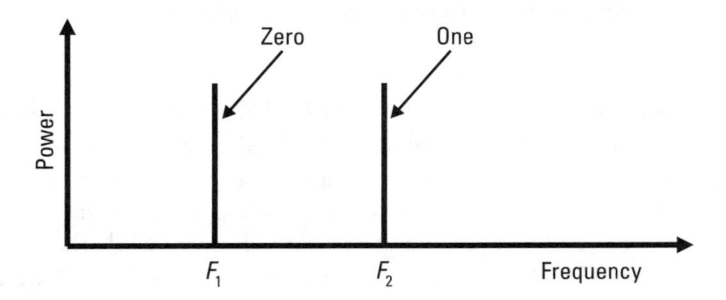

Figure 2.13 Frequency shift keyed signal in frequency domain.

coherent FSK signal which provides performance advantages over noncoherent FSK.

Figure 2.14 shows a type of phase modulation–binary phase shift keying (BPSK). It is shown in the time domain. Any phase modulation is necessarily coherent in order to allow the determination of the phase of the signal during each baud by comparison to a single oscillator in the receiver. As shown in the figure, the signal has one phase during a one and changes phase by 180 degrees during a zero. Thus one bit of digital data is carried during each baud.

A second type of phase modulation is shown in Figure 2.15. This is quadrature phase shift keying (QPSK). This signal has four possible phases for each baud: 0°, 90°, 180°, and 270° shifted relative to a reference oscillator. As shown in the figure, this allows the signal to carry two bits of digital information in each transmitted baud. Thus, when the receiver detects a particular phase in the received signal, it outputs two bits of digital data.

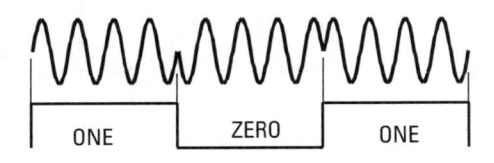

Figure 2.14 Binary phase shift keyed signal in time domain.

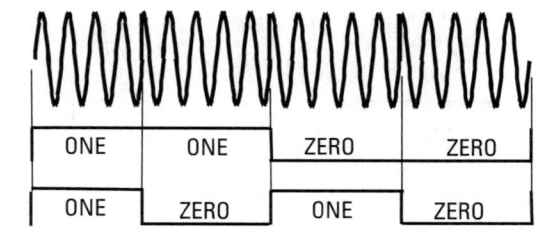

Figure 2.15 Quadriture phase shift keyed signal in time domain.

Another way to consider BPSK and QPSK signals is shown in Figure 2.16. In this type of phase diagram, the fixed power signal rotates one complete cycle (counterclockwise) during each RF cycle. It is common to show the in-phase condition as a vector pointing to the right. In Figure 2.16(a), the "one" is shown as a nonphase-shifted signal, while the "zero" has 180° phase shift. The four phase conditions of the QPSK signal are clearly shown along with the two bits represented by each signal phase in Figure 2.16(b).

There can be more defined phase values in a "nary" phase shift keyed signal. One quite common case is a "32ary" phase shift keyed signal. Each of its 32-phase positions represents five bits of digital data.

Figure 2.17 shows a quadrature amplitude modulated (QAM) signal. This case is a signal with 16 states. Each state has a unique amplitude and phase combination. Each can be considered to have an "I" (in phase) and "Q" (quadrature or 90°-phase delayed) component. Since there are 16 states,

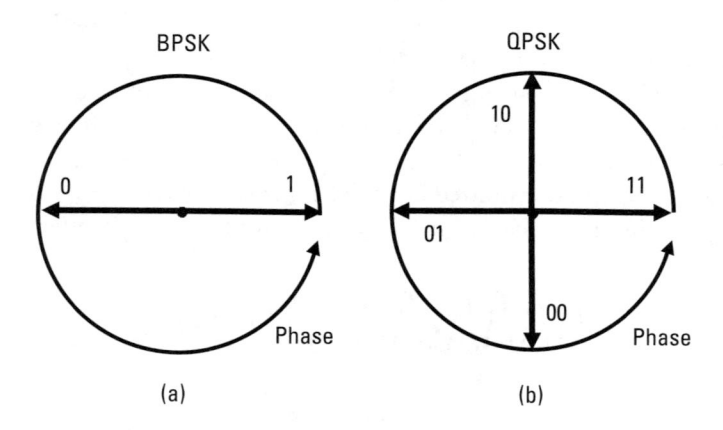

Figure 2.16 Phase diagrams for (a) BPSK, and (b) QPSK modulated digital signals.

Communications Signals 23

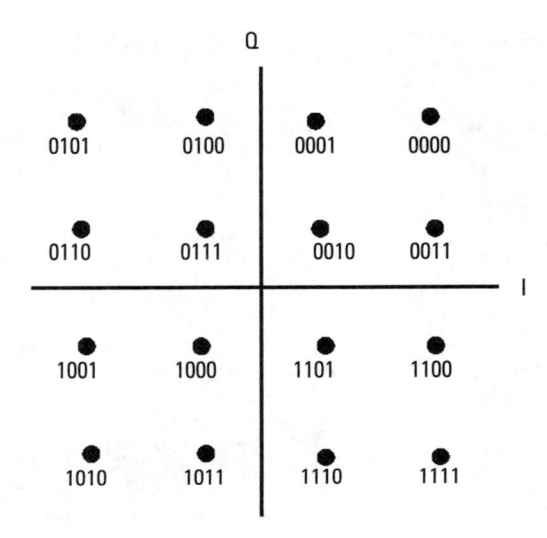

```
              Q
              |
  ●      ●    |   ●      ●
 0101   0100  |  0001   0000
              |
  ●      ●    |   ●      ●
 0110   0111  |  0010   0011
──────────────┼────────────── I
  ●      ●    |   ●      ●
 1001   1000  |  1101   1100
              |
  ●      ●    |   ●      ●
 1010   1011  |  1110   1111
              |
```

Figure 2.17 A 4-bit per baud I & Q modulated digital signal.

each state present during a baud represents four bits of digital data. Complex modulations can have many more states in their "constellations," hence more bits per state.

Modulations which provide multiple bits per baud are considered high efficiency modulations because they allow more digital data to be transmitted in any given RF bandwidth.

2.2.4 Bandwidth of Digital Signals

Figure 2.18 shows what a digital signal looks like on a spectrum analyzer. The fuzzy lump is the main lobe of the transmitted spectrum. The transmit frequency (the carrier) is at the center, and there are distinct nulls on either side. Since the digital data has pseudo-random characteristics, the signal looks noise-like. If you watch the spectrum analyzer screen in real time, the signal within the lobe will be constantly undulating.

Figure 2.19 shows the parts of the frequency response. The null-to-null bandwidth of the main lobe is twice the clock rate. That is, each null is the baud rate from the carrier (i.e., if 1 million bauds per second are sent, the lobe is 2 MHz wide). The side lobes are each one-clock rate wide. The drawing shows only the first side lobes, but they continue, with the side lobe peaks diminishing with frequency from the carrier.

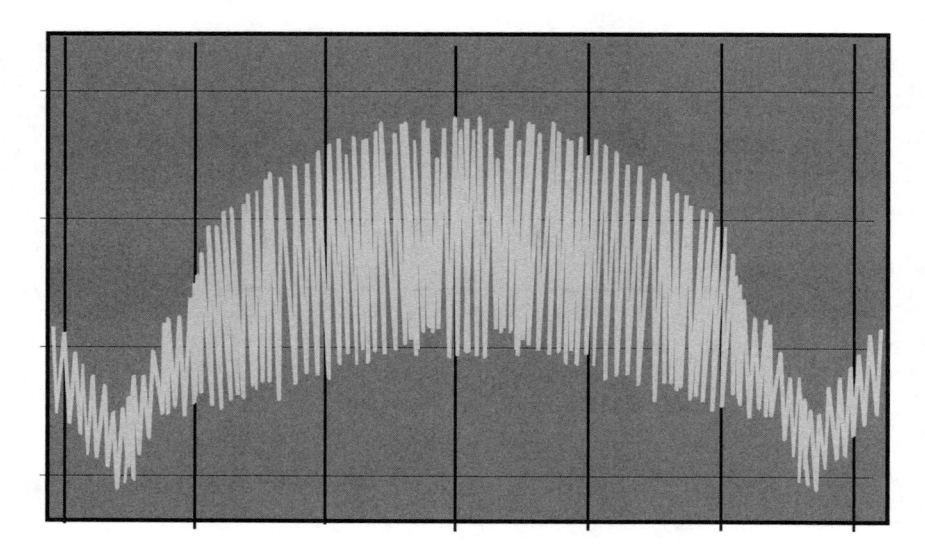

Figure 2.18 A spectrum analyzer display of a digital signal.

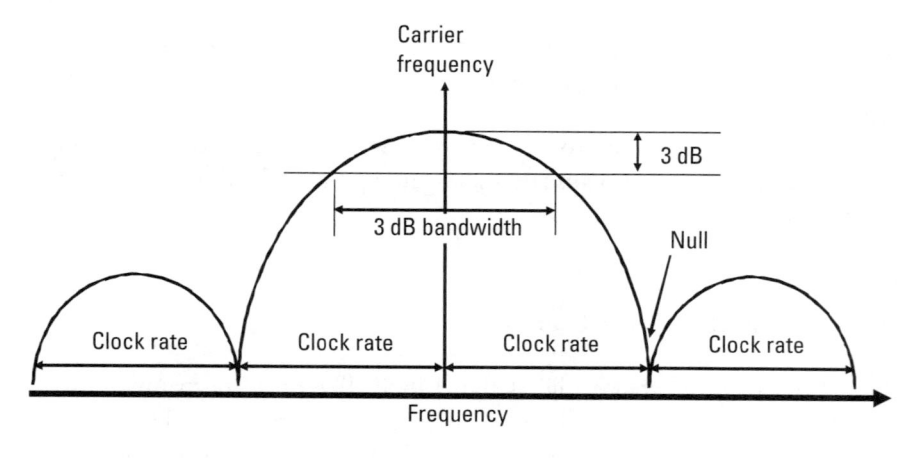

Figure 2.19 Characteristics of the frequency spectrum of a digitally modulated signal.

The 3-dB beamwidth of the main lobe is often taken as the required transmission bandwidth. Note that the whole main lobe (null-to-null) contains 90% of the transmitted energy.

There are some digital modulations which change states using waveforms which minimize the frequency spreading of the transmitted signal. One of these, sinusoidal shift keying, is shown in the time domain in Figure 2.20. The RF signal moves between "one" and "zero" states of its modulated

Sinusoidal shift keyed waveform

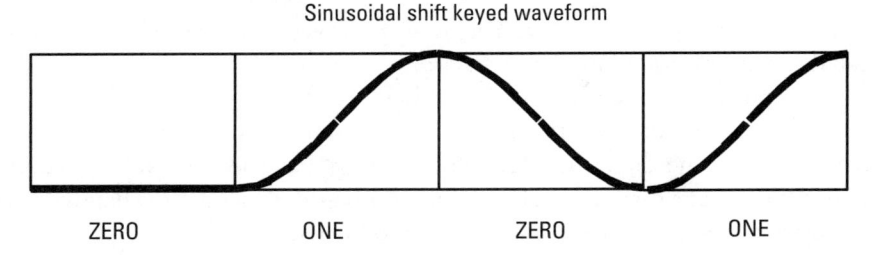

Figure 2.20 Sinusoidal shift keyed waveform.

signal sinusoidally. When the basic modulation is frequency shift keying, this waveform is called sinusoidal frequency shift keying (SFSK).

A widely used efficient modulation is called minimum shift keying (MSK) because it moves between modulation states in an energy-efficient pattern. It has many variations. Classic MSK changes states as shown in Figure 2.21. Dixon's *Spread Spectrum Systems* book (ISBN 0-471-59342-7) clearly explains the details of this family of modulations and how they are generated. The important point about MSK for our current discussion is that it requires significantly less bandwidth than modulation which moves directly between states (described in the literature as having "square pulses").

Table 2.1 shows the effects of modulation types on required bandwidth. Note that MSK requires significantly less bandwidth and has

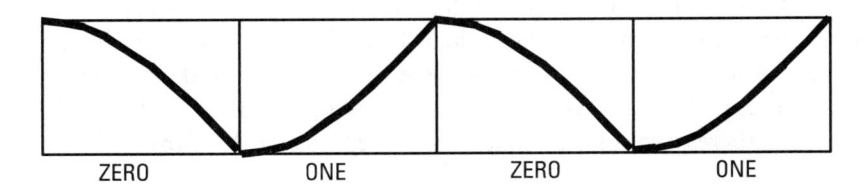

Figure 2.21 Minimum shift keyed waveform.

Table 2.1
Bandwidth Versus Clock Rate Versus Modulation

Waveform	Null-to-Null Main Lobe BW	3-dB BW	First Side Lobe	Roll-Off Rate
PAM, BPSK, QPSK	2 × code clock	0.88 × code clock	−13 dB	6 dB/octave
MSK	1.5 × code clock	0.66 × code clock	−23 dB	12 dB/octave

significantly reduced side lobes. This is why it is very widely used in sophisticated communications applications.

Important points from this table are that MSK requires only three-fourths the bandwidth of the other modulations, and that the 3-dB bandwidth of the digital modulation is 0.88 × the code clock. This latter item means that a digital signal carrying 1 M baud/sec has a 3-dB bandwidth of 880 MHz—or 660 MHz if it is MSK modulated.

2.2.5 Digital Signal Structure

As shown in Figure 2.22, a transmitted digital signal includes more than just the binary data for the information being carried. It also has sychnronization bits, address bits, and parity or error-correction bits. These extra bits are called overhead, and can require from about 10% to over 100% additional bits to be added to those carrying the information.

The synchronization bits are required to allow the receiving system to properly place received digital bits into the correct position in output registers. At the output of the receiver which returns a received signal to ones and zeros, the digital data is either converted back to analog form or the information it carries passed to the appropriate system destination. Either way, the bits of the digital data must be put into the right stages of a register. This requires that the beginning of the digital "frame" be detected. The synchronization bits have a unique pattern that can be automatically detected to perform that important function.

Address bits, when present, allow the digital data to go to the right location in equipment using the digitally carried information. For example, if the digital signal is sent to control some remote equipment, the address could cause the digital data to be switched to the intended controlled chassis. Alternately, if there are multiple users of digitized analog data, the address bits

Synchronization bits	Address bits	Information bits	Parity or EDC bits
~10% of total			Less than 10% to over 100% of info bits

Transmitted digital signal

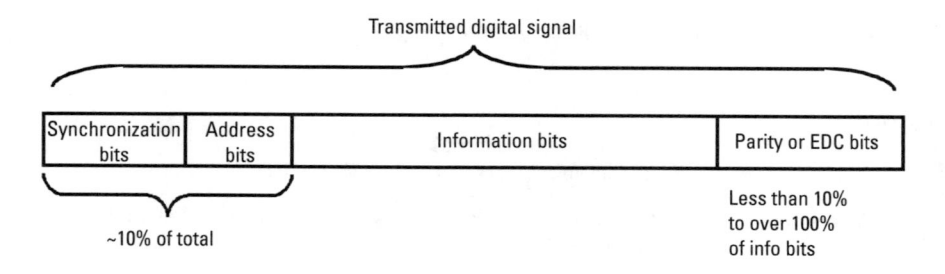

Figure 2.22 Bit structure of a transmitted digital signal.

would be used to switch received signals to the intended user for digital-to-analog conversion.

Bit errors are incorrectly received bits. That is, a transmitted one that is received as a zero or vice versa. These errors can be caused by interference in the transmission environment, by deliberate interference from a jammer, or by noise. The bit error rate of a received digital signal is the number of bit errors divided by the total number of bits received. As we will discuss in Chapter 4, the signal-to-noise ratio, which impacts the bit error rate, is a function of the signal strength of the received signal.

The final section of the transmitted signal structure in Figure 2.22 is for parity or error-correction code bits. Parity bits are added to allow the detection of the presence of bit errors. If all of the parity bits are appropriate, the received word contains no errors. If this section contains an error detection and correction (EDC) code, it is possible to correct any bit errors in the received signal up to some performance limit. There will be more about this in Section 2.5.

2.3 Noise

Signal-to-noise ratio (SNR) is the way that we quantify the quality of a received signal. In practice, any random distortion of a received signal is often called noise, even though it may actually be interference or quantization errors. We will be discussing the impact of signal-to-noise ratio on the sensitivity of receivers in Chapter 4. Our purpose here is to clarify the way that we talk about noise to support later discussions of modulations and electronic warfare operations.

Figure 2.23 shows a received signal with noise as it would appear on an oscilloscope. Note that the noise seems to be the thickness of the line, while the sinusoidal line is the signal. The irregular thickness of the line is often called "the grass," because of its irregular shape. This is a drawing of a signal which has fairly high signal-to-noise ratio. The noise is approximately one-tenth the height of the signal. This means that the signal-to-noise ratio is approximately 20 dB. The vertical deflection of an oscilloscope is proportional to the voltage of its input signal.

You will recall (from Chapter 1) that the voltage ratio is converted to dB by multiplying the logarithm of the ratio by 20. Figure 2.24 shows the same signal as it might appear on a spectrum analyzer. If the spectrum analyzer is set to a power scale, the vertical axis shows the power of the input signal versus frequency. In this diagram, you can see the signal and the noise

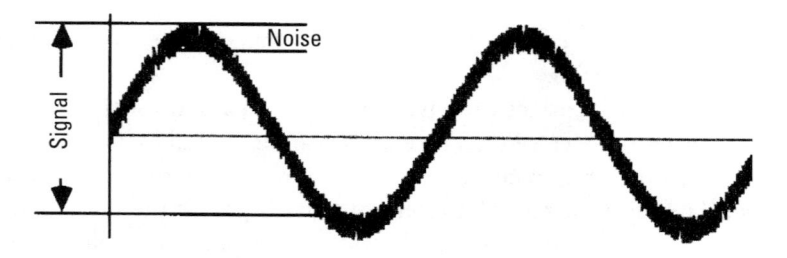

Figure 2.23 Signal with noise in time domain.

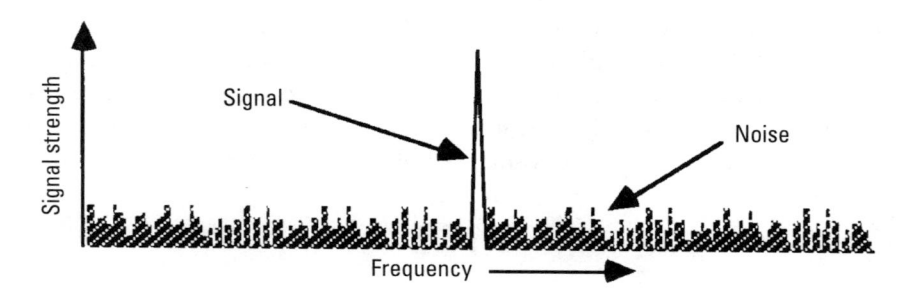

Figure 2.24 Signal with noise in frequency domain.

which is present. You will note that the noise is fairly constant versus frequency.

A characteristic of noise is that it is proportional to the bandwidth of the receiver system in which it is received. If the receiver bandwidth is doubled, the measured noise will double (i.e., increase 3 dB). If the receiver bandwidth is increased by a factor of 10, the noise increases 10 dB, and so forth.

Figure 2.25 shows the time domain display of a signal with much lower signal-to-noise ratio. Note that the thickness of the line (the noise) is several times greater than the amplitude of the signal. This means that we have a negative signal-to-noise ratio (in dB). Figure 2.26 shows a low signal-to-noise ratio signal in the frequency domain. This diagram is of a signal with a low but positive signal-to-noise ratio. Note that the signal is slightly above the noise.

When we talk about noise in a receiver, we normally mean the thermal noise generated inside the receiver. However, two other effects that look (and sound) like noise are also called noise. One is the quantization inaccuracy

Figure 2.25 Low SNR signal with noise in time domain.

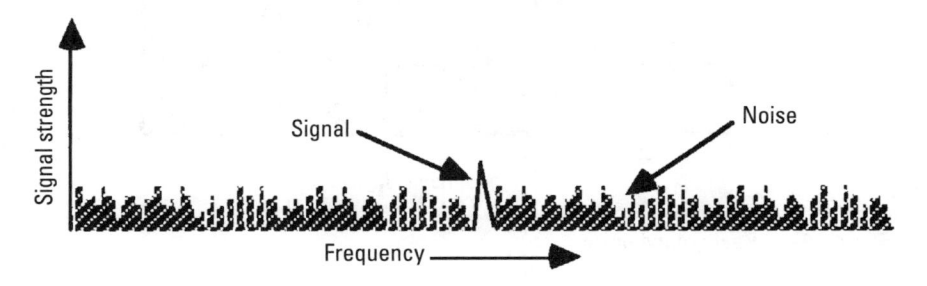

Figure 2.26 Low SNR signal with noise in frequency domain.

introduced in the digitization process. This would more accurately be called "quantization noise," but is often just called noise.

The second effect is environmental noise. This noise enters the receiver through the antenna, so can be thought of as external noise. In very quiet rural settings, this is the combination of random signals generated by the many stars in our galaxy. In urban or suburban settings, the environmental noise is the combination of many small interfering signals such as: street cars, automotive ignition interference, electric motors, and others. Figure 2.27 is a commonly used diagram which describes the external noise level in terms of the frequency and the density of human activity in the area. The assumption of this chart is that the receiver is connected to an isotropic antenna (described in Chapter 3). The curves are independent of receiver bandwidth in that the level of background noise is given as the number of dB above "kTB." kTB refers to the internal noise level in an ideal receiver system. The "B" in kTB is the bandwidth of the receiver system. This term will be defined in detail and its significance discussed along with receiver system sensitivity

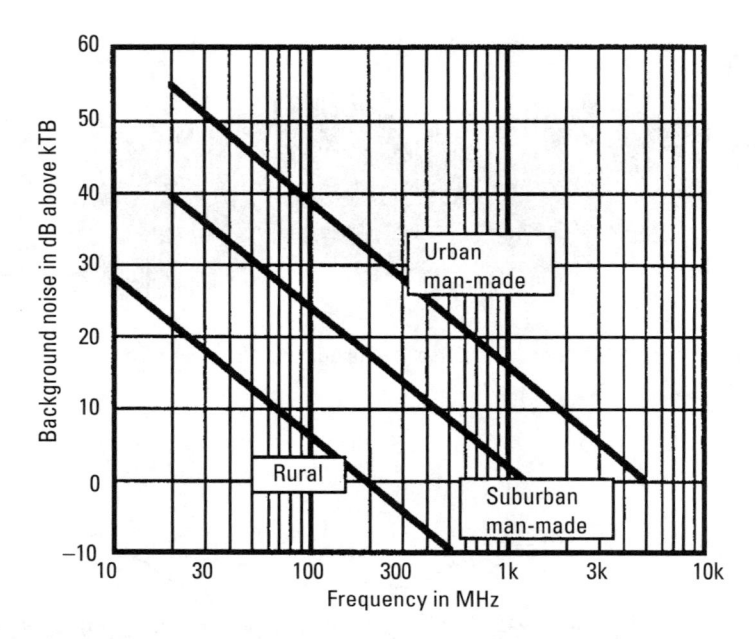

Figure 2.27 Background noise.

in Chapter 4. The environmental noise level is often taken to be the "suburban manmade" level for tactical military applications.

The SNR in a receiver system is given in dB. It is the way we quantify the quality of the received signal. In a low SNR signal, the noise overwhelms the signal and it is very difficult to recover the information carried by that signal.

2.4 LPI Signals

Low probability of intercept (LPI) communication signals have special modulations which spread their spectrum over a very wide bandwidth. LPI communication signals are also called spread spectrum signals, and are used for selective addressing, hiding transmitted signals, and rejecting interfering signals. There are three general types of LPI signals:

- Frequency hopping;
- Chirp;
- Direct sequence.

Each of these will be discussed later in this chapter. Note that LPI techniques provide transmission security—that is, they make it difficult for an enemy to detect the presence of a signal or to intercept or jam that signal. This is different from message security. Message security is provided by encryption. As you will learn in later chapters, there are ways in which spread spectrum techniques can be penetrated by sophisticated systems and techniques and in some intercept geometries. Thus, LPI techniques should ordinarily not be assumed to provide message security. Encryption must be added if that feature is required.

For each of the above listed LPI techniques, the spreading is done using pseudo-random codes which can either be public codes or carefully guarded military secrets. The code that is used to spread the signal is required to collapse the signal back to its prespread bandwidth. As shown in Figure 2.28, there is a synchronization scheme common between transmitters and their intended receivers that allows the same code to be used and to keep the transmitter and receiver codes in phase with each other. Once the signal is despread, it can be demodulated in the appropriate, narrow bandwidth in which it has a high signal-to-noise ratio. Thus, the spreading process is transparent to the intended receiver. In public code transmission situations, anyone can design a receiver system to despread the signal. With nonpublic codes, a hostile receiver does not have the synchronization scheme and thus has no access to the spreading code.

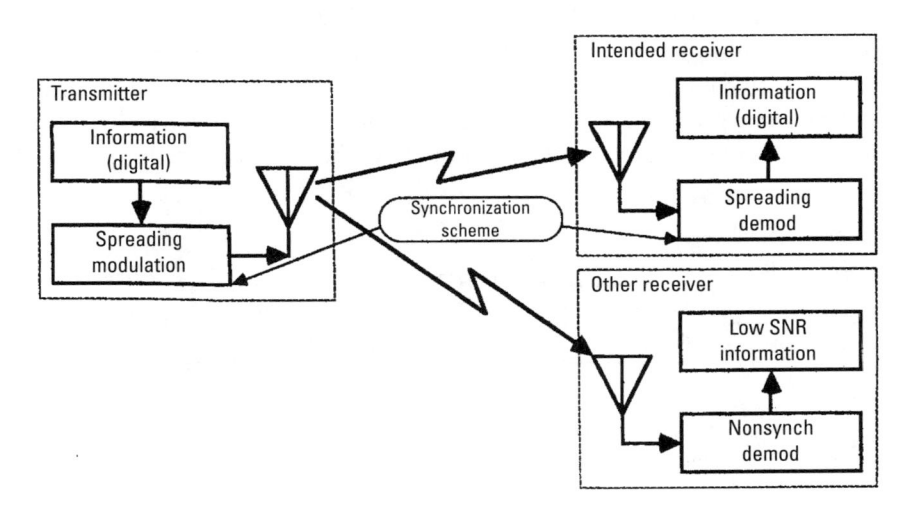

Figure 2.28 Spread spectrum signal operation.

As shown in Figure 2.29, the information to be passed by the communication system is input as a narrowband (i.e., high signal-to-noise ratio) signal modulation to a spreading modulator. This spreading modulator applies a second modulation to the signal which is only to spread its frequency spectrum—by one of the three techniques listed earlier. Then, the frequency spread signal is transmitted. In the intended receiver, a spreading demodulator removes the secondary modulation. The transmitting and receiving pseudo-random codes are many bits long, but the synchronization scheme allows the codes in the two locations to be perfectly aligned.

Remember that the noise power in the receiver is proportional to the receiver bandwidth, thus when the signal is artificially spread in frequency, the receiver must have more bandwidth to receive it, significantly increasing the noise power. This causes the SNR to be reduced. In practice the SNR can be reduced 20 to 40 dB by spread spectrum techniques. When the spreading modulation is removed, the receiver need only have enough bandwidth to receive the despread signal modulation so the SNR is back at its original level. For convenience, we will call the despread bandwidth the "information bandwidth" and the spread bandwidth the "transmission bandwidth."

Figure 2.30 is a diagram showing the effects of the spreading demodulator. The frequency spread signal is input to the demodulator. Since its power is spread across a wide frequency bandwidth, its power at any one frequency is reduced. If the demodulator is set to the same code as the transmitter that generated the received signal, the spreading modulation will be removed and the signal will be output with all of its power concentrated in the information bandwidth. If the demodulator is not set to the matching code, the signal out of the demodulator is not despread. Since the following circuitry expects an information bandwidth signal, it may not even be able to detect the presence of the signal. On the other hand, if a normal, nonspread

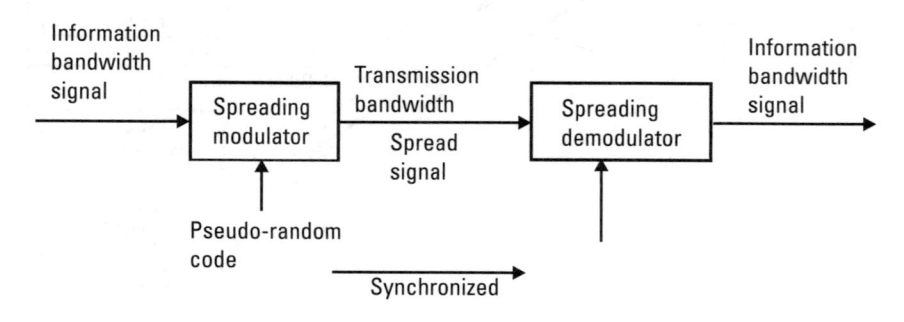

Figure 2.29 Generation of LPI signal.

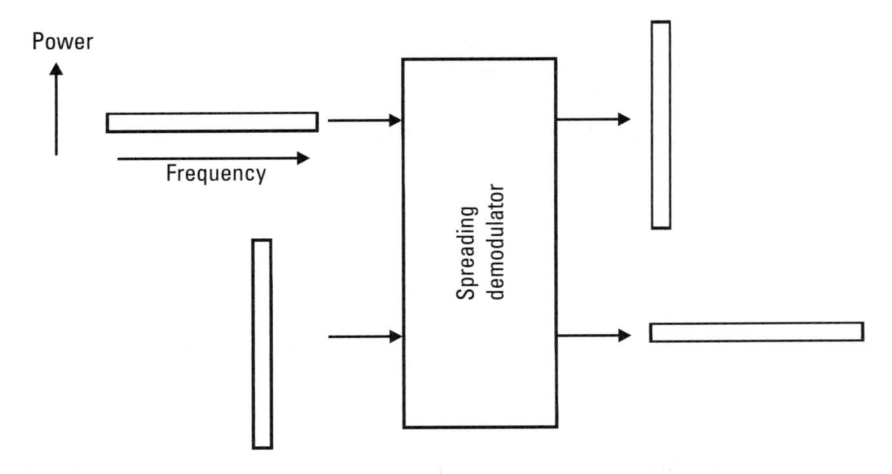

Figure 2.30 Frequency spectrum of LPI signals.

signal is input to the demodulator, the demodulator will spread it and the following circuitry will see a signal so low that it may not even be detectable. This explains the ability of spread spectrum communication systems to operate in the presence of significant interference from other signals in the environment.

2.4.1 Pseudo-Random Codes

Pseudo-random codes are used in encryption, the generation of frequency hopping sequences, the control of pseudo-random synchronization, and as direct sequence spreading bit streams. These codes are generated using shift registers which can be implemented in hardware or software. The more stages in the shift register, the longer the code will be. The code length is 2^n-1 where n is the number of stages in the shift register. Table 2.2 shows a few values for the length of the code as a function of the number of shift register stages.

The characteristics of codes are that they appear to be randomly generated and that the number of ones and zeros are close to equal. When the code is correlated with itself the number of agreements are equal to the number of bits. However, if the code is compared to an unsynchronized version of itself, the bit agreements minus the bit disagreements are equal to −1. Another, practical, way to consider this is that a pseudo-random digital bit stream has 50% autocorrelation when it is not synchronized and 100% autocorrelation

Table 2.2
Shift Register Length Versus Code Length

Stages	Code Length
3	7
4	15
5	31
6	63
7	127
31	2, 147, 483, 647

when it is synchronized. This is sometimes called the "thumb tack correlation" characteristic of digital signals.

A shift register passes the state (one or zero) in one stage on to the next stage each time the system clock cycles. If feedback loops are added to the shift register, the sequence of ones and zeros is modified. Figure 2.31 is a three-stage shift register with a simple feedback loop. The state of stage one is added (modulo 2) to the state of stage three and input to stage three. Figure 2.32 is a timing diagram of the states of each of the three stages and the modulo-2 adder at each clock cycle. Note that the shift register needs to start with one in each stage. The initial output of the modulo-2 adder is zero because one + one equals zero in modulo-2 addition. After the first clock cycle, the zero from the modulo-2 adder is shifted into stage 3, the one in stage 3 is shifted into stage 2, the one in stage 2 is shifted into stage 1, and the shift register outputs the second bit of its code as a one. The output code is the series of states of stage 1 of the shift register:

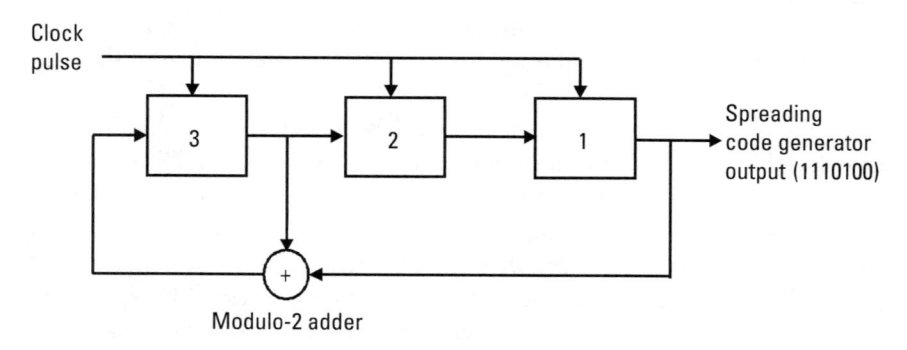

Figure 2.31 Shift register for generation of a pseudo-random code.

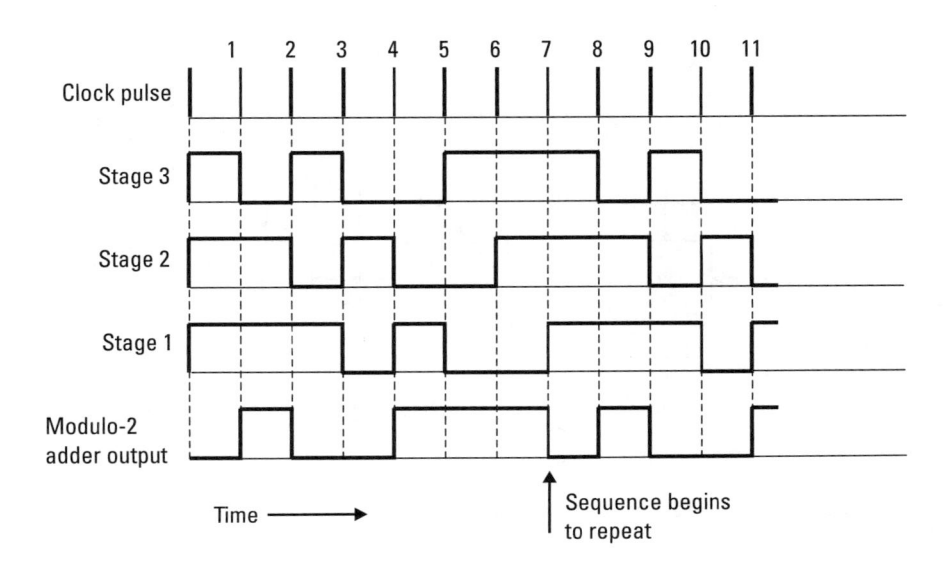

Figure 2.32 Shift register timing diagram.

$$1, 1, 1, 0, 0, 1, 0, 0$$

then the code repeats. Thus, as stated in Table 2.2, the three-stage shift register produces a seven-bit code.

There are two classes of codes, linear and nonlinear. Linear codes are appropriate for applications in which security is not required. There can be any reasonable number of shift register stages and feedback loops, but all of the feedback loops use modulo-2 adders. It is fairly easy to recreate the shift register configuration for the short, linear codes typical of commercial applications.

Nonlinear codes are appropriate for applications in which transmission security is important. This is typically the case for military applications, which also can be expected to use long codes that may not repeat for many days operation. Nonlinear codes employ more complex operations, such as AND and OR gates in the feedback loops. It is, of course, much more difficult and time consuming to recreate nonlinear codes.

Figure 2.33 is the same shift register shown in Figure 2.31, however the status of all three stages are brought out in parallel to form a binary numerical value. As shown in Table 2.3, the states of the three stages cycle over seven clock pulses to form binary values for a series of numbers which pseudo-randomly cover the seven values from 1 to 7. A longer shift register would, of course, generate random numbers covering a larger range.

Parallel code outputs

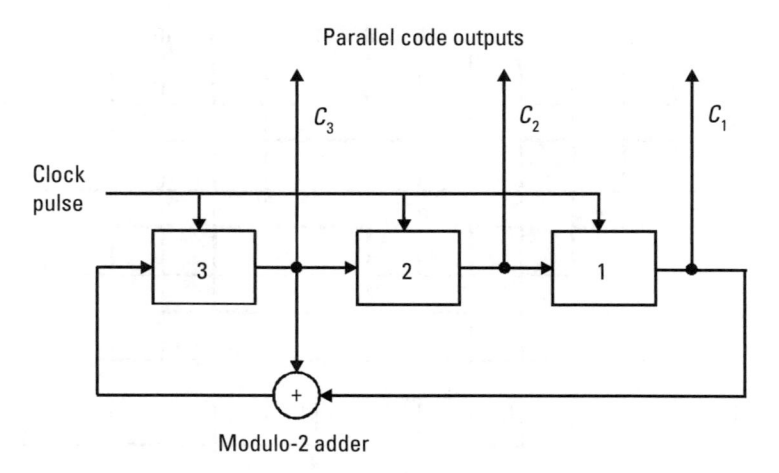

Figure 2.33 Hopping sequence generator.

Table 2.3
Random Numbers Generated by Shift Register

Clock Pulse	Binary Code			Number
	C_3	C_2	C_1	
1	1	1	1	7
2	0	1	1	3
3	1	0	1	5
4	0	1	0	2
5	0	0	1	1
6	1	0	0	4
7	1	1	0	6
...	Code Repeats			

As we talk about the specific types of LPI signals, you will see how these forms of codes are applied.

2.4.2 Frequency Hopping Signals

Frequency hopping (FH) signals are an extremely important electronic warfare consideration because they are widely used in military systems and because the conventional detection, intercept, emitter location, and jamming techniques are not effective against them. One advantage of frequency hopping over the other two types to be discussed here is that it can achieve significant frequency diversity.

Frequency Versus Time Characteristic

As shown in Figure 2.34, a frequency hopping signal remains at a single frequency for a short time, then it "hops" to a different frequency. The hopping frequencies are normally spaced at regular intervals (for example, 25 kHz) and cover a very wide frequency range (for example, 30 to 88 MHz). In this example, there are 2,320 different frequencies which the signal might occupy. The time that the signal remains at one frequency is called the "hop period" or the "hop time." The rate at which it changes frequency is called the "hop rate."

For reasons explained later, frequency hopping signals carry their information in digital form, so there is a data rate (the bit rate of the information signal) and a hop rate. Signals are described either as "slow hoppers" or "fast hoppers." By definition, a slow hopper is a signal in which the data rate is faster than the hop rate, and a fast hopper has a hop rate faster than the bit rate of the information transmitted. See Figure 2.35. However, many people speak of a signal with a hop rate of about 100 hops per second as a slow hopper and a signal with a significantly higher hop rate as a fast hopper. Most tactical FH systems are slow hoppers. Since they put all of their power onto one frequency at a time, FH signals are easier to detect using sophisticated receivers than are the other types of LPI signals. This is particularly true of

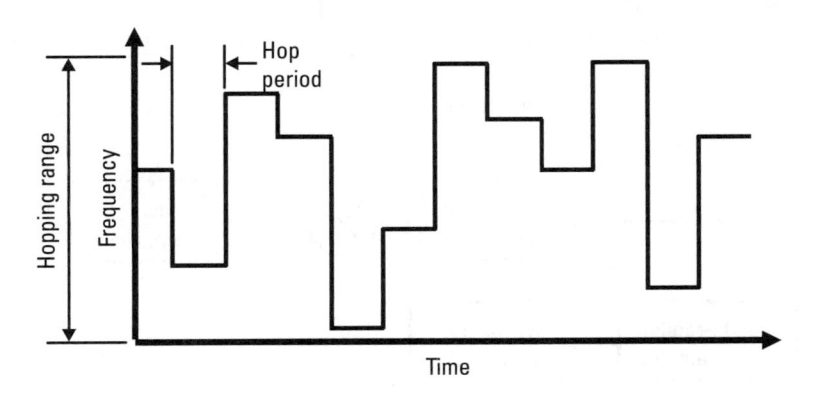

Figure 2.34 Frequency hopping signal.

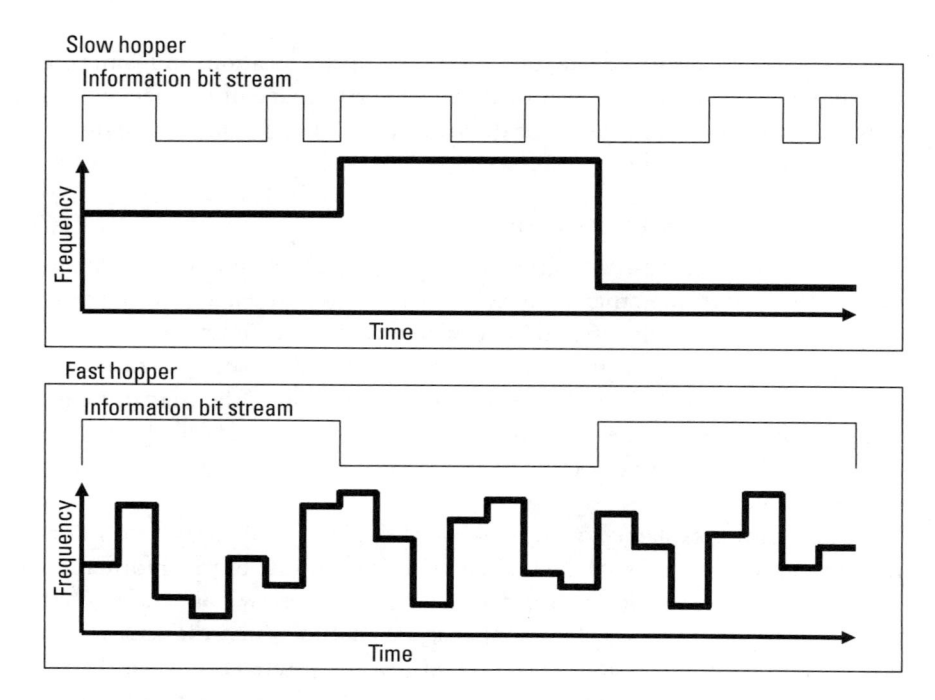

Figure 2.35 Slow versus fast hop.

slow hoppers, since they generally dwell at one frequency for multiple milliseconds.

Frequency Hopping Transmitter

Figure 2.36 shows a very general block diagram of a frequency hopping transmitter. First, it generates a signal which carries the information in its

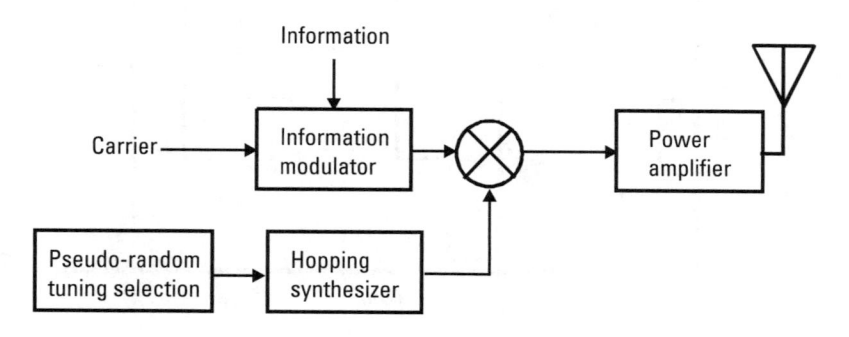

Figure 2.36 Frequency hopping transmitter.

modulation. Then, the modulated signal is heterodyned to the transmission frequency by a local oscillator which is a fairly fast synthesizer. For each hop, the synthesizer is tuned to a frequency selected by a pseudo-random process. The parallel output of the shift register in Figure 2.33 is a simple example. This means that although a hostile listener has no way to predict the next tuning frequency, there is a method by which a cooperating receiver can be synchronized to the transmitter. The design of the cooperative receiver is very similar to the transmitter with the hopping synthesizer controlled to the same code. When synchronized, the cooperative receiver tunes with the transmitter, so it can receive the signal almost continuously. Thus, the transmitter and receiver hop to the same frequencies at the same times.

Figure 2.37 is a simplified block diagram of a phase-lock-loop synthesizer. Note that the loop bandwidth is a design parameter. The wider the loop bandwidth, the faster the synthesizer can tune; but the narrower the loop, the purer the output signal. The loop bandwidth for slow hop frequency synthesizers is normally assumed to be such that it can settle onto a new frequency in a small percentage (typically 15%) of the hop period, as shown in Figure 2.38. Thus, there is a period of time at the beginning of each hop during which no data can be transmitted. This settling time is the reason that the information must be transmitted in digital form. Input data can be fed into a shift register and clocked out at a higher rate during the post settling part of the hop. Then at the receiver, another shift register can be used to slow the received data to the original bit rate. This will allow a continuous output signal at the receiver, so the human ear won't have to deal with signal drop-outs during the hopping transitions.

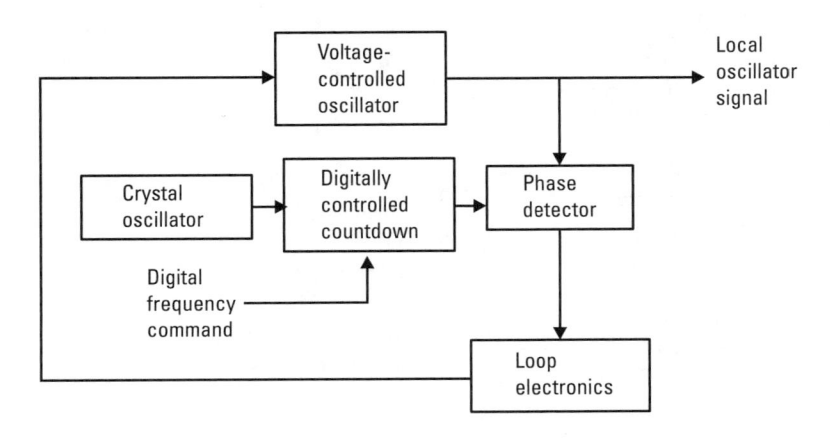

Figure 2.37 Phase-lock loop synthesizer.

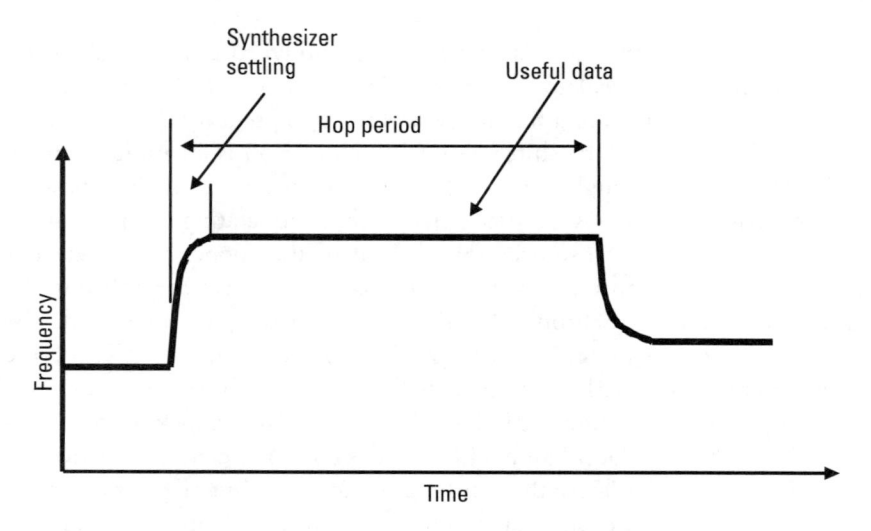

Figure 2.38 Settling of frequency hopped signal.

Fast hopping FH transmitters create a much higher level of transmission security than slow hoppers, but they can be expected to require much faster hopping synthesizers. At extremely high tuning speeds, "direct synthesizers" may be required. They are more complex than phase lock loop synthesizers. A simple example of direct synthesizer is shown in Figure 2.39. Selected oscillators are very rapidly switched into a mixer to generate the

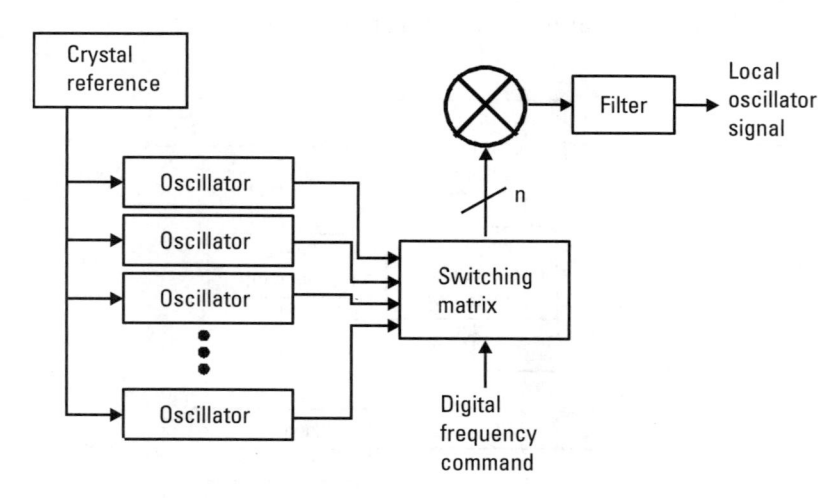

Figure 2.39 Direct synthesizer.

desired output frequency. The mixer outputs the input frequencies and the sums and differences and harmonics of all input frequencies. When properly filtered, a single, selectable output frequency is generated. Direct synthesizers are very fast, but are quite complex if many available frequencies are required.

Low Probability of Intercept

The frequency hopper is a LPI signal because the amount of time that it occupies a frequency is too short for an operator to hear that a signal is present. Taking the earlier example, the signal would be expected to be at any given frequency only 0.04% of the time, so its received power (over time) is significantly reduced, even though its full power is present at a single frequency for the hop period.

2.4.3 Chirp Signals

A second type of LPI signal is the "chirped" signal. When a swept frequency modulation is applied to communication or data signals, we say they are "chirped" because when the signal is received it sometimes sounds like a bird call. The purpose of the sweeping is to prevent detection, intercept or jamming of the signal, or the location of the transmitter.

Frequency Versus Time Characteristic

As shown in Figure 2.40, a chirped signal is rapidly swept across a relatively large frequency range at a relatively high sweep rate. It is not necessary that the sweep waveform be linear as shown in the diagram. It is important to vulnerability minimization that it be difficult for a hostile receiver to predict when the signal will be at any specific frequency. This can be accomplished by varying the sweep rate (or shape of the tuning curve) in some random way, or by causing the sweeps to have pseudo-random start times, or both.

Chirp Transmitter

Figure 2.41 shows a very general block diagram of a chirp signal transmitter. First, it generates a signal which carries the information in its modulation. Then, the modulated signal is heterodyned to the transmission frequency by a local oscillator which is swept at a high rate. The receiver will have a sweeping oscillator synchronized to the transmitter sweep. This oscillator will be used to reconvert the received signal to a fixed frequency. This allows the receiver to process the received signal in the information bandwidth, making the chirp process "transparent" to the receiver. Like the frequency hopping

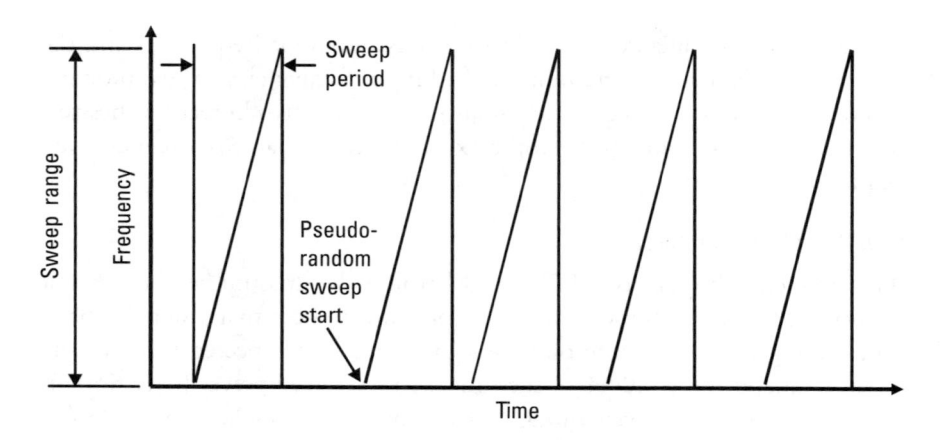

Figure 2.40 Chirp signal.

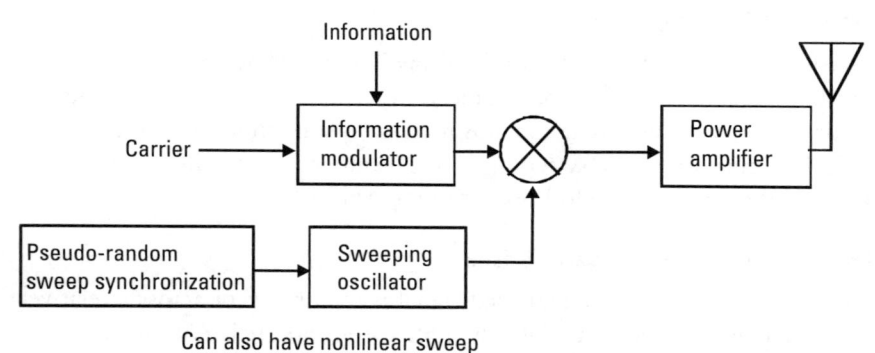

Figure 2.41 Chirp signal transmitter.

LPI scheme, the data transmitted can be expected to be digital so that data blocks can be synchronized to the sweeps and then reorganized into a continuous data stream in the receiver. This is particularly true if the sweep start times are delayed by a pseudo-randomly selected factor.

Low Probability of Intercept

The LPI qualities of the chirped signal have to do with the way receivers are designed. A receiver typically has a bandwidth approximately equal to the frequency occupancy of the signal it is designed to receive. This provides optimum sensitivity (as described in Chapter 4). To maximize transmission efficiency, the signal modulation bandwidth is approximately equal to the bandwidth of the information it is carrying (or varies by some fixed, reversible factor caused by the modulation).

A signal must remain within a receiver's bandwidth for a time equal to "one divided by its bandwidth" for the receiver to detect the signal with its full sensitivity. (For example, a 10-kHz bandwidth requires the signal to be present for 1/10,000 Hz = 100 microseconds.) This will be discussed in detail in Chapter 6. As shown in Figure 2.42, the chirped signal is present in the bandwidth of an information-bandwidth receiver for only a small fraction of the required time.

For an example, assume that the information bandwidth is 10 kHz and that the signal is chirped across 10 MHz at a rate of one linear sweep per millisecond. The swept signal remains within any 10-kHz segment of its 10-MHz sweep range for only 1 microsecond—only 1% of the duration required to adequately receive the signal.

2.4.4 Direct Sequence Spread Spectrum Signals

A third type of spread spectrum signals is direct sequence, commonly called direct sequence spread spectrum (DSSS). This type of signal most closely meets the definition of a spread spectrum signal since it is literally spread in frequency rather than being rapidly tuned across a wide frequency range. DSSS has many military and civil applications, since it can protect against both intended and unintended interference and can also provide multiple use of a frequency band.

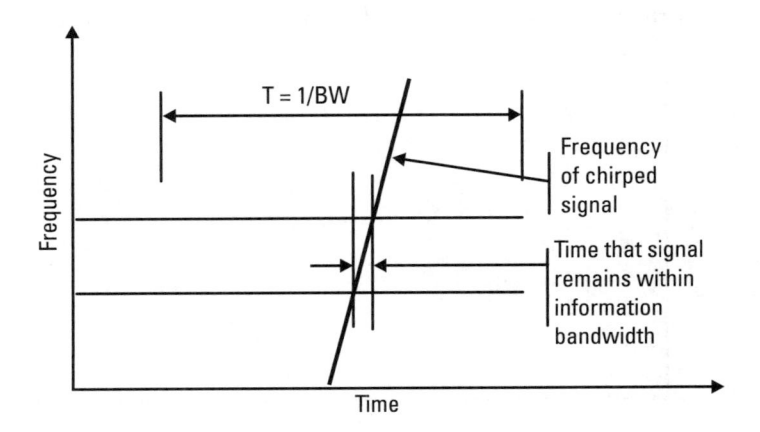

Figure 2.42 Detection of chirp signal.

Frequency Versus Time Characteristic

As shown in Figure 2.43, a direct sequence spread spectrum signal continuously occupies a wide frequency range. Since the DSSS signal power is distributed over this extended range, the amount of power transmitted within the information bandwidth of the signal (i.e., its bandwidth before it was spread) is reduced by the spreading factor. In Chapter 4, we will cover a formula for the amount of noise power (kTB) in any given receiver bandwidth. In a typical application, the amount of signal power from a DSSS spread spectrum signal will be much less than this amount of noise power. Figure 2.42 is a little misleading, since it implies that the power of the DSSS signal is evenly spread across frequency. Actually, the signal power is spread with the waveform shown in Figure 2.44 because the spreading is caused by adding a high-rate digital modulation to the signal.

Direct Sequence Spread Spectrum Transmitter

Figure 2.45 shows a very general block diagram of a direct sequence spread spectrum transmitter. First, it generates a signal which carries the information in its modulation. This signal has adequate bandwidth to carry the transmitted information, thus we say it is an "information bandwidth" signal. Then, the modulated signal is modulated a second time with a high bit-rate digital signal. One of several phase modulation schemes is used for this second modulation step. The digital modulating signal has a bit rate (called the "chip rate") which is one or more orders of magnitude higher than the maximum information signal frequency, and it has a pseudo-random bit pattern. The pseudo-random nature of the modulation causes the frequency spectrum of the output signal to spread evenly over a wide frequency range.

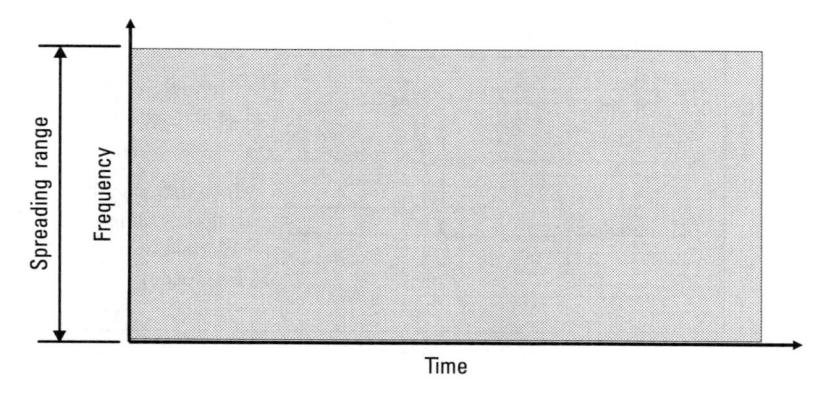

Figure 2.43 Direct sequence spread spectrum signal.

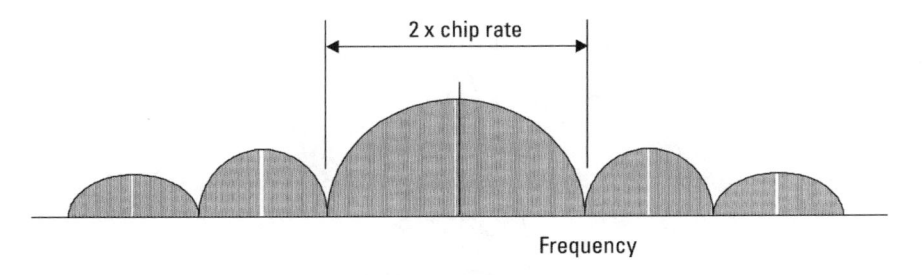

Figure 2.44 Spectrum of direct sequence signal.

The power distribution characteristic varies with the type of modulation used, but the effective bandwidth is of the order of magnitude of one divided by the chip rate.

The effect of adding a second, much higher rate digital modulation to a signal is shown in Figure 2.46. Note that the area of the top and bottom waveforms are the same. By increasing the bit rate in the bottom figure, the total power is spread over a greater frequency range and thus reduced in amplitude at any one frequency. This figure shows only a 5 to 1 increase in bit rate. Consider that DSSS systems typically increase the bit rate by a factor of 100 or 1,000, and you will understand how much the amplitude of the spread signal is reduced. If the frequency spectrum of the spread signal is significantly less than the noise spectrum, the signal is said to be "below the noise," and is thus extremely difficult to receive.

Direct Sequence Spread Spectrum Signal Receiver

A receiver designed for direct sequence spread spectrum signal reception has a despreading demodulator which applies the same pseudo-random signal that was applied by the transmitter. Since the signal is pseudo-random, it has

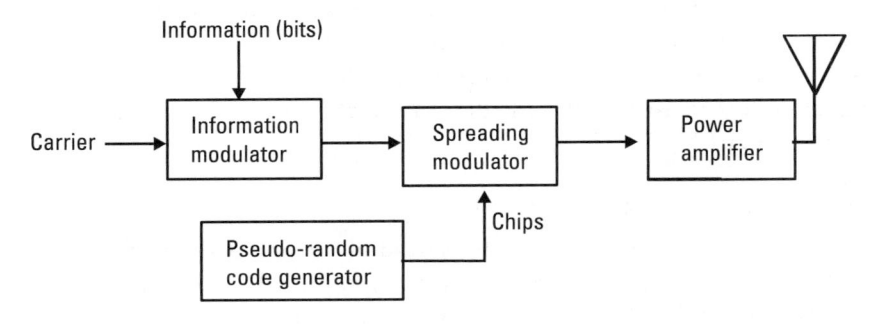

Figure 2.45 DSSS transmitter.

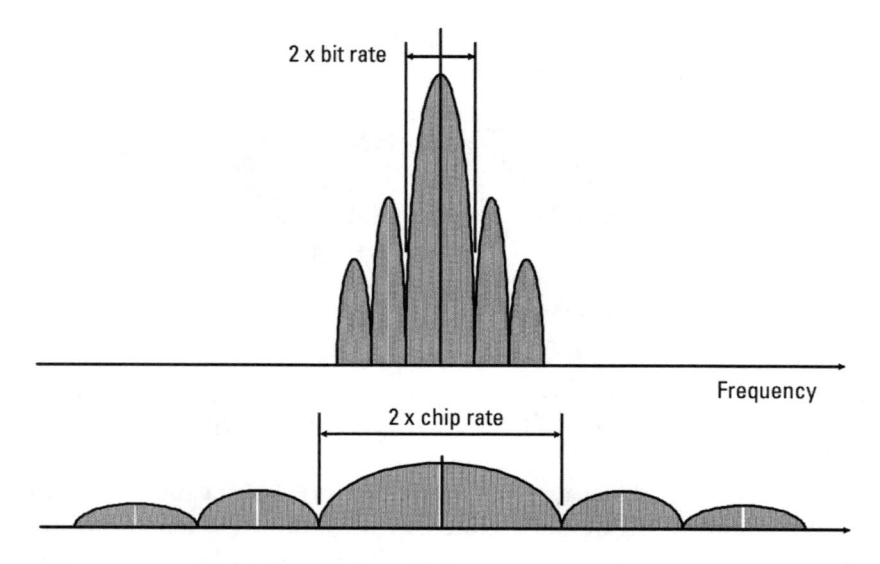

Figure 2.46 Frequency spreading versus bit rate.

the statistical characteristics of a random signal, but it can be recreated. A synchronization process allows the receiver code to be brought in phase with the code on the received signal. When this occurs, the received signal is collapsed back down to the information bandwidth, recreating the signal that was input to the spreading modulator in the transmitter.

Low Probability of Intercept

The low probability of intercept of the DSSS signal comes from the fact that any noncompatible receiver wide enough to receive the signal will have so much noise in that bandwidth that the signal-to-noise ratio of the intercepted signal will be extremely low.

Since, in military applications, the spreading code is closely guarded—just as the pseudo-random codes used in encryption are controlled—an enemy trying to intercept the DSSS signal cannot collapse the signal and must thus deal with the very low power density of the spread transmission.

Despreading Nonspread Signals

A very useful characteristic of the spreading demodulator is that signals which do not contain the correct code are spread by the same factor that the properly coded signal is despread as shown in Figure 2.30. This means that a narrow band signal (i.e., from a regular transmitter) that is received by the

DS receiver will be spread in frequency, and will thus have significantly reduced impact on the desired (despread) signal. Since most of the interfering signals encountered in almost any application have narrow bandwidth, a DSSS link can provide excellent communication in a cluttered environment. This gives the technique significant commercial as well as military applications.

Another reason for the use of DS spread spectrum is to allow multiple use of the same signal spectrum through code division multiple access or CDMA. There are sets of codes which are designed to be mutually "orthogonal." That is, the cross correlation of any two of the sets is very low. This orthogonality is expressed as a dB ratio—the output of the discriminator is reduced by so many dB if the correct code of the set is not selected.

A Great Unclassified Example

The GPS global navigation system transmits its signals with direct sequence spectrum spreading. In the first level of spreading, each of the satellites broadcasts one of a set of known codes that are about 40-dB orthogonal. This means that the tiny GPS receiver/processors (some of them very inexpensive as such things go) can use simple fixed-tuned receivers. The receiver applies various codes in the set, and the correct code causes the signal to increase by 40 dB, identifying the received satellite. (Actually, that is a shameless simplification: the processors keep track of which satellites are in view and operate in a much more sophisticated manner.)

For authorized military applications, another level of spreading is applied. This does not use public codes, and thus restricts the use to authorized receivers as well as providing another level of jamming protection.

2.4.5 Combined Technique LPI Signals

More than one type of spectrum spreading can be applied to an LPI signal. This is done to provided an additional level of transmission security. It can also be practical to simplify the synchronization process if two shorter synchronization schemes are used for the two spreading processes. As we will see in Chapter 9, jamming communication protected by multiple spreading techniques can be very challenging. The LPI technique combinations are:

- Direct sequence spread spectrum and frequency hopping;
- Chirp and direct sequence spread spectrum;
- Frequency hopping and chirp.

There are also "time hopping" techniques, although these are not typically for transmission security. They provide multiple time slots so that multiple users can independently communicate over the same communication network.

The most common combination LPI approach is direct sequence with frequency hopping. Figure 2.47 shows the frequency spectrum of this type of signal. Note that each of the round "lumps" in the waveform is like the digital signal spectrum shown in Figure 2.44. The peak of each "lump" is one of the randomly selected hopping frequencies. This type of signal is generated by a transmitter with the general block diagram shown in Figure 2.48. The spacing of the frequency hopping channels will typically be less than the null-to-null bandwidth shown for digital signal in Figure 2.19. That is, the spacing of the hop channels will be less than the inverse of the chip rate used to direct sequence spread the signal.

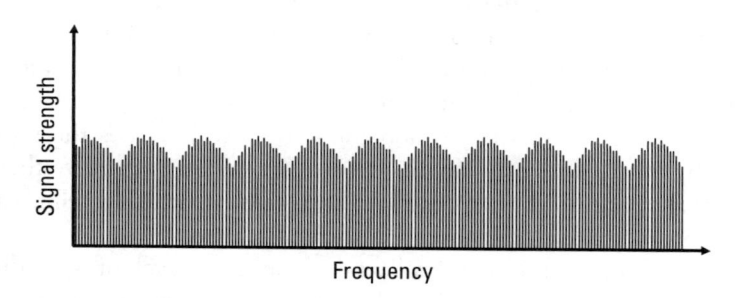

Figure 2.47 Spectrum of hopped direct sequence signal.

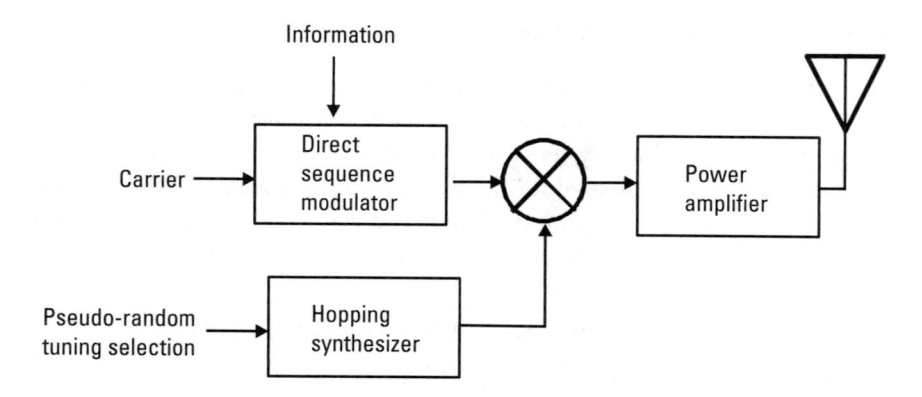

Figure 2.48 Hopped direct sequence transmitter.

2.4.6 Cell Phone Signals

Cell phones have become a very important threat signals because of their wide use by enemy combatants in asymmetrical warfare. Figure 2.49 shows a cell phone system. Cell phones are connected by radio links to cell towers, each of which has a base station (BS). The base stations are controlled by a mobile switching center (MSC). The MSC and the base stations are interconnected by wideband lines or microwave links. The MSC is also connected to the public switched telephone network so that wired phones and cell phones can communicate. The cells in the system are areas in which cell phones can interconnect with cell towers. The cells overlap and a cell phone can move from cell to cell in the system. Cell phones operate in several bands, including 450 MHz, 800 MHz, 900 MHz, 1,800 MHz, and 1,900 MHz. These are the approximate frequencies used. Specific frequency allocations are assigned by appropriate authorities. Each system can have a large number of RF channels, all time shared by many users under control of the MSC. All cell phone systems provide full duplex operation (i.e., instantaneous two-way conversation) so there is an uplink and a downlink (at different frequencies) between the cell tower and the cell phone.

The MSC authorizes access to the cell system, controls the timing for access, and controls the movement of signals through the system. It also controls the power output of each connected cell phone to minimize its power

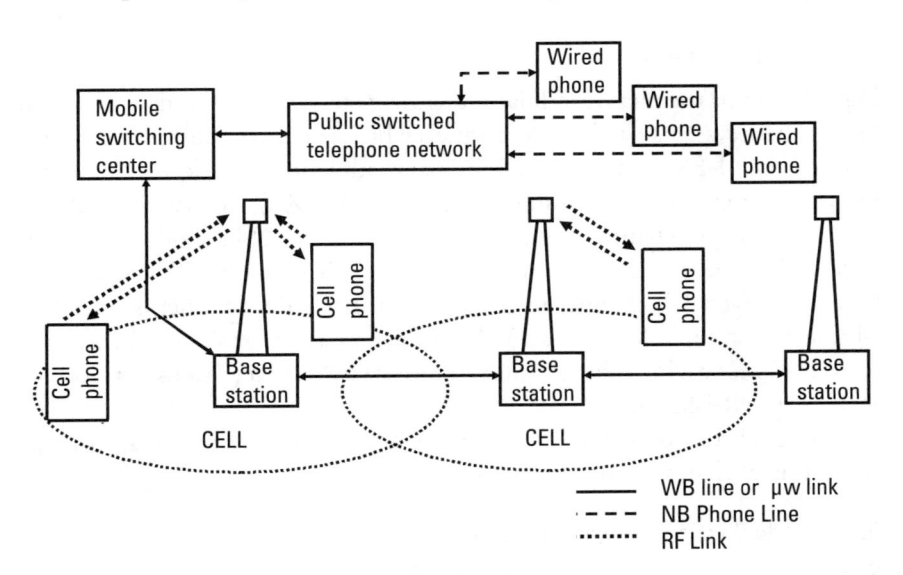

Figure 2.49 Cell phone system.

draw. Cell phone systems can carry conversations and data in either analog or digital form, but all use digital control channels through which the MSC controls the system and the connected cell phones.

The cell towers output 10 to 50 watts of effective radiated power, while the cell phones operate at significantly lower power. The amount of radiated power and the height of cell towers determines the size of cells. When cell towers are widely separated, for example in Nordic Mobile Telephone (NMT) systems in Scandinavia, cell phones can output as much as 15 watts. However, in most systems, the cells are smaller and the maximum cell phone power is typically 1 to 6 watts. Since this much power is not usually required to communicate with the nearest cell, the MSC commands the cell phone power down to the required level which can be as low as 6 milliwatts.

Analog Systems

Analog systems carry one conversation per RF channel using frequency modulation. The channels are wide enough (10 to 30 kHz) to carry the FM signals. These systems normally require a separate control channel over which the cell phones and the MSC communicate with digital messages. Multiple signals require separate RF channels, so the system is said to use frequency division multiple access (FDMA). In general, analog systems are being phased out in favor of digital systems, but there are still many analog systems in operation.

Digital TDMA Systems

Digital systems use phase modulation to carry conversations or data in digital form. These systems have wider channels which can carry multiple conversations on each RF channel using additional modulation techniques. Some systems use time division multiple access (TDMA) to carry multiple conversations per channel. The Global System for Mobile Communication (GSM) is a very widely employed approach which uses TDMA. GSM timeshares each RF channel in 8 time slots. Each RF channel is 200 kHz wide and each time slot carries 33 kbps of data. Each time slot carries a single conversation, or with voice encoders can carry two conversations per slot, for 16 users per RF channel.

Digital control data is carried in one or more of the time slots, so a separate control channel is not required.

Digital Code Division Multiple Access Systems

Some systems use a different technique for carrying multiple conversations on each frequency channel. These systems use code division multiple access

(CDMA), which is the same basic process used in direct sequence spread spectrum (DSSS) described in Section 2.4.4. A unique code is applied to each signal on a channel to spread the transmitted signal energy, and a matched decoder at the receiving end of the link removes the code, dispreading the signal. Unmatched codes are not decoded, so other signals are received at levels low enough to prevent interference with the match-coded signal. By applying a different code to each conversation, 64 conversations can be carried on each RF channel in a 1.23-MHz bandwidth.

Like TDMA systems, CDMA systems carry control data in one or more coded paths, so a separate control channel is not required.

Cell Phone Operation

When a cell phone is activated, it searches a set of control channels and chooses the strongest signal (presumably the closest cell tower). It listens for messages from the MSC for system identification and setup information. Then it enters the idle mode. Access time is deliberately randomized by the MSC to optimize the availability of the system to a large number of users. In idle mode, the cell phone waits for an incoming call or the user dials an outgoing call. In either case, the MSC assigns the cell phone an RF channel and (if it is digital) a time slot or code that is available.

Then the voice or data information is passed. The MSC can move the cell phone to different access channels to optimize system throughput, particularly if the user is moving from cell to cell.

An important aspect of cell phone communication is that the MSC contains information about which users are on the system and generally where each is located. At least the MSC knows which cell each mobile user is using. In many cellular systems, all newly activated cell phones are required to have GPS receivers. This means that the MSC will at some point have complete access to the location of each cell phone by reading its location directly from the GPS feature.

2.5 Error-Correction Codes

Back in Section 2.2.5, we discussed the structure of transmitted digital signals which include parity and error-correction bits along with the actual information bits. If parity bits are added to the data, they allow detection of bit errors in the received signals. With several parity bits, almost any received signal containing bit errors will be rejected. However, if an error-detection and correction (EDC) code is used, received errors can be corrected (up to a limit set by the power of the code). The more EDC bits that are added to the data, the higher percentage of bit errors can be corrected. The use of EDC is also called "forward error correction."

There are two classes of EDC codes: convolutional codes and block codes. Convolutional codes correct errors bit by bit, while block codes correct whole bytes (for example, 8 bits). Block codes don't care if one bit in a byte is bad or if all of them are bad—it corrects the whole byte. In general, if the errors are evenly spread, convolutional codes are better. However, if there is some mechanism that causes groups of errors, the block codes are more efficient.

An important application for block codes is for frequency hopping communication. If the signal is hopped to a frequency occupied by another signal, all of the bits transmitted during that hop will be erroneous.

The power of a convolutional code is stated as (n, k). This means that n bits must be sent to protect k information bits. The power of a block code is stated as (n/k) meaning that n code symbols (bytes) must be sent to protect k information symbols.

Example of an EDC Code

In this example, we will use a simple (7,4) Hamming code. For this code, there are four information bits and three error-correction bits per transmitted character. Figure 2.50 shows the encoding scheme. The message sent is four bits, "1010." If the first bit is a one, the encoder puts the first seven-bit word from the generator into a register. If the second bit were a one, it would enter the second line of the generator—however, since the second bit is actually a zero, all zeros are entered. The third bit being a one, the third line of the generator is put into the register. Since the last message bit is a zero, all zeros are added in the fourth position of the register.

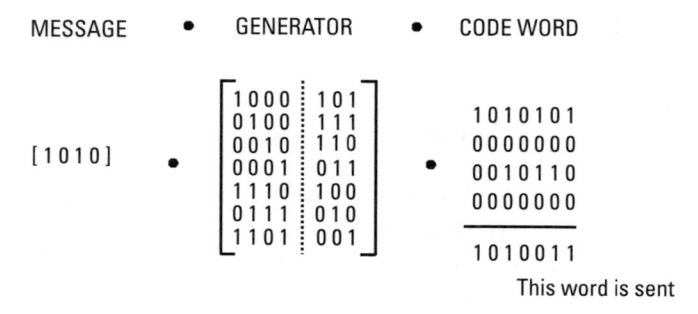

$$
\text{MESSAGE} \bullet \text{GENERATOR} \bullet \text{CODE WORD}
$$

$$
[1010] \bullet \begin{bmatrix} 1000 & 101 \\ 0100 & 111 \\ 0010 & 110 \\ 0001 & 011 \\ 1110 & 100 \\ 0111 & 010 \\ 1101 & 001 \end{bmatrix} \bullet \begin{array}{c} 1010101 \\ 0000000 \\ 0010110 \\ 0000000 \\ \hline 1010011 \end{array}
$$

This word is sent

Figure 2.50 (7,4) Hamming code generator.

Communications Signals

Now, the four lines in the output register are added to form the transmitted code word, "1010011." Seven bits are transmitted to carry the four-bit message. To make it interesting, we are adding an error to the received signal. The error signal is, "0010000." This causes the receiver to get the erroneous message, "1011011."

The receiver has the decoder as shown in Figure 2.51. Like the code generator, the decoder adds the first line of its decoder to an output register if the first bit is a one, and all zeros if it is a zero. All seven of the bits of the received character are processed and the seven binary numbers in the register are added to form the three-bit word 011. Looking back into the decoder, you will see that "011" is in the fourth position. Therefore, we need to add a 1 to the fourth bit of the received message to correct it.

It is left as an exercise for the reader to run the correct transmitted signal through the decoder. You will see that the seven digits in the register add to "000" indicating that there were no errors.

Example of a Block Code Application

Link 16 (JTDS), a widely used military link, uses a (31/15) Reed Solomon EDC code. It sends 31 bytes in each block including 15 information-carrying bytes. The code can correct up to 8 bad bytes of the 31. Naturally, the extra bytes require that the receiver bandwidth be more than doubled. Link 16 frequency hops (31 bytes sent per hop), and if a hop frequency is occupied or jammed all 31 bytes will be lost.

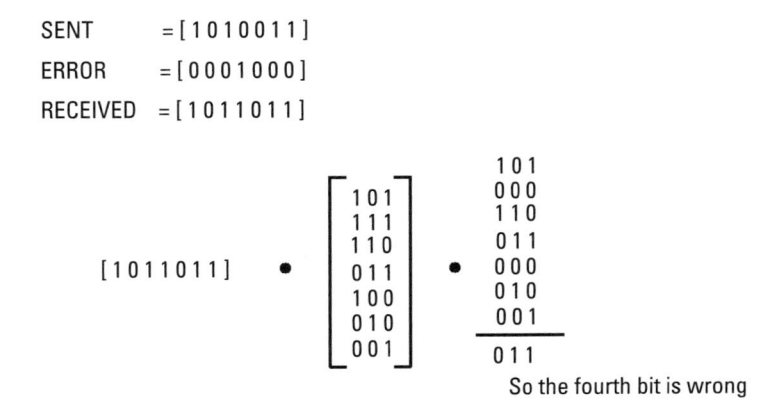

```
SENT      = [1 0 1 0 0 1 1]
ERROR     = [0 0 0 1 0 0 0]
RECEIVED  = [1 0 1 1 0 1 1]
```

$$[1011011] \bullet \begin{bmatrix} 101 \\ 111 \\ 110 \\ 011 \\ 100 \\ 010 \\ 001 \end{bmatrix} \bullet \begin{matrix} 101 \\ 000 \\ 110 \\ 011 \\ 000 \\ 010 \\ 001 \\ \hline 011 \end{matrix}$$

So the fourth bit is wrong

Error is corrected by adding 0 0 0 1 0 0 to received word

Figure 2.51 (7,4) Hamming code decoder.

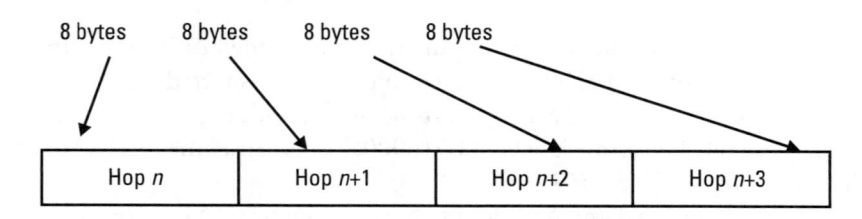

Figure 2.52 Simple interleaving scheme.

Since the code can correct only 8 bad bytes, the data is interleaved to transmit no more than 8 of 31 bytes in a single hop. Figure 2.52 shows a simplified interleaving scheme; actually, the placement of the bytes in a modern communication system will be pseudo-random.

3

Communication Antennas

An antenna is any device which converts electronic signals (i.e., signals in cables) to electromagnetic waves (i.e., signals out in the atmosphere or in space)—or vice versa. They come in a huge range of sizes and designs, depending on the frequency of the signals they handle and their operating parameters. Since this book is about communications electronic warfare, more attention will be given to the types of antennas commonly associated with communications. However, it is important to remember that an antenna does not care about the modulation on signals it handles—so some coverage of many types of antennas is included.

Functionally, any antenna can either transmit or receive signals, however antennas designed for high-power transmission must be capable of handling large amounts of power.

3.1 Antenna Parameters

Common antenna performance parameters are shown in Table 3.1. Each of these terms will be covered in context later in this chapter.

The antennas discussed in this chapter include both threat and EW antennas. This chapter covers the parameters and common applications for various types of antennas, a guide for matching the type of antenna to the job it must do, and some simple formulas for the trade-off of various antenna parameters.

Table 3.1
Commonly Used Antenna Performance Parameters

Term	Definition
Gain	The increase in signal strength (commonly stated in dB) as the signal is processed by the antenna. (Note that the gain can be either positive or negative, and that an "isotropic antenna" has unity gain, which is also stated as 0 dB gain.)
Frequency Coverage	The frequency range over which the antenna can transmit or receive signals and provide the appropriate parametric performance.
Bandwidth	The frequency range of the antenna in units of frequency. This is often stated in terms of the percentage bandwidth [100% × (maximum frequency − minimum frequency)/ average frequency].
Polarization	The orientation of the E waves transmitted or received. Mainly vertical, horizontal, or right- or left-hand circular—can also be slant linear (any angle) or elliptical.
Beamwidth	The angular coverage of the antenna, usually in degrees.
Efficiency	The ratio of the gain to the directivity.

3.1.1 Types of Antennas

There are many types of antennas used in electronic warfare applications. They vary in their angular coverage, in the amount of gain they provide, in their polarization, and in their physical size and shape characteristics. Selection of the best type of antenna is highly application dependent, often requiring difficult performance trade-offs with significant impact on other system design parameters.

Selecting an Antenna to Do the Job

To do what is required for any specific EW application, an antenna must provide the required angular coverage, polarization and frequency bandwidth. Table 3.2 provides an antenna selection guide in terms of those general performance parameters. In this table, the angular coverage is just divided between "360 degree azimuth" and "directional." Antennas with 360° azimuth coverage are often called "omni-directional," which is not true. An omni-directional antenna would provide consistent spherical coverage, whereas these types of antennas provide only limited elevation coverage (some more limited than others). Still, they are "omni" enough for most applications in which signals from "any direction" must be received at any instant—or for which it is desirable (or acceptable) that signals be

Table 3.2
Antenna Selection Guide

Angular Coverage	Polarization	Bandwidth	Antenna Type
360° Azimuth	Linear	Narrow	Whip, dipole, or loop
		Wide	Biconical or swastika
	Circular	Narrow	Normal mode helix
		Wide	Lindenblad or 4-arm conical spiral
Directional	Linear	Narrow	Yagi, array with dipole elements or dish with horn feed
		Wide	Log periodic, horn or dish with log periodic feed
	Circular	Narrow	Axial mode helix or horn with polarizer or dish with crossed dipole feed
		Wide	Cavity-backed spiral, conical spiral or dish with spiral feed

transmitted in all directions. Directional antennas provide limited coverage in both azimuth and elevation. Although they must be pointed toward the desired transmitter or receiver location, they typically provide more gain than the 360° types. Another advantage of directional antennas for EW applications is that they significantly reduce the level at which undesired signals are received—or alternately, the effective radiated power transmitted to hostile receivers.

Next, the table differentiates by polarization, and finally by bandwidth (only narrow or wide). In most EW applications, "wide" bandwidth means an octave or more (sometimes much more).

3.1.2 General Characteristics of Various Types of Antennas

Table 3.3 is a convenient summary of parameters of the various types of antennas used in electronic warfare applications. For each antenna type, the left-hand column shows a rough sketch of the antenna's physical characteristics. The center column shows a very general elevation and azimuth gain pattern for that antenna type. Only the general shape of these curves is useful—the actual gain pattern for a specific antenna of that type will be determined by its design The right-hand column is a summary of the typical specifications to be expected. Typical is an important word here, since the

Table 3.3
Characteristics of Types of Antennas

Antenna Type	Pattern	Typical Specifications
Dipole	El, Az	Polarization: Vertical Beamwidth: 80° x 360° Gain: 2 dB Bandwidth: 10% Frequency Range: zero through μw
Whip	El, Az	Polarization: Vertical Beamwidth: 45° x 360° Gain: 0 dB Bandwidth: 10% Frequency Range: HF through UHF
Loop	El, Az	Polarization: Horizontal Beamwidth: 80° x 360° Gain: -2 dB Bandwidth: 10% Frequency Range: HF through UHF
Normal Mode Helix	El, Az	Polarization: Horizontal Beamwidth: 45° x 360° Gain: 0 dB Bandwidth: 10% Frequency Range: HF through UHF
Axial Mode Helix	Az & El	Polarization: Circular Beamwidth: 50° x 50° Gain: 10 dB Bandwidth: 70% Frequency Range: UHF through lowμw
Biconical	El, Az	Polarization: Vertical Beamwidth: 20° x 360° Gain: 0 to 4 dB Bandwidth: 4 to 1 Frequency Range: UHF through mmw
Lindenblad	El, Az	Polarization: Circular Beamwidth: 80° x 360° Gain: -1 dB Bandwidth: 2 to 1 Frequency Range: UHF through μw
Swastika	El, Az	Polarization: Horizontal Beamwidth: 80° x 360° Gain: -1 dB Bandwidth: 2 to 1 Frequency Range: UHF through μw
Yagi	El, Az	Polarization: Horizontal Beamwidth: 90° x 50° Gain: 5 to 15 dB Bandwidth: 5% Frequency Range: VHF through UHF
Log Periodic	El, Az	Polarization: Vertical or Horizontal Beamwidth: 80° x 60° Gain: 6 to 8 dB Bandwidth: 10 to 1 Frequency Range: HF through μw
Cavity Backed Spiral	Az & El	Polarization: R & L Horizontal Beamwidth: 60° x 60° Gain: -15 dB (min freq) +3 dB (max freq) Bandwidth: 9 to 1 Frequency Range: μw
Conical Spiral	Az & El	Polarization: Circular Beamwidth: 60° x 60° Gain: 5 to 8 dB Bandwidth: 4 to 1 Frequency Range: UHF through μw
4 Arm Conical Spiral	El, Az	Polarization: Circular Beamwidth: 50° x 360° Gain: 0 dB Bandwidth: 4 to 1 Frequency Range: UHF through μw
Horn	El, Az	Polarization: Linear Beamwidth: 40° x 40° Gain: 5 to 10 dB Bandwidth: 4 to 1 Frequency Range: VHF through mmw
Horn with Polarizer	El, Az	Polarization: Circular Beamwidth: 40° x 40° Gain: 4 to 10 dB Bandwidth: 3 to 1 Frequency Range: μw
Parabolic Dish	Az & El	Polarization: Depends on Feed Beamwidth: 5° to 30° Gain: 10 to 55 dB Bandwidth: Depends on Feed Frequency Range: UHF to μw
Phased Array	El, Az	Polarization: Depends on Elements Beamwidth: 5° to 30° Gain: 10 to 40 dB Bandwidth: Depends on Elements Frequency Range: VHF to μw

possible range of parameters is much wider—for example, it is theoretically possible to use any antenna type in any frequency range. However, practical considerations of physical size, mounting, and appropriate applications cause a particular antenna type to be used in that typical frequency range.

3.2 Important Types of Communication Antennas

For tactical communication, the whip antenna is the most widely used type. It has the advantage of 360° azimuth coverage, so the transmitter does not need to know the direction to the receiver, or the receiver the direction to the transmitter in order to communicate. One interesting characteristic of a whip antenna is that its effective height is the height of the bottom of the whip. This is an important input to propagation equations to be discussed in Chapter 5. Monopole antennas on aircraft have the same favorable characteristics

The log periodic antenna is an excellent choice for communication jamming from the ground since it can be designed to cover any part of the tactical communication range. It also covers a wide frequency range and can be designed with a significant amount of gain.

Parabolic dish antennas are widely used for data links and for communication with satellites. These links typically operate at microwave frequencies.

Dipole antennas are very widely used in direction finding arrays at tactical communication frequencies. When operated over narrow frequency ranges, they provide reasonable gain. However, many direction-finding systems must operate over several octaves. As you will see in Chapter 7, a 5-to-1 frequency range is often covered by a single array. Thus, array dipoles are used with matching networks that significantly reduce their gain. Typical gain values can be as low as –20 dB at the low end of the frequency range.

3.3 The Antenna Beam

One of the most important (and misstated) areas in the whole electronic warfare field has to do with the various parameters defining an antenna beam. Several antenna beam definitions can be described from Figure 3.1, which is the amplitude pattern (in one plane) of an antenna. This can be either the horizontal pattern or the vertical pattern or the pattern in any other plane which includes the antenna. This type of pattern is made in an anechoic chamber designed not to reflect signals from its walls. The subject antenna is rotated in one plane while receiving signals from a fixed-test antenna, and the

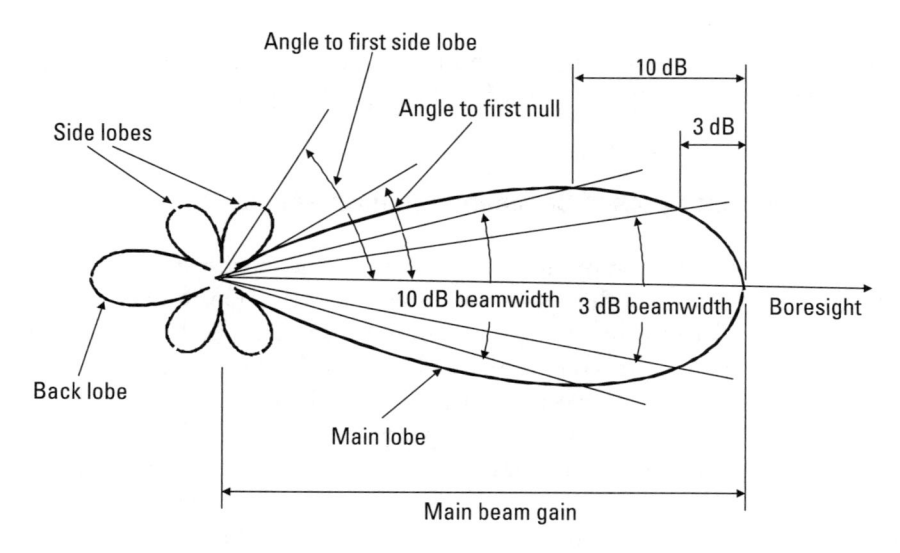

Figure 3.1 Antenna amplitude pattern.

received power is recorded as a function of the antenna's orientation relative to the test antenna.

- *Boresight:* The boresight is the direction the antenna is designed to point. This is usually the direction of maximum gain, and the other angular parameters are typically defined relative to the boresight.

- *Main lobe:* The primary or maximum-gain beam of the antenna. The shape of this beam is defined in terms of its gain versus angle from boresight. Remember that the main lobe has both vertical and horizontal shapes; they can be the same, or significantly different.

- *Beamwidth:* This is the width of the beam (usually in degrees). It is defined in terms of the angle from boresight at which the gain is reduced by some amount. If no other information is given, "beamwidth" usually refers to the 3-dB beamwidth. The horizontal beamwidth is often called the "azimuth" beamwidth, and the vertical beamwidth is often called the "elevation" beamwidth.

- *3-dB beamwidth:* The two-sided angle (in one plane) between the angles at which the antenna gain is reduced to half of the gain at the boresight (i.e., 3-dB gain reduction). Note that all beamwidths are two-sided values. For example, in an antenna with a 3-dB beamwidth of 10°, the gain is 3-dB down 5° from the boresight, so the two 3-dB points are 10° apart. If just "beamwidth" is stated, the implication is that it refers to the 3-dB beamwidth.

- *"n" dB beamwidth:* The beamwidth can be defined for any level of gain reduction. The 10-dB beamwidth is shown in the figure.

- *Side lobes:* Antennas have other than intended beams as shown in the figure. The back lobe is in the opposite direction from the main beam, and the side lobes are at other angles.

- *Angle to the first side lobe:* This is the angle from the boresight of the main beam to the maximum gain direction of the first side lobe. Note that this is a single-sided value. (It makes people crazy the first time they see a table in which the angle to the first side lobe is less than the main beam beamwidth—before they realize that beamwidth is two-sided and the angle to the side lobe is single-sided).

- *Angle to the first null:* This is the angle from the boresight to the minimum gain point between the main beam and the first side lobe. It is also a single-sided value.

- *Side-lobe gain:* This is usually given in terms of the gain relative to the main beam boresight gain (a large negative number of dB). Antennas are not designed for some specific side-lobe level—the side lobes are considered bad, and thus certified by the manufacturer to be below some specified level. However, from an EW or reconnaissance point of view, it is important to know the side-lobe level of the transmitting antennas for signals you want to intercept. EW receiving systems are often designed to receive "0-dB side lobes"—which is to say that the side lobes are down from the main lobe gain by the amount of that gain. For example, 0-dB side lobes from a 40-dB gain antenna would transmit with 40-dB less power than observed if the antenna boresight is pointed directly at your receiving antenna.

Antenna Efficiency

Antenna efficiency refers to the ratio of the gain of the antenna to its directivity. The maximum efficiency obtainable in a narrow frequency coverage parabolic dish antenna is 55%. Narrow frequency means less than 10% bandwidth. When an antenna is operated over a wider bandwidth, the efficiency will be less. For example, a typical EW antenna operating over the frequency range 2 to 18 GHz can be expected to have about 30% efficiency at the lowest frequency and 40% efficiency at the highest frequency for an average efficiency of 35%.

3.4 More About Antenna Gain

In order to just add the antenna gain to a received signal's strength, we need to state signal strength out in the "ether waves" in dBm—which is not really true. dBm is really a logarithmic representation of power in milliwatts—which only occurs in a circuit. The strength of a transmitted signal is more accurately stated in microvolts per meter (μv/m) of field strength, and the sensitivity of receivers with an integral antennas are often stated in μv/m. Convenient formulas for the conversion between dBm and μv/m are given in Section 5.2.

3.5 Polarization

From an EW point of view, the most important effect of polarization is that the power received in an antenna is reduced if it does not match the polarization of the received signal. In general (but not always) linearly polarized antennas have geometry which is linear in the polarization orientation (e.g., vertically polarized antennas tend to be vertical). Circularly polarized antennas tend to be round or crossed, they can be either right-hand or left-hand circular (LHC or RHC). The gain reduction from various polarization matches is shown in Figure 3.2. The polarization loss between right-hand and left-hand circular polarized antennas is shown as approximately 25 dB. This is an important consideration, because instrumentation antennas, like those used in anechoic chambers, can be expected to have close to 25-dB cross-polarization loss. The small cavity-backed spiral antennas used for aircraft radar warning receiver systems have closer to 10-dB isolation. Narrow bandwidth circularly polarized antennas for space-to-ground links are carefully designed for polarization isolation, and can have up to 33-dB cross-polarization loss.

An important EW polarization trick is to use a circularly polarized antenna to receive a linearly polarized signal of unknown orientation. You always lose 3 dB, but avoid the full loss that would occur if you were cross-polarized. When the received signal can have any polarization (i.e., any linear or either circular), it is common practice to make quick measurements with LHC and RHC antennas and chose the stronger signal.

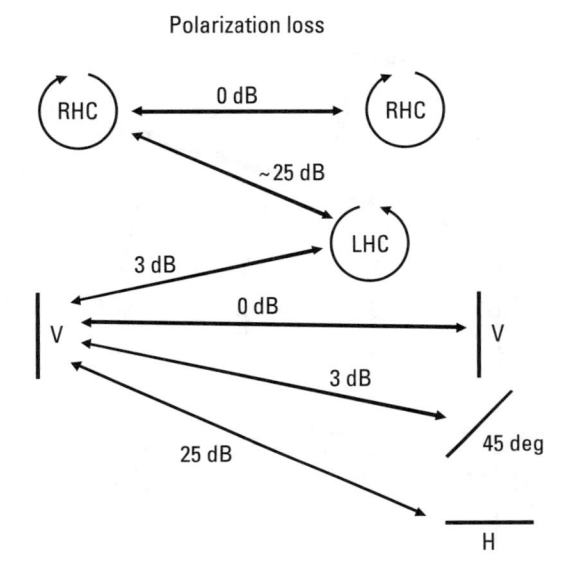

Figure 3.2 Cross-polarization losses.

3.6 Phased Arrays

For a number of extremely practical reasons, phased array antennas are becoming increasingly important in microwave communication links. They have the advantage of very quick electronic steering and/or the ability to null-out jamming signals. An additional advantage of a phased array is that it can be made conformal to the skin of a vehicle. This is particularly valuable in an airborne platform, since a parabolic antenna (the other primary way to create a narrow antenna beam) must be mounted in a radome which protrudes significantly from the skin of the aircraft in order to allow steering over a wide angle. Anything protruding from the skin of an aircraft causes significant aerodynamic problems.

A phased array is a group of small antennas which is used together to act like a large antenna. Figure 3.3 illustrates how a phased array works. This figure shows a linear array of antennas and a received signal arriving at the array. Since the signal is arriving from a distant transmitter, the phase of the signal arriving at the antenna location is very close to equal along a line perpendicular to the direction of arrival. This is shown as a line of constant phase, often called the wavefront.

A phase shifter follows each antenna. The phase shifts are set so that the signals from all of the antennas are in-phase when they reach the signal

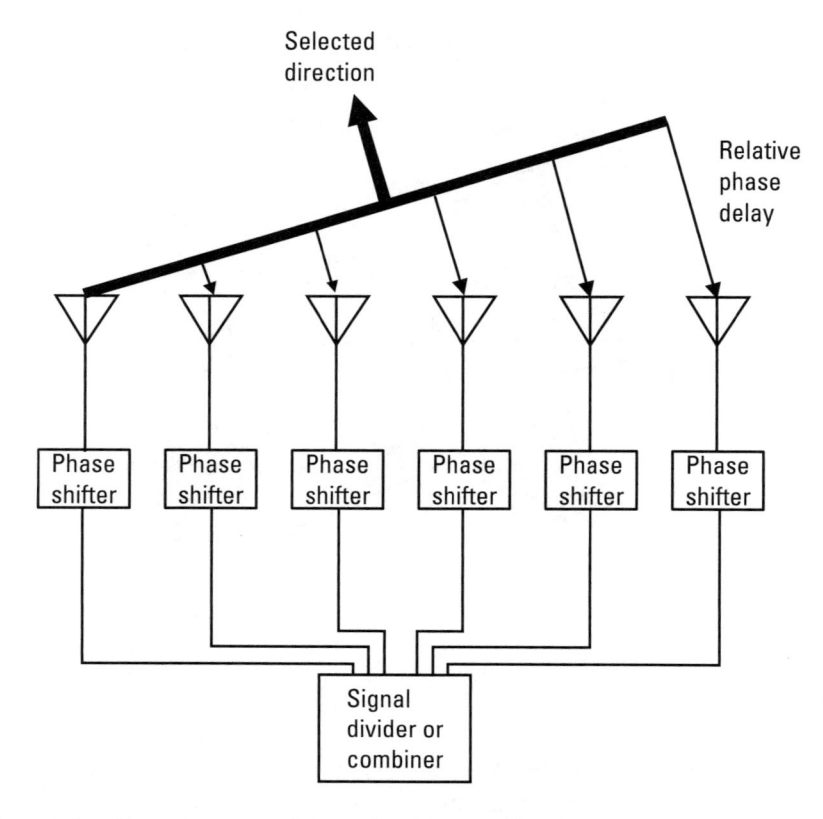

Figure 3.3 Phased array receiving a signal from a selected direction.

combiner—if the direction to the transmitter is as shown. Thus, signals arriving from the desired direction add constructively, while signals from other directions do not. This forms an antenna beam.

Since the phased array can act as either a transmitting or receiving antenna, the phase shifters are shown connected to a block called signal divider or combiner.

Phased arrays can be linear, as shown in the figure, or can be planar. A linear array would create a narrow beam in only one dimension, while a planar array narrows the beam in two dimensions. The array need not be flat, but can be curved around the skin of the vehicle on which it is mounted (i.e., conformal to the vehicle).

The phase shifters can be either fixed or settable. If the phases are fixed, the direction of the beam will be fixed relative to the orientation of the array. In this case, the beam is steered by moving the array. If the phases are set electronically, the beam can be moved to any desired direction by properly resetting the delays. This is called an electronically steered array.

In electronically steered arrays, if the antennas in the array are one-half wave-length apart as in Figure 3.4, the array can be steered over a full 180° without forming false lobes called "grating lobes." Most phased arrays have their antennas spaced more than a half wavelength, and avoid grating lobes by limiting the steering angle to less than ±90°.

3.6.1 Phased Array Beamwidth and Gain

This discussion is based on a planar array with the elements spaced at a half wavelength. The beamwidth of a phased array of dipole antennas is determined from the formula:

$$\text{Beamwidth} = 102/N$$

N is the number of elements in the array and the beamwidth is the 3-dB beamwidth in degrees. This is, of course only for one dimension of the array. For example, a 10-element linear horizontal array of dipoles would produce an antenna beam with a horizontal beamwidth of 10.2°.

If higher gain antennas are used in the array, the beamwidth is the element beamwidth divided by the number of antennas in the array.

These calculations apply when the antenna is pointed perpendicular to the array. This beamwidth increases by the reciprocal of the cosine of the angle from boresight as shown in Figure 3.5. Thus, if the phase shifters are adjusted to steer the beam 45° from the boresight, the antenna beamwidth would be 14.4°.

The gain of a phased array is given by the formula:

$$G = 10 \log_{10}(N) + G_e$$

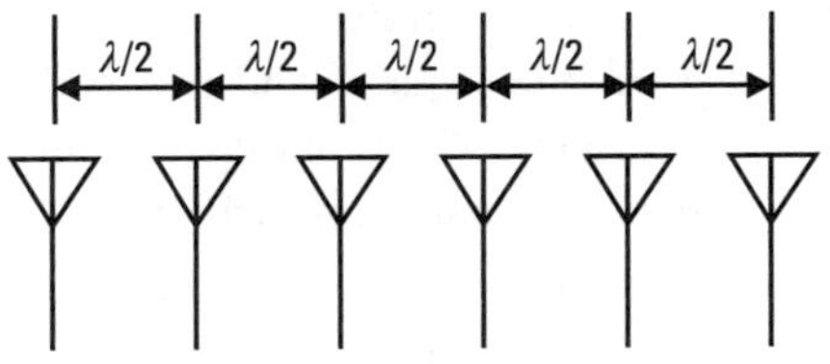

Figure 3.4 Phased array with half-wavelength element spacing.

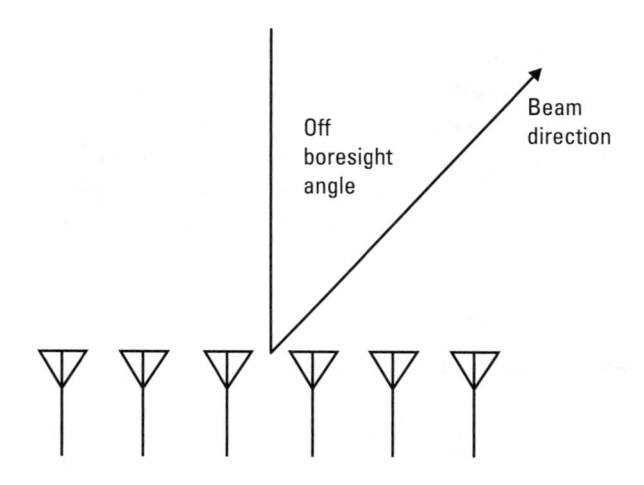

Figure 3.5 Beam directed away from perpendicular to array.

G is the maximum gain in dB, N is the number of elements in the array, and G_e is the gain of one element. Thus if there are 10 elements in the array and each element has 6-dB gain, the array gain would be 16 dB.

If the phase shifters are set to steer the array 45° away from the array boresight, the gain is reduced by at least the cosine of the angle from boresight.

3.7 Parabolic Dish Antennas

One of the most flexible types of antennas used in electronic warfare applications (and many others) is the parabolic dish. The definition of a parabolic curve is such that it reflects rays from a single point (the focus) to parallel lines. By placing a transmitting antenna (called a feed) at the focus of a parabolic dish, we can direct all of the signal power which hits the dish in the same direction (theoretically). A parabola is a curve which reflects all rays from its focus to the same direction, however a parabola is an infinite curve. A practical reflector is a parabolic section, and the feed antenna at the focus of the parabola transmits about 90% of its energy into the reflector. This causes the antenna to generate a main lobe which rolls off in angle, a back lobe, and side lobes. (See Figure 3.1.)

There is a relationship between the size of an antenna's reflector, the operating frequency, the efficiency, the effective antenna area, and the gain. This relationship is presented later in several useful forms.

Gain Versus Beamwidth

Figure 3.6 shows the gain versus the beamwidth of a parabolic antenna with 55% efficiency. This efficiency is what you expect in a commercially available antenna which covers a relatively small frequency bandwidth (about 10%). For the wide bandwidth antennas often used in EW and reconnaissance applications (an octave or more) the efficiency will be less than 55%. The beam is assumed to be symmetrical in azimuth and elevation. To use the table, draw a line from the antenna beamwidth up to the line and then left to the gain in dB.

Effective Antenna Area

Figure 3.7 is a nomograph of the relationship between the operating frequency, the antenna boresight gain, and the effective antenna area. The line on the figure is for an isotropic antenna (0-dB gain) with a one-square-meter effective area. You can see that this occurs at approximately 85 MHz. The equation for this nomograph is:

$$A = 38.6 + G - 20 \log(F)$$

where
 A is the area in dBsm (i.e., dB relative to one square meter)

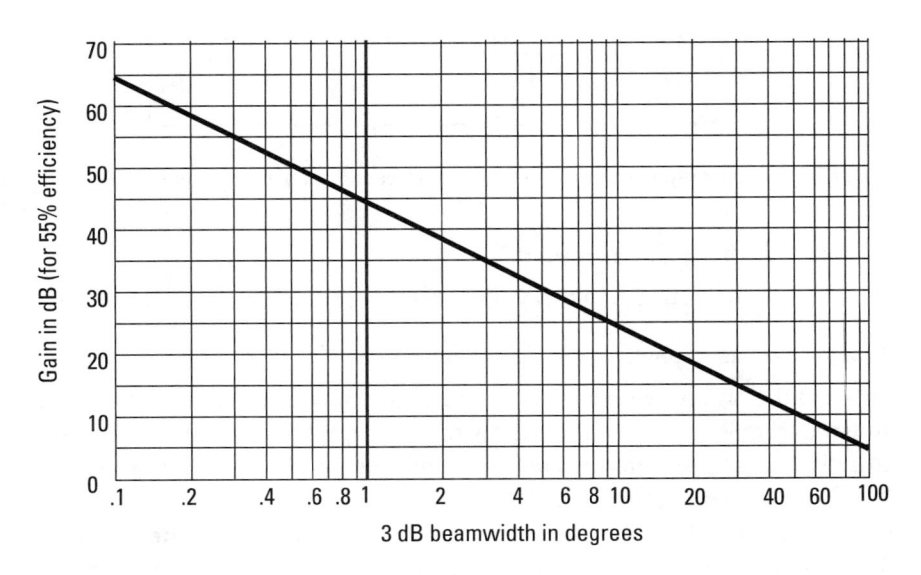

Figure 3.6 Gain versus 3-dB beamwidth of 55% efficient parabolic antenna.

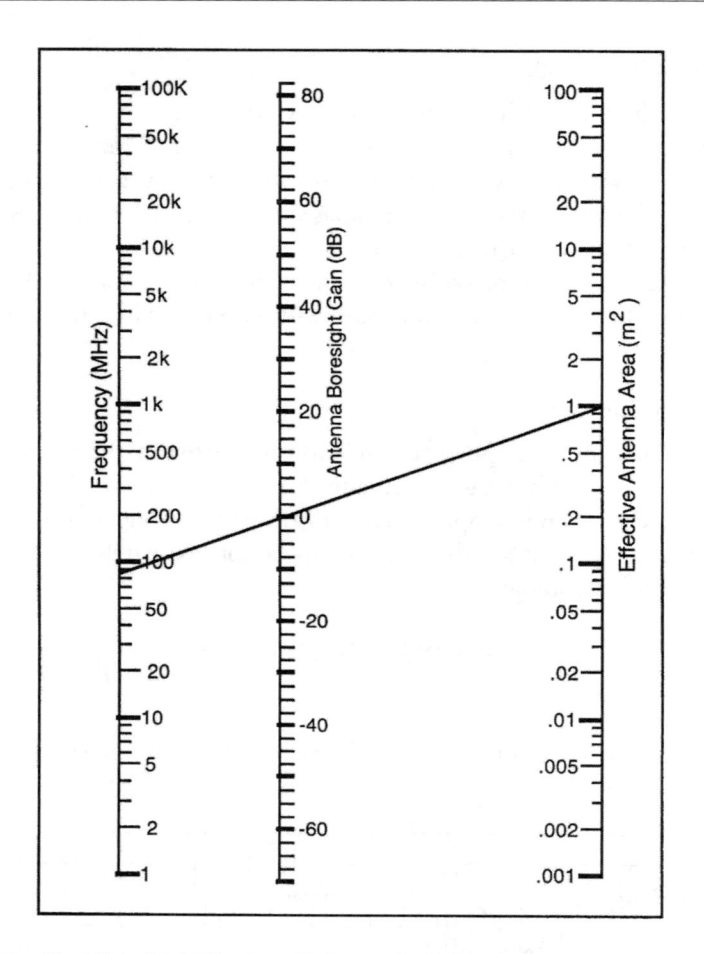

Figure 3.7 Nomograph of effective antenna area versus gain and frequency.

G is the boresight gain (in dB)

F is the operating frequency (in MHz)

Antenna Gain as a Function of Diameter and Frequency

Figure 3.8 is a nomograph which can be used to determine the gain of an antenna from its diameter and the operating frequency. Note that this is specifically for a 55% efficient antenna. The line on the figure shows that a 0.5-meter diameter antenna with 55% efficiency will have a gain of approximately 32 dB at 10 GHz. This nomograph assumes that the surface of the dish is a parabola accurate to a small part of a wavelength at the operating

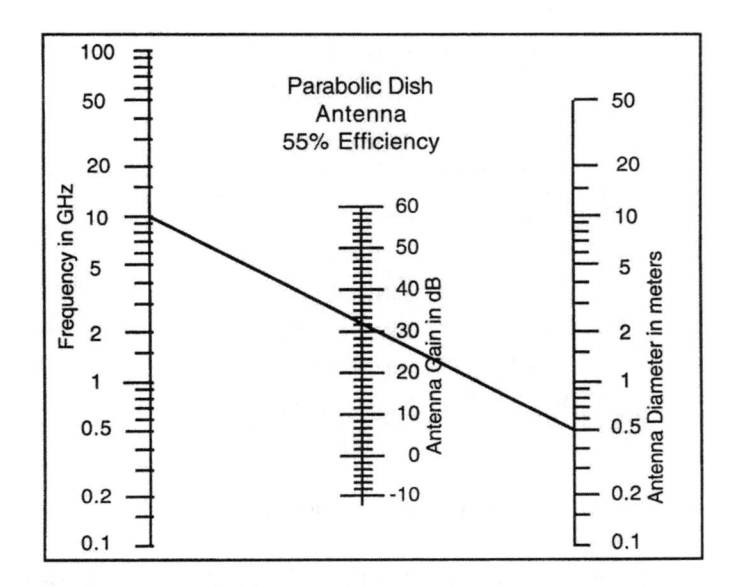

Figure 3.8 Nomograph of parabolic antenna gain versus antenna diameter and operating frequency at 55% efficiency.

frequency; otherwise, there is a gain reduction. The equation for this nomograph is:

$$G = -42.2 + 20\log(D) + 20\log(F)$$

where
  G is the antenna gain in dB
  D is the reflector diameter in meters
  F is the operating frequency in MHz

Table 3.4 shows adjustment of gain as function of antenna efficiency. Since both Figures 3.7 and 3.8 assume 55% efficiency, this table is very useful in adjusting the determined gain numbers for other values of efficiency.

Gain of Nonsymmetrical Antenna

The earlier discussion assumes that the antenna beam is symmetrical (i.e., azimuth and elevation of the beam are equal). The gain of a 55% efficient parabolic dish with a nonsymmetrical pattern can be determined from the equation:

Table 3.4
Gain Adjustment for Efficiency

Antenna Efficiency	Adjustment to Gain (Versus 55%)
60%	Add 0.4 dB
50%	Subtract 0.4 dB
45%	Subtract 0.9 dB
40%	Subtract 1.4 dB
35%	Subtract 2 dB
30%	Subtract 2.6 dB

$$\text{Gain (not in dB)} \cong [29{,}000/(\theta_1 \times \theta_2)]$$

where θ_1 and θ_2 are the 3-dB beamwidth angles (in degrees) in two-orthogonal directions (e.g., vertical and horizontal).

Naturally, this is converted to the gain in dB by taking 10 × the log of the value on the right side of the equation.

This equation can be derived by assuming that the gain is equivalent to the energy concentration within the 3-dB beamwidth. Thus the gain is equivalent to the ratio of the surface of a sphere to the surface area inside an ellipse on the sphere's surface with major and minor axes (dimensioned in spherocentric degrees) equal to the two angles describing the antenna beam coverage (remember the 55% efficiency factor). (Note: If you actually do that derivation, you will calculate the factor in the gain equation to 28,889, but 29,000 is commonly used.)

Antenna Scales on Slide Rule

Figure 3.9 shows side 1 of the antenna and propagation slide rule provided with this book—with the antenna scales highlighted. These scales are for parabolic antennas. They will allow you to calculate the gain and beamwidth of an antenna from its reflector diameter, its operating frequency, and its efficiency.

Figure 3.10 is a close-up of the antenna scales. The top window shows frequencies on the slide, and the diameter in feet on the body of the rule. By moving the slide, you can align the operating frequency to the antenna diameter. In Figure 3.10, 2 GHz is aligned with 10-feet diameter at point A. This is a good time to point out that right below the heavy line on side 1 of the

Communication Antennas

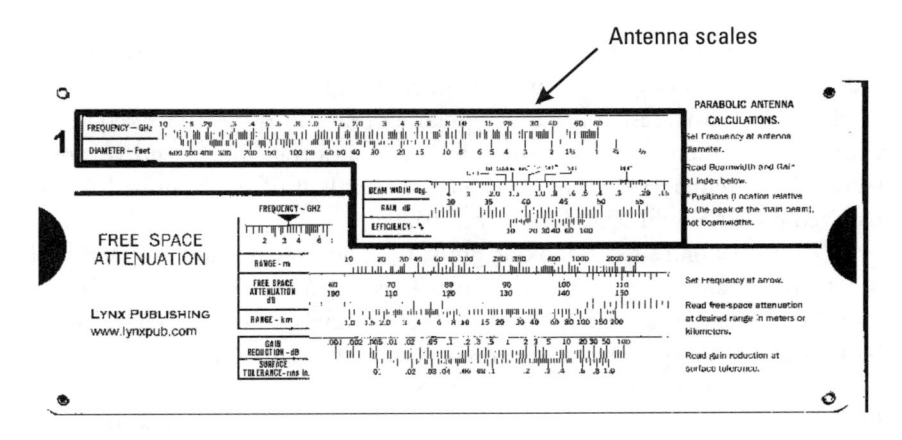

Figure 3.9 Slide rule with antenna scales highlighted.

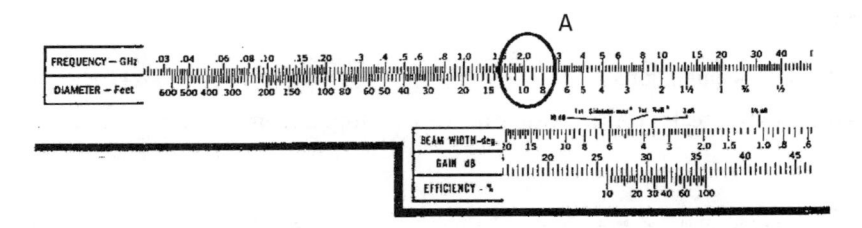

Figure 3.10 Operating frequency aligned with antenna diameter.

slide rule there is a frequency window. This window is not used for antenna and beamwidth calculations. If the antenna diameter is given in meters, you will need to convert to feet in order to use this slide rule: 1 foot = 0.305 meters; 1 meter = 3.28 feet.

Now look at point B on the slide rule (Figure 3.11). Note that there is a dark line by 55% efficiency. If the antenna efficiency is 55%, the gain at the

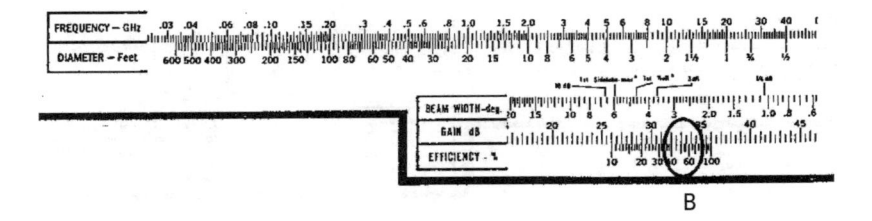

Figure 3.11 Read antenna gain at antenna efficiency.

antenna boresight is 33.4 dB. Now look at point C on Figure 3.12. The 3-dB line on the body of the rule is aligned with the 3-dB beam-width value on the slide. In this case it is 3.6°. Remember that this is the two-sided beamwidth. If the antenna had its boresight aimed at a transmitter and you redirected the antenna 1.8° away from the emitter (half of the beamwidth), the received power would be reduced by 3 dB.

Also on Figure 3.12, look at point D. The 10-dB line is aligned with the 10-dB beamwidth of the antenna. This is the angle between the two points on the antenna pattern at which the antenna gain is reduced by 10 dB. Note that the ¼ degree beamwidth is shown at the right end of this window.

On the same scale, you can find two more antenna gain parameters (see Figure 3.13). At point E, the first null line aligns with the angle from the boresight to the first null in the antenna pattern. This is one-sided (versus two-sided values for the beamwidths). The "1st side-lobe max" line aligns with the angle from the boresight to the peak of the first antenna side lobe at point F.

Slide Rule Assumptions

The antenna scales on the slide rule are developed from the equation:

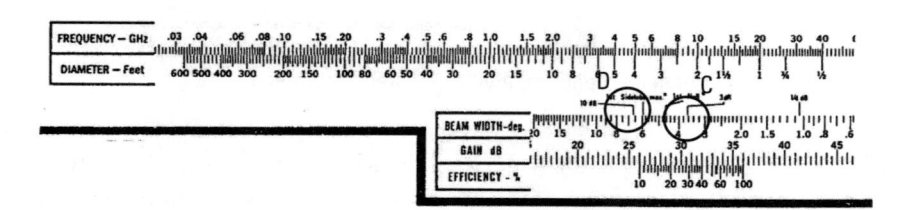

Figure 3.12 3-dB and 10-dB antenna beamwidth.

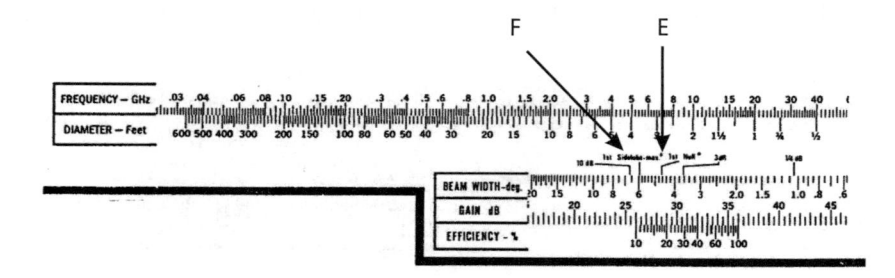

Figure 3.13 First null and first side-lobe peak of antenna gain pattern.

$$\text{Gain} = \text{Boresight gain} \times \left[\sin(\text{offset angle})/\text{offset angle}\right]$$

where "offset angle" refers to the angle from the boresight to the direction in which the antenna gain is being predicted or measured
This is often called the "sinx/x" function.
Other assumptions built into the slide rule are:

- The vertical and horizontal antenna patters are the same.

- 90% of the energy to or from the feed antenna is directed into or received from the reflector.

- The reflector has a perfect parabolic surface.

On side 1 of the slide rule, there is a scale that deals with the situation in which the reflector is not a perfect parabola. Figure 3.14 highlights the frequency and gain reduction windows. Figure 3.15 is a close-up of the appropriate scales. To use this feature, move the slide until the operating frequency is aligned with the arrow at point G. It is set to 2 GHz in the figure. Now the bottom window shows the gain reduction as a function of the surface tolerance. If the surface has an RMS error from a perfect of parabola of 0.1 inch

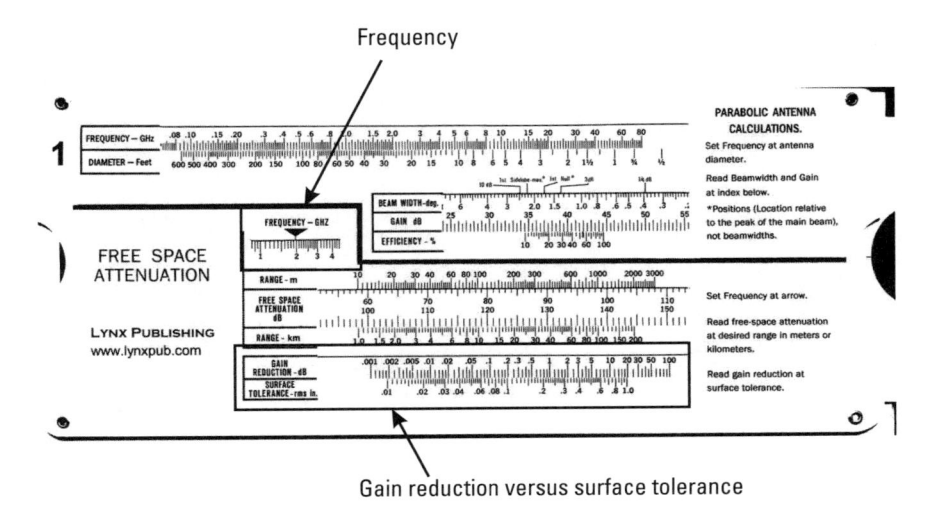

Figure 3.14 Slide rule with scales for gain reduction versus surface tolerance highlighted.

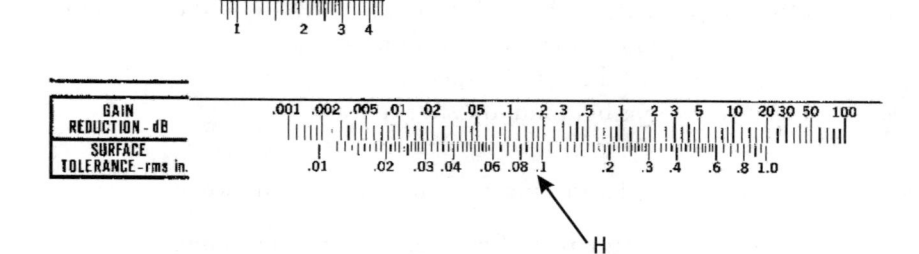

Figure 3.15 Gain reduction at 2 GHz and 0.1-inch RMS tolerance.

(point H), the boresight gain of the antenna is actually 0.2 dB less than shown at point B in Figure 3.11.

4
Communications Receivers

This chapter covers the types of receivers used in electronic warfare, with emphasis on those important to communications EW applications. There will be particular emphasis on digital receivers which are being incorporated into almost all new EW systems.

The chapter will also cover the calculation of sensitivity and dynamic range for receiver systems and common multiple receiver systems important to communication EW applications.

4.1 Types of Receivers

There are several types of receivers that are used in electronic support (ES) and reconnaissance systems. Each type of receiver has specific advantages and disadvantages, thus most ES systems have multiple receiver types which are computer controlled in response to the types of threat signals encountered. Table 4.1 shows the most common receiver types used in electronic warfare along with typical sensitivity, applications, and impact on system performance. Since some of these receiver types apply primarily to EW systems targeted against radar (rather than communication) threats, they will be covered at a more cursory level.

Table 4.1
Common Types of EW Receivers

Receiver Type	Typical Sensitivity	Typical EW/ Recon Application	System Performance Impact
Crystal Video	Low	RWR	Wide-frequency coverage, fast response time, one signal at a time.
Instantaneous Frequency Measurement	Low	RWR	Up to octave frequency coverage, one signal at a time, measures only frequency.
Superheterodyne	Medium to High	RWR, ELINT, COMINT, Comm ES, Targeting	Selects one of multiple signals, recovers any modulation.
Tuned Radio Frequency	High	Early method of dealing with multiple signal environment.	Tuning is complex process in true TRF receiver.
Fixed Tuned	Medium to High	CDMA signals, time reference, and so forth	Receives and demodulates one signal in complex environment.
Channelized	Medium to High	RWR, EW and recon systems for complex signals	Multiple simultaneous signals; Can recover modulation.
Bragg Cell	Medium to High	Seldom used	Measures only frequency, multiple simultaneous signals, very small dynamic range.
Compressive	Medium to High	ELINT, COMINT, supervisor position	Measures frequency of multiple simultaneous signals.
Digital	Medium to High	Every type of EW and recon application	Highly flexible, supports normal or complex signal processing, can be very fast.

4.1.1 Pulse Receivers

The receivers in this section have performance characteristics that limit their use, in most cases, to relatively low duty cycle pulsed signals. Therefore, there are few circumstances in which they are useful for EW communication systems.

Crystal Video Receiver

Crystal video receivers (CVR) were the first commercial broadcast receivers, and were widely used in early 1960s reconnaissance systems, but are most commonly found now in radar warning receivers (RWRs), where high probability of intercept is essential and low sensitivity can sometimes be tolerated.

This receiver type is principally suited to the recovery of fairly strong pulse signals.

Early commercial broadcasts used amplitude modulation, which could be detected with a piece of crystal with a "cat's whisker": a thin piece of spring wire touching it. This was, in effect, a point contact diode. The detected amplitude modulation was low-pass filtered to produce the audio signal—which could be heard through earphones. If those earphones were placed in a thin china bowl, the whole family could gather close around to listen to the broadcast.

In the 1950s, special low-noise diodes capable of operation at microwave frequencies became available. These, when connected between an antenna and a video amplifier could detect radar signals over a relatively wide frequency range. This made the modern CVR practical in time for the development of radar warning receivers that played such an important role in the Vietnam War.

As shown in Figure 4.1, a diode detector (made from crystal) detects amplitude modulation to produce an audio or video signal which is amplified and band limited to produce the output signal. Since reception of most hostile signals requires a high dynamic range, the diode detector output is passed to a logarithmic amplifier.

The low level of received RF signals cause the diode detector to operate in the square law region, which limits the sensitivity of the CVR. Typical sensitivity for a nonpreamplified CVR is about −40 dBm. This can be improved to about −65 dBm with a preamplifier. Sensitivity will be discussed in detail in Section 4.5.

The big advantage of the CVR is its high probability of intercept. One CVR can continuously cover an extremely wide frequency range—up to several GHz. Thus, it will recover the modulation of any AM signal that is present anywhere in the band. The problem is that it demodulates all of the signals present in the band. In environments where there are only low duty cycle pulses or a few high duty cycle signals, this causes relatively few

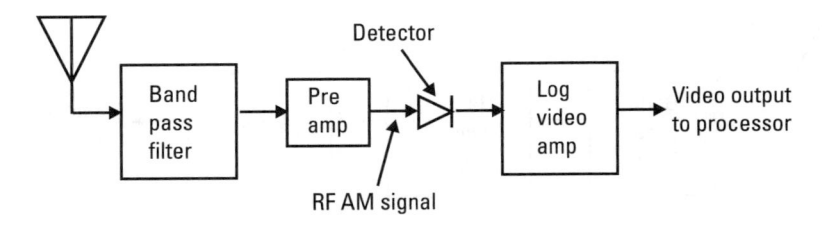

Figure 4.1 Crystal video receiver.

problems. When a few continuous or high duty cycle signals are present, they can be removed using band-stop filters. However, in the HF, VHF, and UHF ranges the high density of high duty cycle signals makes the CVR useful only in very special circumstances. Another problem with the CVR is that it recovers only the modulation and cannot determine the actual frequency of any received signals within the band it covers.

Instantaneous Frequency Measurement Receiver

The instantaneous frequency measurement receiver (IFM) measures the frequency of RF signals in real time and provides a digital output on parallel lines. The process is so fast that the frequency of an individual pulse down to 50-ns duration can be measured. The sensitivity of the IFM is approximately equal to that of the CVR. It is very common for an IFM to be included in systems with CVRs.

In the IFM circuit, RF signals are sampled before and after a delay line. The ratio of the two samples varies as a function of the frequency of the RF signal. The ratio also varies with signal strength. Thus, the digitized output is a function of both frequency and signal strength. In the late 1960s, when IFMs were first available, this variation limited the usefulness of the IFM for EW applications.

As shown in Figure 4.2, the IFM circuit is now used downstream of a hard-limiting RF amplifier. Since the input to the IFM circuit now has constant signal strength, the IFM output is only a function of frequency. Hard-limiting RF amplifiers became commercially available in the mid 1970s, so the IFM was practical, and came into wide use in EW receiving systems.

An IFM can cover up to an octave of bandwidth, and provides frequency resolution of about a tenth of a percent of bandwidth. Thus, an IFM covering a 4-GHz frequency range would provide frequency resolution to 4 MHz.

The IFM provides a useful output only when a single signal is present. When two or more signals are received simultaneously, the digital output

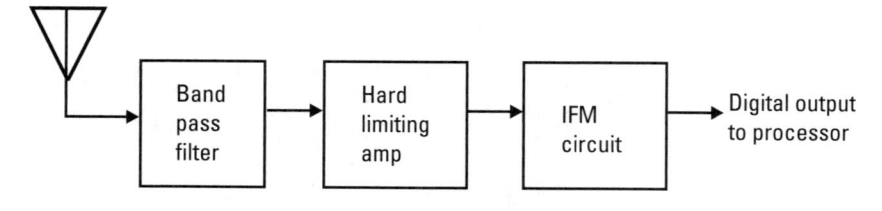

Figure 4.2 IFM receiver.

from the IFM is a jumble of seemingly random bits. If only low duty cycle pulsed signals are present, the speed of the IFM process allows frequency measurement for all pulses except for the brief periods when two pulses overlap. However, when a single strong CW signal is present at the IFM input, no frequency information can be provided for any pulsed or continuous signals present. Like the CVR, the IFM is often installed downstream a tunable band-stop filter which allows it to operate effectively in the presence of a very small number of continuous signals. The single-signal restriction severely limits the usefulness of the IFM in HF, VHF, and UHF bands which normally contain very large numbers of continuous signals. Thus, the IFM is useful in communication EW systems only in very special (low-signal density) situations.

4.1.2 Superheterodyne Receiver

The superheterodyne receiver (SHR) is the most commonly used type of receiver in communications and commercial broadcast receivers as well as EW and reconnaissance systems. It allows reception of one of many signals in a dense environment, can recover any type of signal modulation, and can provide very good sensitivity.

Heterodyning occurs when a local oscillator is input to a nonlinear device along with a received signal to convert the received signal to a different frequency. The nonlinear device is a mixer. Its output signals include the two input signals, signals at their sum and difference frequencies, and signals at all of their multiples and at the sum and difference frequencies of those multiples.

In this receiver (see Figure 4.3), a tunable preselector (bandpass filter) is set to a frequency band of interest. The local oscillator (LO) is set to a fixed

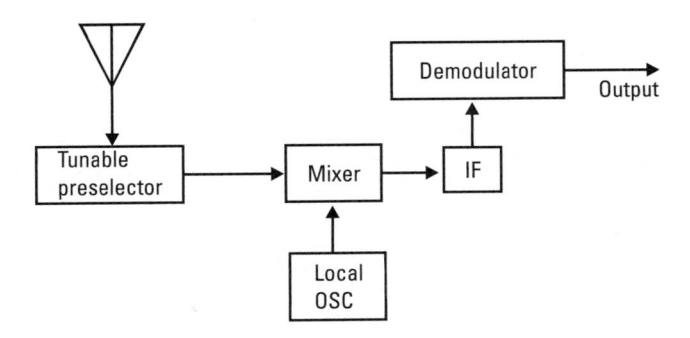

Figure 4.3 Superheterodyne receiver.

frequency offset above or below this band of interest, and the output of the mixer is passed to an intermediate frequency (IF) amplifier centered at a frequency equal to the offset of the local oscillator from the center of the preselector frequency band. For example, if the band of interest is centered at 100 MHz and the IF is centered at 21.4 MHz, the LO could be set to 121.4 MHz or 78.6 MHz. If the LO is at 121.4 MHz (above the band of interest), we say the receiver uses "high side" conversion, if the LO is below the signal of interest (i.e., 78.6 MHz in this case), the receiver uses "low side" conversion. High side conversion is more common.

The IF provides very good filtering to remove all (or most) of the output signals from the mixer other than the signal at the difference frequency between the preselector band and the local oscillator. All of the other mixing products are considered spurious signals or "spurs." Common IF frequencies are 455 kHz, 10.7 MHz, 21.4 MHz, 60 MHz, and 160 MHz. The receiver can be designed with any desired intermediate frequency, but commercial IF amplifiers with excellent stability and filtering are readily available at these frequencies. The IF amplifier provides sufficient gain to raise signal level from its received level to about 10 milliwatts, which is required for input to a demodulator.

In communication EW systems, it is often necessary to cover very wide frequency ranges with a single receiver. When the frequency range is greater than an octave (i.e., highest frequency twice the lowest frequency) it is difficult to filter out the significant spurious signals. This is solved by performing two or more frequency conversions and filtering off the spurious signals at each stage as shown in Figure 4.4. The first IF is at a relatively high frequency. This means that the difference frequency between the LO and the received signal is large—causing the spurious output frequencies to be more widely spread and thus easier to filter away. After filtering (in the first IF), the

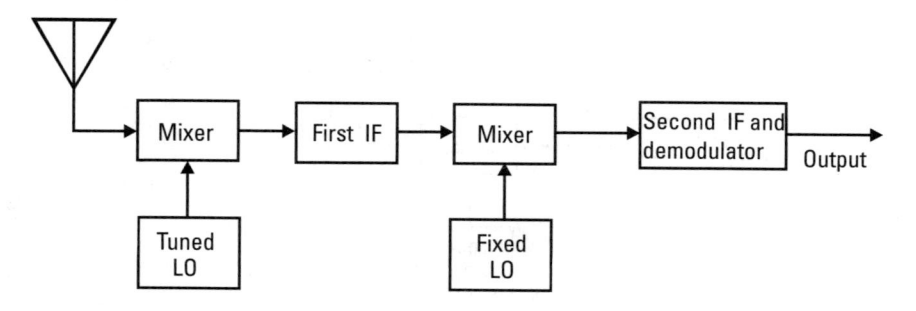

Figure 4.4 Double-conversion receiver.

converted signals are passed to a second conversion stage and the second IF amplifier/filter.

The output of the second IF amplifier goes to demodulator circuits to recover the modulation from received signals. It can also be passed to a digitizer, to provide input signals to a computer in which a digital receiver is implemented.

Frequency Converters

When the more complex types of receivers discussed later in this chapter are incorporated into receiver systems, it is often advantageous to cover a large frequency range incrementally. This is accomplished with a frequency converter which converts the large frequency range to a single frequency range as shown in Figure 4.5. Then, the complex receiver can operate over a reasonable frequency range for the technology involved (normally less than an octave and often at a lower center frequency). The frequency converter uses the heterodyning process, and often uses high-side and low-side conversion so that the same LO can handle multiple channels. The design of the converter must include careful consideration of spurious signal frequencies and how they are to be filtered off.

4.1.3 Tuned Radio Frequency Receiver

Some early receivers had multiple stages of amplification and filtering—each of which had to be tuned to change the receiver to different signal frequencies. These were tuned radio frequency (TRF) receivers. Because it is much easier to achieve good performance at a single frequency, the TRF was functionally replaced by superheterodyne receivers (as discussed earlier).

During the 1950s and early 1960s, there was a significant problem with the intercept of microwave signals in a multiple-signal environment. Transistor amplifiers and oscillators (for practical superhet receivers) were not yet commercially available at microwave frequencies, so large traveling

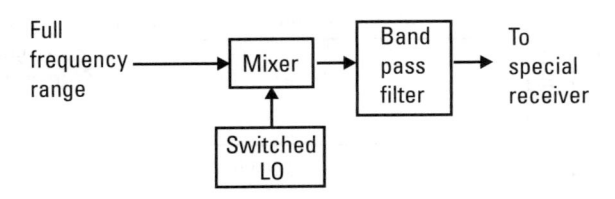

Figure 4.5 Frequency converter.

wave tube (TWT) amplifiers were required (about 13 cm in diameter and 30 cm long, and very heavy). Electronically tunable yttrium iron garnet (YIG) bandpass filters were also available, so they were used in reconnaissance systems to tune to one of multiple simultaneous signals. There were receivers configured with a TWT preamplifier, a YIG filter before and after the TWT amplifier, and a crystal video receiver to form what was called a "TRF receiver." They were predictably called "YIG, TWT, YIG" receivers (pronounced "yig, twit, yig").

4.1.4 Fixed-Tuned Receiver

If only one signal frequency is to be covered by a receiver, it can be fixed tuned. Examples are receivers to monitor time broadcasts or emergency channels. Other important examples include GPS receivers. All 24-GPS satellites broadcast on the same frequency—but with spreading codes that allow selection of one satellite's signal by a receiver. Thus GPS receivers are fixed tuned

4.1.5 Channelized Receiver

If multiple fixed-tuned receivers are fed by a bank of fixed filters tuned to adjacent frequencies, a channelized receiver is formed, as shown in Figure 4.6. The filters are typically designed so that the half power (3 dB) points of adjacent filters are at the same frequency.

This allows multiple simultaneous signals to be received with high quality. Channelized receivers can be designed for any frequency range, with the number of channels determined by the task to be performed and the size/weight/power available. Modern digital and RF circuit miniaturization techniques have made channelized receivers more and more practical.

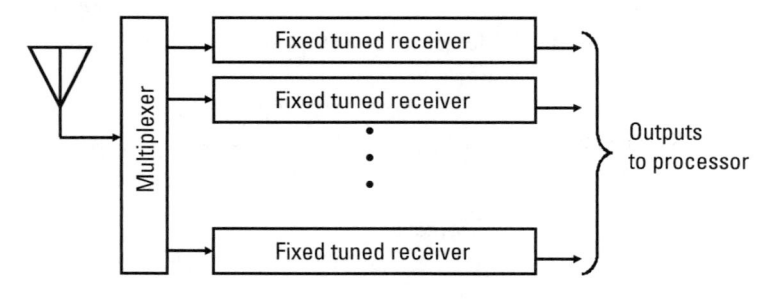

Figure 4.6 Channelized receiver.

Each of the channel receivers need not be a complete receiver—each can be only a receiver front end. Then, some of the outputs can be switched to a smaller number of channels (when energy from a signal is detected in the selected channel). Then, further down selection can be made based on the analysis of some quality (perhaps modulation) of the signals in the second set of channels. Finally, the selected signals can be passed to recorder channels or to a computer for signal analysis.

Figure 4.7 shows a typical channelized receiver configuration. There are a relatively large number of "front ends," which include local oscillators, mixers, and a stage or two of IF amplification and filtering. They cover the range of all possible anticipated signal frequencies. Each front end measures received signal level. When signal energy is detected in any front-end channel, the processor applies priorities to choose whether or not to assign an IF and discriminator channel to that front end. Since it is assumed that there are fewer signals present than front-end channels, the smaller number of IF and discriminator channels are applied to all front ends showing signal energy. If there are more signals than available output channels, the signals serviced can be chosen either by setting frequency priorities, or by passing some additional information about the signals to the processor. For example, there may be some clues as to the type of modulation present on the signals or the general location of the emitter.

There are, of course, many ways to set up a channelized receiver. There are often multiple down-selection phases, and the outputs can also be sorted into appropriate storage locations to collect similar signal types on to one recording channel.

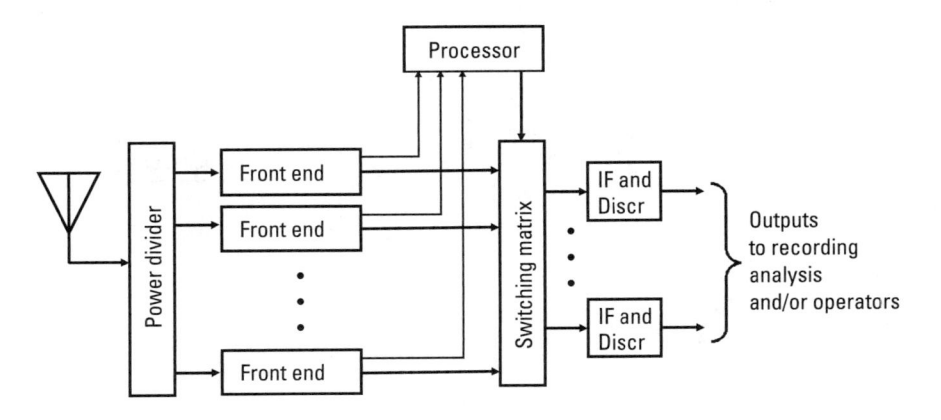

Figure 4.7 Typical channelized receiver application.

4.1.6 Bragg Cell Receiver

Electro-optical (or Bragg cell) receivers were considered the ultimate answer to rapidly handling multiple simultaneous signals in the 1960s. A Bragg cell receiver would determine the frequency of each signal present within its bandwidth (typically an octave). Then a narrow band receiver could be tuned to each signal of interest.

As shown in Figure 4.8, a laser is passed through a crystal (lithium niobate) Bragg cell. A received signal band is highly amplified and applied to the cell. Each signal present causes compression lines across the cell. These compression lines act as diffraction gratings—with spacing corresponding to the frequency of each signal present. Since the spacing of the grating determines the angle at which the light beam will be diffracted, the Bragg cell breaks the laser beam into separate beams for each signal. A detector array (or a set of arrays) allows the determination of the angles at which laser energy is present—hence the frequencies which have active signals. Narrow analysis receivers can then be very rapidly tuned to each signal to perform intercept and/or analysis functions.

This capability was so powerful that Bragg cell receivers were written into many important programs. However, there was a significant problem—limited dynamic range. EW and reconnaissance receivers must be able to detect weak signals in the presence of strong in-band signals, so 60 to 120 dB of dynamic range is typically required. With about 1 kilowatt of prime power, early Bragg cells could achieve about 20 dB of dynamic range. With less prime power, the dynamic range was significantly less. The dynamic range problem proved so intractable that the Bragg cell receiver is functionally constrained to applications in which a receiver need only deal with the

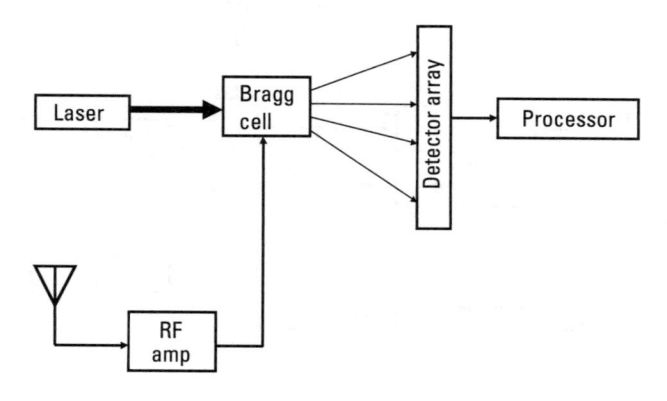

Figure 4.8 Bragg cell receiver.

strongest signal present. Mean while, other techniques have been developed to solve the multiple simultaneous signal problem. The usefulness of the Bragg cell receiver is limited to cases in which the system need only determine the frequency of the strongest signal present, and that signal is perhaps 10-dB stronger than any other in-band signals.

4.1.7 Compressive Receiver

The compressive (or micro-scan) receiver can determine the frequencies of multiple simultaneous signals with good sensitivity and dynamic range. Only the frequency is determined. The frequency of each signal can then be digitally supplied for tuning of set-on receivers and/or an operator display. In one important application, a supervisor manages a large number of intercept operators. A compressive receiver drives a display on which the supervisor can see which frequencies are active, and can assign operators to signals of interest.

In another application, a 1-MHz bandwidth receiver is set onto a signal anywhere in the 2- to 4-GHz band within two microseconds.

Figure 4.9 shows the block diagram of a compressive receiver. A wide intermediate frequency (IF) amplifier is swept across a signal band at a very high sweep rate using the heterodyne principle. The output of the wide IF is passed to a compressive filter which has a delay that varies as a function of frequency. The key to compressive operation is that the delay time versus frequency slope is the same as the frequency versus time slope of the local oscillator converting input signals to the IF frequency. See Figure 4.10. When considering this figure, remember that a positive tuning slope for the IF will cause the signal to enter the filter at maximum frequency—causing the

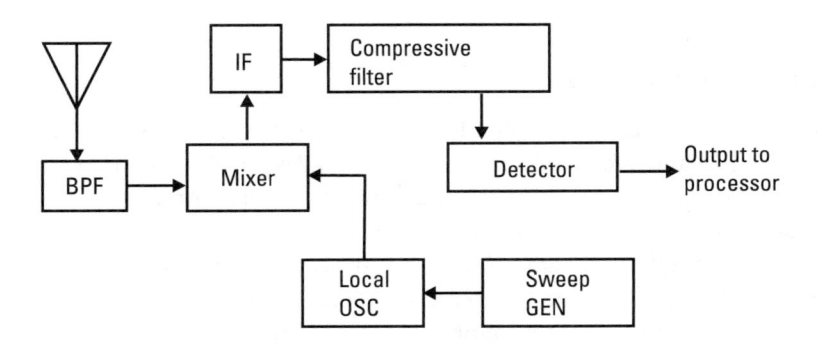

Figure 4.9 Compressive receiver.

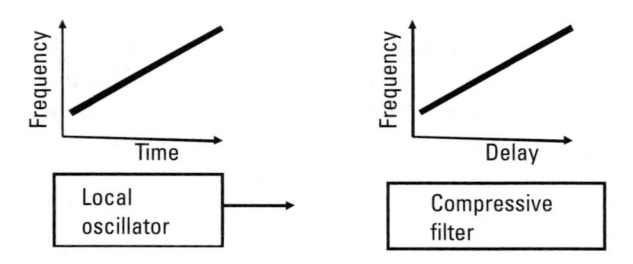

Figure 4.10 Delay versus frequency and tuning rate.

maximum delay. Then, as the IF tunes through the signal, it will be at a progressively lower frequency in the filter—causing reduced delay.

The output bandwidth of the compressive filter is much narrower than the IF amplifier, which allows the compressive receiver to provide reasonably good sensitivity. Consider the operation of the compressive filter as the IF is tuned through a signal: As the edge of the IF (and compressive filter) reaches the signal frequency, the signal is delayed to the time that the IF will leave the signal frequency. A moment later, the IF has moved in frequency so that the signal is no longer at the filter edge—the signal is delayed by a lesser amount—just enough delay to the time when the IF leaves the signal frequency. This process continues as the IF is tuned past the signal frequency. Thus, the signal remains in the output of the compressive filter for an extended time.

Ordinarily, a signal must remain in an IF for a time equal to the inverse of the IF bandwidth (i.e., 1-MHz bandwidth requires 1-microsecond dwell time) to allow the signal to be detected. The IF is tuned at a much faster rate than one bandwidth in time = 1/bandwidth for the compressive filter output. However, because of the compression, the signal dwells long enough to be detected in the output.

4.1.8 The Digital Receiver

The digital receiver is arguably the most common type of receiver employed in new EW and recon systems and in upgrades to older systems. This is intended only as an overview of a much more complex subject. There is significantly more detail available from textbooks on digital receivers; this and other relevant references are described in Appendix B. In this chapter, we will discuss the digital receiver concept, sampling, digitization, factors limiting performance, and a few important application issues.

A digital receiver digitizes analog signals and then performs all of the receiver functions in software. It provides significant flexibility in EW and reconnaissance applications and can perform some functions that are highly impractical to implement in hardware. A digital receiver cannot typically operate directly on received signals; some signal conditioning is required before digitization can be performed. As shown in Figure 4.11, a digital receiver has an analog front end, a digitizer, and a processor.

The analog front end can be significantly more complex than the simple IF translator shown in the figure. It typically covers a relatively wide frequency range in steps using a frequency converter as discussed earlier. The output of the frequency converter is a wide IF output in a frequency range and at a signal strength that can be input to the digitizer. An early digitizer required input from a so-called "zero IF" that had a very low bottom frequency, requiring very complex conversion schemes to mitigate spurious responses. However, most modern digitizers operate directly at a higher intermediate frequency. Received signals are normally –60 to –120 dBm, far too low for input to the digitizer. Thus the analog front end must supply significant signal gain, and must have a bandwidth compatible with the digitizing rate.

Digital Receiver Applications

There are several reasons that digital receivers are becoming so widely used in electronic warfare systems:

- The state of the art in digitizers, processors, and software is quite mature, so that digital receivers are reliable and cost effective.

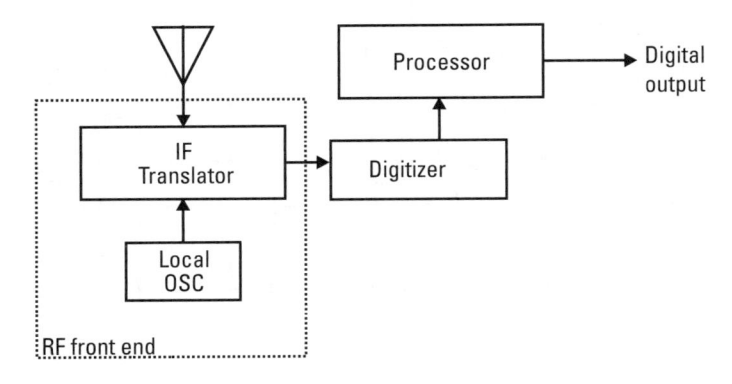

Figure 4.11 Digital receiver.

- The miniaturization in digital components and computers allows significant size and weight reduction as compared to analog receiver techniques in most cases.
- Digital receivers allow significant operational flexibility.
- There are significant processing advantages in digital receivers that allow previously impractical tasks such as chip detection and following a frequency hopper (both of which are described later).

4.2 Digitization

The digitizer is an analog-to-digital converter (ADC). As described in Section 4.2.2, it accepts a band of analog signals and produces a digital output which captures the amplitude of the combined analog signals in the input band. All of the signals in the analog input band form a single complex waveform. This combined waveform is digitized as described in Section 2.2.2 and as shown in Figure 4.12.

Digitizers can operate up to GHz sample rates, and can have up to 18 or more bits—but not in the same devices. Very fast digitizers have very few bits per sample. The state of the art in digitizer's bits versus sample rate is

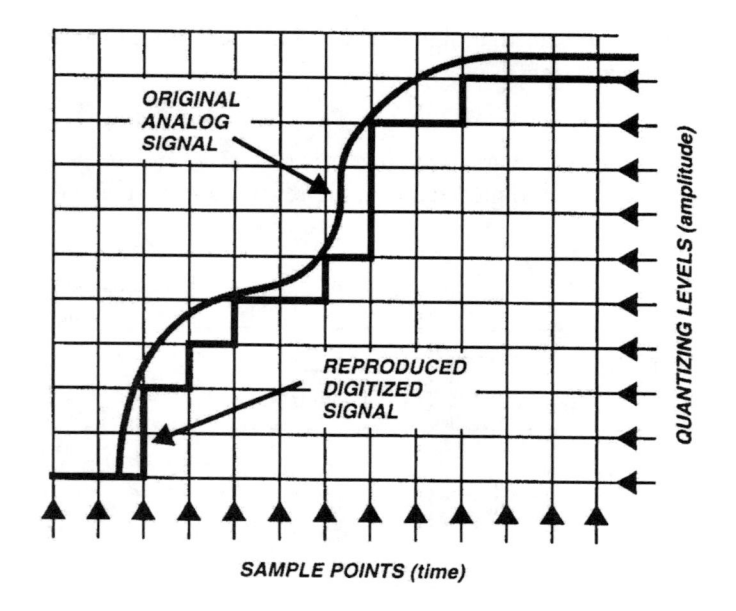

Figure 4.12 PCM digitization.

constantly improving in response to significant investment by both military and commercial users.

Processors include many types of computers. The computer accepts the digital characterization of the input analog band of signals. Then, using software, it performs filtering, demodulation, and analysis tasks to produce the necessary outputs (digital) to provide the information that the receiver is specified to recover from the signals in the covered band.

4.2.1 Sampling Rates

The required sampling rates are often stated in terms of the Nyquist rate. The Nyquist sampling rate is twice the highest frequency of the signal being digitized. Modern A/D converters can sample at twice the bandwidth of the input (IF) band being digitized as long as the IF frequency is not at too high. Theoretically, the signal can be reproduced from samples at the Nyquist rate—however, people who make digital RF memories (DRFM) (which reproduce modified versions of the original analog signals) say that the sampling rate needs to be about 2.5 times the bandwidth to allow adequate reproduction of the signal.

4.2.2 Digital Waveforms

The output of the digitization shown in Figure 4.12 is pulse code modulation (PCM), in which binary words represent the instantaneous value of the analog waveform at the sample times. These value levels can be equally spaced, or can be "companded" (higher levels are more widely spaced than lower levels) to reduce the number of bits required.

A second type of digitization is shown in Figure 4.13. This is delta modulation, in which the digital output represents the change or slope of the

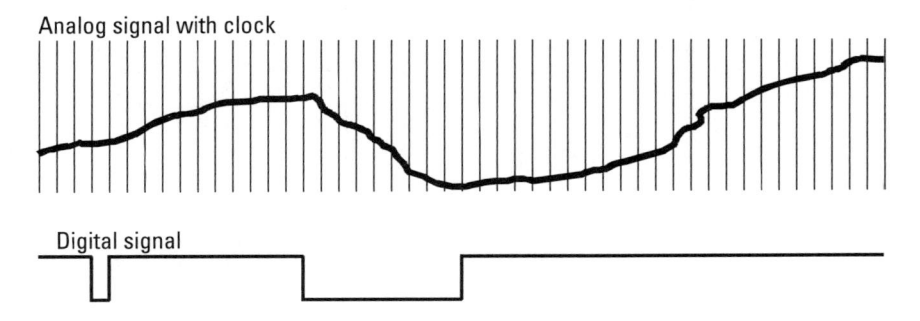

Figure 4.13 Delta modulation.

analog signal waveform rather than its absolute value. A "1" shows that the slope is rising and a "0" shows it is falling. The required bit rate is determined by the maximum significant rate of change of the signal.

4.2.3 Digitizing Techniques

There are several digitization techniques used. We will discuss only two here: successive approximation A/D and flash A/D. Variations of these two techniques are sometimes required for specific applications.

Successive approximation is used when relatively slow conversion times (10 to 300 μs) are acceptable. In this technique, the analog input waveform is sequentially compared to reference voltages—one for each bit level (related in the ratios 1, 2, 4, 8, and so forth). Thus an output digital code of n bits is generated in n steps with minimum hardware complexity.

A flash A/D provides very fast conversion times (10 to 50 ns). It uses parallel comparators at each analog value to be considered as shown in Figure 4.14. That is, an 8-bit digital code requires 255 comparators. While this technique allows extremely high sample rates, it requires significantly increased hardware complexity.

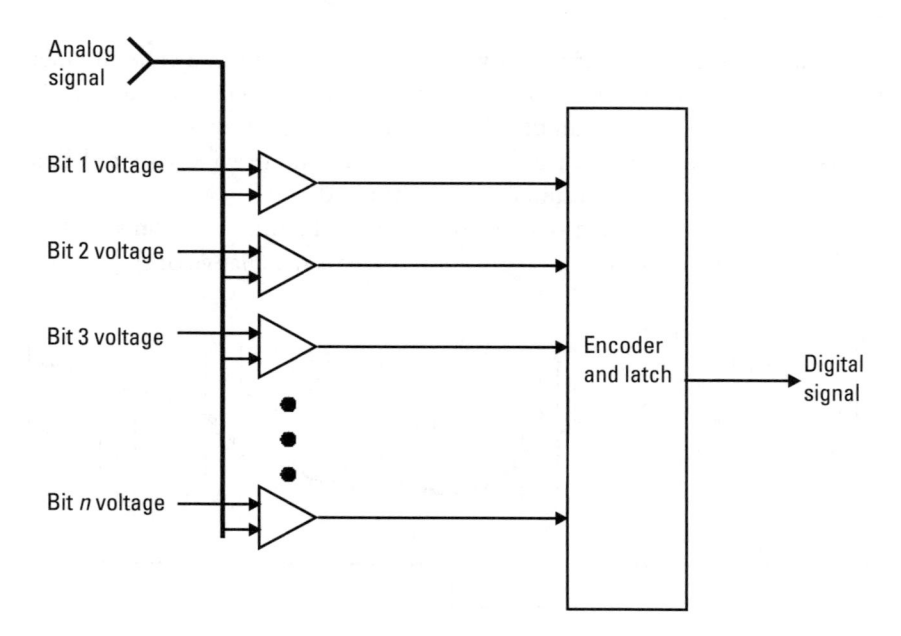

Figure 4.14 Flash encoder.

4.2.4 I & Q Digitization

In many applications it is necessary to preserve the phase of the digitized signal. In addition to some digital receivers, these applications include many DRFMs and several types of radars. Phase is preserved by performing "in-phase and quadrature" (I & Q) digitization. As shown in Figure 4.15, this involves digitizing the signal twice per cycle with the second measurement delayed approximately a quarter cycle (i.e., 90 degrees in phase) relative to the first.

4.3 Digitized Signal Quality Issues

Since the digital part of the receiver accepts its input from the analog RF front end, the sensitivity of the front end must be adequate to receive the desired signals. Sensitivity is discussed in Section 4.4, and you will note that the effective receiver bandwidth is one of the elements of system sensitivity. In a digital receiver, it is often the processing that determines the effective bandwidth. However, the system noise figure (another element of sensitivity) is mainly influenced by the receiver front-end design.

Processing in Digital Receivers

Once a signal is digitized, it is passed to a processor where all of the normal receiver functions are accomplished in software. This includes filtering to set the effective bandwidth, demodulation to recover the modulation from the

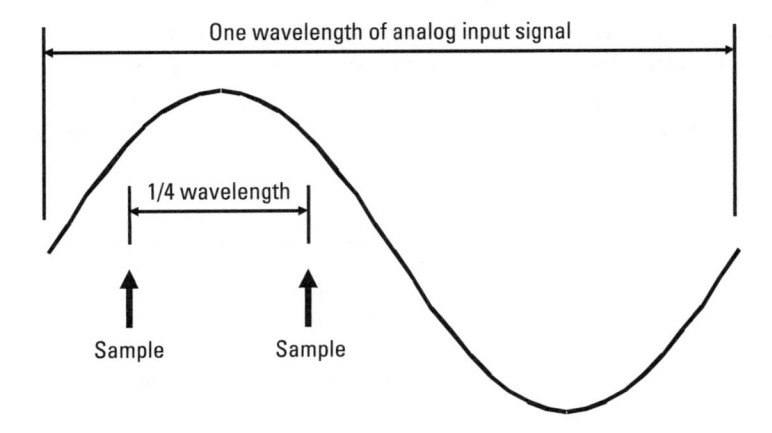

Figure 4.15 I & Q digitization.

received signal, and several other functions that are difficult to accomplish in analog receivers. The following discussion covers two processing tasks uniquely suited to digital receivers. Both of these examples relate to low probability of intercept signals—which are discussed in Chapter 2.

4.3.1 Chip Detection

Direct sequence spread spectrum signals are generated by (binary) addition of a high bit-rate pseudo-random digital waveform to a much lower bit-rate digital signal before transmission. The bits of the spreading waveform are called "chips." This causes the transmitted signal to spread over such a wide frequency range (perhaps 1,000 times as wide as that of the unspread signal) that it may not even be detectable by a normal receiver. Since the signal frequency is spread by 1,000, its power per Hz is reduced by 30 dB over the period of the sweep, as described in Chapter 2.

For the spread signal to be recovered by a cooperative receiver, the high bit-rate signal (synchronized with that in the transmitter) must be added to the received signal. Synchronization requires that the spreading waveform have a stable chip rate. If you were to view the spreading signal on an oscilloscope which is synchronized on the waveform, you would see a scope picture as shown in Figure 4.16. Note the square block transitions are at the chip rate. Since the number of sequential ones and zeros is highly variable, the first few transitions will show somewhat lighter on the scope face.

If the signal is processed with time collapsing, the chip energy can be integrated to the point that "chip detection" is possible. This technique creates the equivalent of a tapped delay line in software with the taps spaced at the chip period. This process overlays a large number of chips so the energy of the chip waveform can be integrated to a level that can be detected. The integration is over the chip period, with a sample near the end of the chip. If this is done in software, the tap spacing can be varied to search for the exact

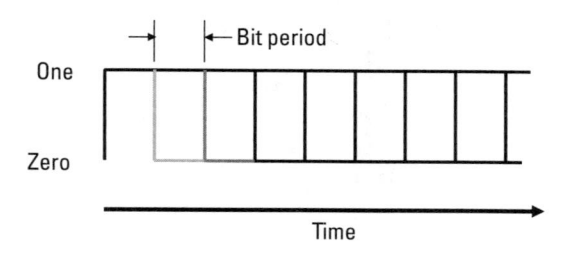

Figure 4.16 DSSS chips on an oscilloscope.

chip rate and phase. This will allow the detection and subsequent location of the direct sequence spread spectrum transmitter.

4.3.2 Catching a Frequency Hopping Signal

A frequency hopping signal changes frequency in a pseudo-random sequence every few milliseconds. The time at each frequency is called the "hop duration." Considering the Jaguar-V VHF frequency hopper as a typical example: the hop period is 10 ms, the channel spacing is 25 kHz, and the maximum hopping range is 30 to 88 MHz. Thus, there are 2,320 transmission frequencies over a 58-MHz range, any one of which might carry the signal during any hop. Figure 4.17 shows the frequency versus time for this type of signal. If we could determine the new frequency of an enemy transmitter within a small part of the hop period, we could very quickly determine the emitter location or effectively jam the enemy's frequency hopper and avoid jamming friendly communication by tuning a jammer to the frequency of each hop. The key is that the frequency measurement must be completed very quickly.

At this point, we will only discuss the determination of the frequency at each hop, leaving discussion of the emitter location and jamming to Chapters 7 and 9, respectively.

The signals are digitized in I & Q A/D converters because the emitter location technique we will be using requires that the phase of the received signal be captured at each of two antennas. Figure 4.18 is a block diagram of the digital receiver configuration for this task. There are two parallel I & Q digitizers so that the phase of the signal received at each of the two antennas can be captured.

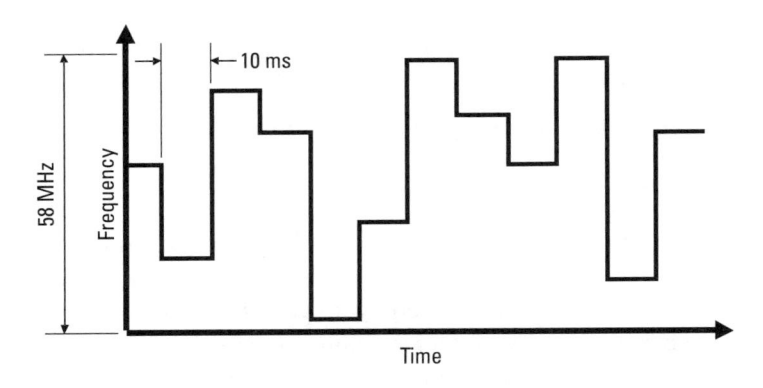

Figure 4.17 Jaguar frequency hopping pattern.

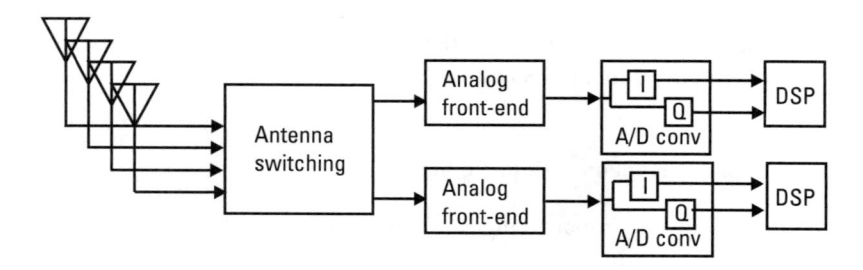

Figure 4.18 Digital receiver to capture phase of hopping signal.

Now, some more assumptions: Let's say you want to design your receiver with standard VME architecture. This restricts you to a 40 MByte data rate—so you can only sample the signal at a 40-MByte rate. This is a sample interval of 25 ns. Using the Nyquist sampling criteria limits the receiver input bandwidth to 20 MHz. If we are trying to catch a VHF hopper, it can hop over as much as a 58-MHz range with 25-kHz channel spacing. Thus, our system can only sample about a third of the range (i.e., we must make three samples to cover all possible hopper channels).

The processing technique used will be fast Fourier transform, to create a software channelized receiver. We need to have a channelizer with less than 25-kHz-wide channels. The number of points of the FFT (i.e., the number of samples processed in a single FFT) must be twice the number of channels of the channelizer created. However, since we are taking parallel I & Q samples, we only need to take the number of samples equal to the number of required channels. So, we need to take 1,000 I & Q samples to get 1,000 channels.

One thousand samples at 25 ns per sample takes 25 μs. Thus, the time to collect all of the data for the whole hopping frequency range is:

$$\text{Data Gathering Time} = 3 \times 25 \,\mu s = 75 \,\mu s$$

FFT Timing

Now, consider the time required to perform the processing. A rule of thumb is that an FFT requires $n \log_2(n)$ complex adds and $n/2 \log_2(n)$ complex multiplies—where n is the number of points in the FFT. For this example, n is 1,000. Combining these two values and converting to 10-base logs calculates the number of floating point operations (FLOPS) for an FFT at:

$$\text{FLOPS per FFT} = [1.5n \log_{10}(n)]/0.30103 = 4.988n \log_{10}(n) \approx 15{,}000 \text{ FLOPS}$$

The speed of a digital signal processor (DSP) is specified in (FLOPS/sec). For this example, we use a SHARC 600 MFLOP/sec data processor. The time required to perform an FFT is: 15,000 FLOPS/600 MFLOPS/sec = 25 μs. If our receiver processes the previous data set while a second date set is being collected, the throughput of the system is optimized because the FFT processing time is the same as the data collection time. Thus, the whole spectral data collection and analysis to determine the frequency to which the signal has hopped requires 75 μs. This is a very small percentage of the 10 ms that the frequency hopper remains at one frequency.

Note that at this point we have only determined the frequency of every signal present in the 30- to 88-MHz range. There are other important aspects of the problem which will be covered in subsequent chapters.

4.4 Receiver System Sensitivity

The sensitivity of a receiver system is most easily defined as the minimum signal strength it can accept and still do the job for which it is designed. For example, a television receiver is to provide an apparently snow-free picture on the screen. An important part of the definition is that this signal strength is determined directly at the output of the system's receiving antenna (as shown in Figure 4.19), so that the antenna gain (in dBi) subtracted from the receiver system sensitivity (in dBm) determine the power density that must arrive at the antenna to provide the required receiver output performance.

The sensitivity of some receiver systems is expressed in terms of that power density. This is quite appropriate when there is an intimate relationship between the antennas and receivers—and the sensitivity includes the antenna gain. However, if the sensitivity of the receiver (without the antenna) is given in power density terms, a zero dB antenna gain is assumed.

Power density has the units of microvolts/meter (μv/m), however, since the received power at the receiver is commonly calculated by the following formula:

$$P_R = P_T + G_T - L + G_R$$

where
P_R = Received power (in dBm)

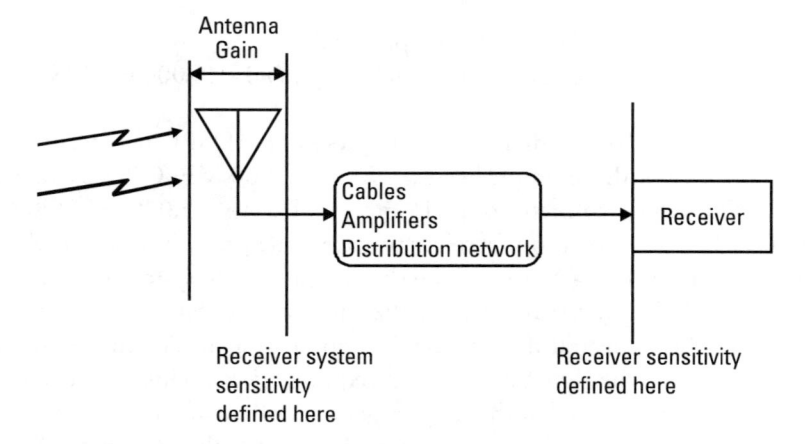

Figure 4.19 Sensitivity definition locations in receiver system.

P_T = Transmitter output power (in dBm)
G_T = Transmit antenna gain (in dB)
L = Propagation loss (in dB)
G_R = Receiving antenna gain (in dB)

It is often most convenient to express the arriving signal in terms of signal strength (in dBm) arriving at the receiving antenna in dBm.

If the arriving signal is stated in terms of its field strength (in μv/m) conversion to signal strength (in dBm) is easily made by use of the formula:

$$P = -77 + 20 \log(E) - 20 \log(F)$$

where

P = the signal strength arriving at the antenna in dBm
E = the arriving field density in μv/m
F = the frequency in MHz

Conversely, the arriving signal strength can be converted to field density by the formula:

$$E = 10^{[P+77+20\log(F)]/20}$$

where

E = field density in μv/m
P = signal strength in dBm

F = frequency in MHz

As a note of interest, these formulas are based on the effective area of an ideal isotropic antenna and the impedance of free space, producing an output power (in dBm) from an input field strength (in μv/m).

Sensitivity Components

It is useful to consider receiver sensitivity in terms of three component parts a shown in Figure 4.20.

Sensitivity is then:

$$S = kTB + NF + RFSNR$$

where

S = sensitivity in dBm

kTB is the internal thermal noise of the receiver

NF is the receiver system noise figure

$RF\ SNR$ is the predetection signal-to-noise ratio

4.4.1 kTB

kTB is thermal noise in the receiver. It is the product of Boltzmann's constant, the temperature of the receiver (in degrees Kelvin) and the effective receiver bandwidth. It is common practice in EW applications to use the standard temperature (290K) to create an expression for kTB only in terms of bandwidth. This expression is:

$$kTB = -114 \text{ dBm} + 10 \log(BW/1 \text{ MHz})$$

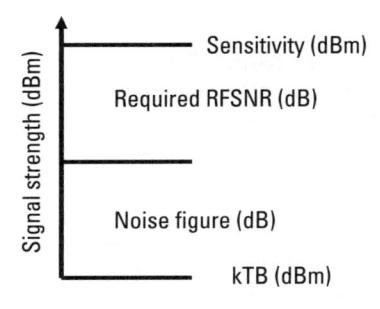

Figure 4.20 Components of sensitivity.

where *BW* is the effective receiver bandwidth.

The right side of this equation is often stated as −114 dBm per megahertz. It is also stated as −174 dBm per hertz (which is exactly the same number).

4.4.2 Noise Figure

Noise figure is the noise above kTB produced by the receiver system —referenced to the input as shown in Figure 4.21. Another way of stating this is: If the receiver produced no such noise, how much noise would have to be injected into the input to produce the output noise observed. The noise figure of an actual receiver is acquired from the receiver manufacturer as a specification, but our concern here is the noise figure of the whole receiver system.

If there are no active components (i.e., amplifiers) between the antenna and the actual receiver, the system noise figure is the sum (in dB) of the receiver noise figure and the losses of all the passive components up stream of the receiver. Examples of passive components are: cables, switches (when they are on), filters, and power dividers.

In order to improve receiver system sensitivity, a low noise amplifier (called a preamplifier) can be placed ahead of as many passive components as practical. Ideally, the preamplifier would be connected directly to the antenna, but it is often inconvenient to provide the power required for the preamplifier to that physical location.

The system noise figure when a preamplifier is included is determined by the following formula:

$$NF = L_1 + N_p + Deg$$

where

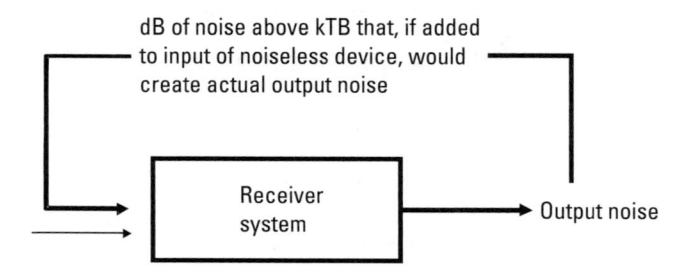

Figure 4.21 Noise figure definition.

NF = the system noise figure (in dB)

L_1 is the loss of all components ahead of the preamplifier (in dB)

N_p = the noise figure of the preamplifier

Deg is a degradation factor determined from Figure 4.23

In Figure 4.22, L_1 is the loss before the preamplifier, G_P is the gain of the preamplifier (in dB), N_P is the noise figure of the preamplifier, L_2 is the loss between the preamplifier and the receiver, and N_R is the noise figure of the receiver. For example, let the preamplifier gain be 20 dB, the preamplifier noise figure be 5 dB, loss L_1 be 2 dB, the loss L_2 be 8 dB, and the receiver noise figure be 12 dB.

The degradation factor can be determined from the chart in Figure 4.23. It is read at the intersection of a horizontal line from the ordinate value determined by $G_P + N_P - L_2$ and a vertical line from N_R on the abscissa. The lines drawn on the chart reflect the component values listed earlier. The horizontal line is at $20 + 5 - 8 = 17$ and the vertical line is at 12. The two lines intersect on the 1-dB degradation curve, so the degradation factor is 1 dB. Thus, the noise figure of the receiving system in the example is:

$$2 \text{ dB} + 5 \text{ dB} + 1 \text{ dB} = 8 \text{ dB}$$

4.4.3 Required Predetection Signal-to-Noise Ratio

The required predetection signal-to-noise ratio is a function of the required output signal quality and of the type of modulation on the signals received.

First, some definitions: The predetection signal-to-noise ratio is the way we quantify the quality of the signal at the input to the receiver system (i.e., at the output of the receiving antenna). In most communication theory texts this is abbreviated CNR" for "carrier-to-noise ratio." This can be confusing, because the received signal contains a carrier signal at the nominal

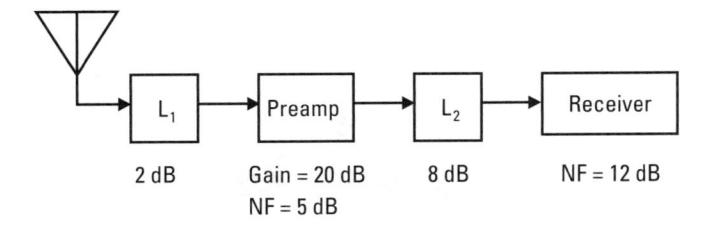

Figure 4.22 Receiver system with preamplifier.

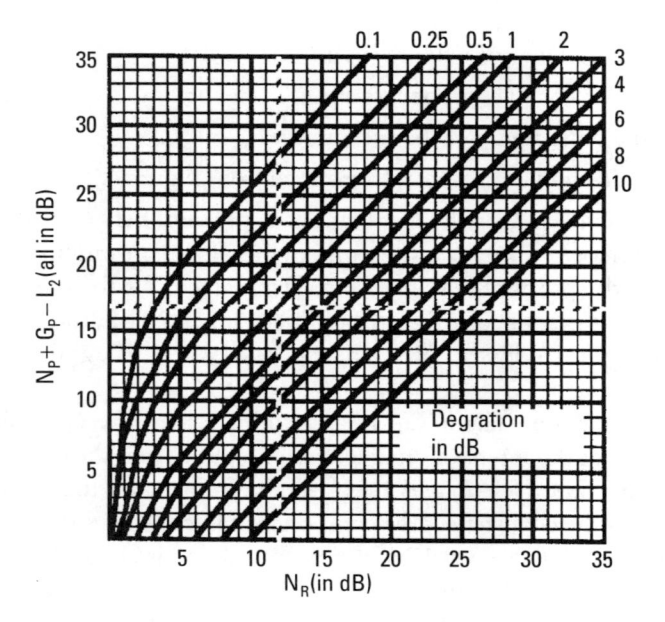

Figure 4.23 Noise figure degradation.

transmission frequency and sidebands which carry the transmitted information. Figure 4.24 shows a diagram of the predetection signal frequency spectrum. The total predetection signal power is divided between the carrier and

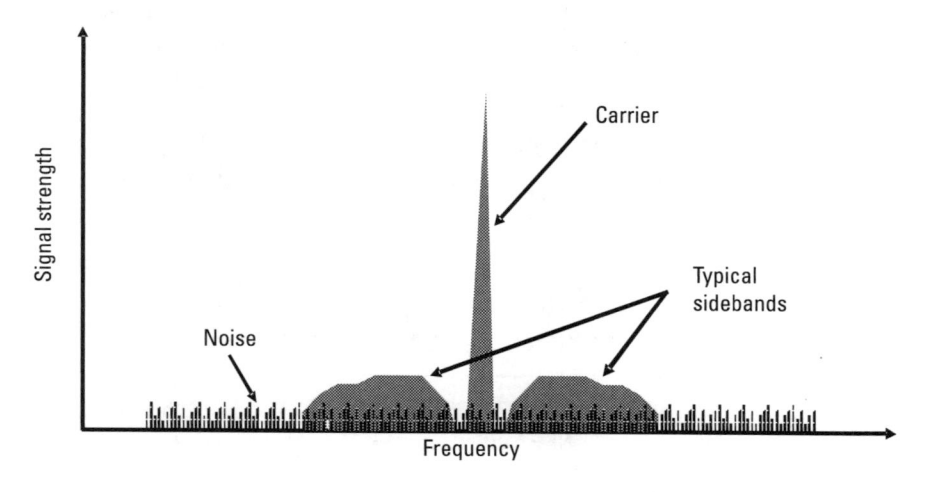

Figure 4.24 Predetection signal frequency spectrum.

the sidebands. Thus the predetection signal-to-noise ratio is the ratio of the power of all signal components to the noise power in the effective receiver bandwidth.

CNR is literally the ratio of the carrier power to noise. To avoid this confusion, this book uses the abbreviation RFSNR (for radio frequency signal-to-noise ratio) when referring to the predetection signal-to-noise ratio.

RFSNR Versus SNR

The SNR is the output signal-to-noise ratio from the receiver system. Thus SNR is the way we quantify the quality of the information recovered from received signals. Although this is what we really care about, we must consider the RFSNR when determining the signal strength that must be present at the receiver system input to produce the desired output signal quality. The relationship between RFSNR and SNR depends on signal modulation, and that relationship can be complex.

Table 4.2 shows typical required values for RFSNR and SNR for various types of modulations and receiver system applications. The first four rows of the table relate to amplitude modulated signals, so the RFSNR and output SNR are the same. For visual displays to be viewed on an oscilloscope by an expert operator, 8 dB is sufficient. If a computer is analyzing the data, 15 dB is normally considered adequate. The video portion of a television signal is amplitude modulated, and a 40-dB SNR will provide a snow-free picture. For AM communication, the signal-to-noise ratio required varies with the application. For example, military command and control communication typically uses rigid formats and vocabulary which allow adequate communication with lower SNR. Thus, 10 to 15 dB is required.

Table 4.2
RFSNR and SNR Versus Modulation and Application

Modulation and Application	RFSNR	SNR
AM signals viewed by expert operator on oscilloscope	8 dB	8 dB
AM signals analyzed by computer	15 dB	15 dB
Television video signal	40 dB	40 dB
AM voice communication	10 to 15 dB	10 to 15 dB
FM signals	4 or 12 dB	15 to 40 dB
Digital signals	10 to 14 dB	SQR

FM signals are subject to a signal-to-noise improvement factor based on the modulation parameters—generally making the output SNR significantly greater than the RFSNR.

The output quality of recovered digital signals is set by the digitization and is only secondarily affected by the RFSNR. The SNR is actually the ratio of the signal to quantization "noise" or SQR. However, the RFSNR determines the bit error rate that will be present in the recovered digital signal.

Required RFSNR for FM Signals

There are two types of FM discriminators, tuned and phase-lock loop. As shown in Figure 4.25, the tuned discriminator acts somewhat like a filter, in that it creates a throughput characteristic of output voltage versus input frequency. However, it is designed with a linear slope of voltage versus frequency. Since the FM signal carries its information as frequency variations, the output voltage is the modulating waveform—hence the information carried on the signal.

Figure 4.26 shows a phase-lock loop FM discriminator. Although it is more complex than the tuned discriminator, it provides a performance advantage. The voltage-controlled oscillator (VCO) is tuned to follow the frequency variations of the input FM signal. The difference in phase between the frequency modulated signal and the output of the VCO (i.e., the phase error) creates a tuning signal for the VCO to lock it to the FM signal. Thus, the voltage controlling the oscillator is the modulation signal.

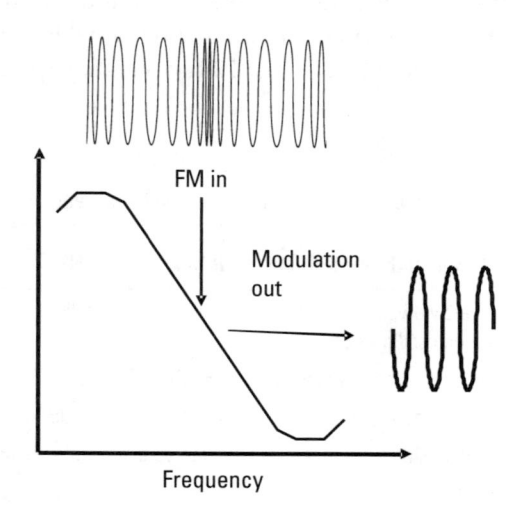

Figure 4.25 Slope detection FM discriminator.

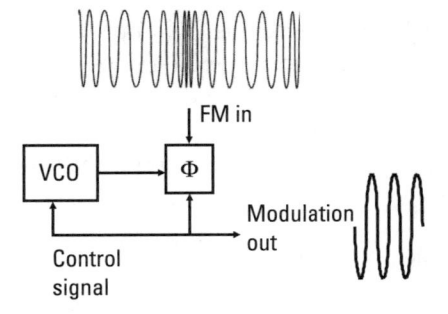

Figure 4.26 Phase-locked loop FM discriminator.

FM Improvement Factor

The output SNR from an FM signal is a function of the RFSNR and the modulation index.

The modulation index is the ratio between the maximum frequency deviation of the FM signal from the carrier frequency divided by the maximum modulating frequency:

$$\beta = \text{Max Deviation}/\text{Max Modulating Frequency}$$

For example, in a commercial FM broadcast, the maximum deviation is 75 kHz and the maximum modulating frequency is 15 kHz, so $\beta = 5$.

The FM improvement is the increase in output SNR above the RFSNR, and can be calculated in dB from the formula:

$$IF_{FM} = 5 + 20\log(\beta)$$

where

IF_{FM} is the FM improvement factor in dB

β is the modulation index

Note that this is a simplified version of the formula which assumes that the predetection bandwidth is matched to the predetection signal and the postdetection bandwidth is matched to the detected signal.

In the example of commercial FM radio, the improvement factor is:

$$IF_{FM} = 5 + 20\log(5) \approx 19 \text{ dB}$$

However, in order to achieve this FM improvement, the RFSNR must be above a required threshold as shown in Figure 4.27. For the tuned discriminator, the threshold is about 12 dB, and for the VCO discriminator the threshold is only about 4 dB.

In our FM broadcast example, the output SNR from a 12-dB RFSNR is 12 + 19 or 31 dB.

Required RFSNR for Digital Signals

For digital signals, SNR is not really signal-to-noise ratio, but quantizing-to-noise ratio The signal-to-quantizing-noise ratio is given by the formula:

$$SQR = 5 + 3(2n - 1)$$

where

SQR is the signal to quantization ratio in dB

n = the number of quantizing bits

Being set by the quantizing, the output SNR is largely independent of the RFSNR. However, the RFSNR has a direct impact on the bit error rate of the recovered digital signal. The digital information must be modulated onto an RF carrier by one of the techniques discussed in Chapter 2. The bit error rate is the number of incorrectly received bits divided by the total bits sent. For each type of digital modulation, there is a curve like the two typical

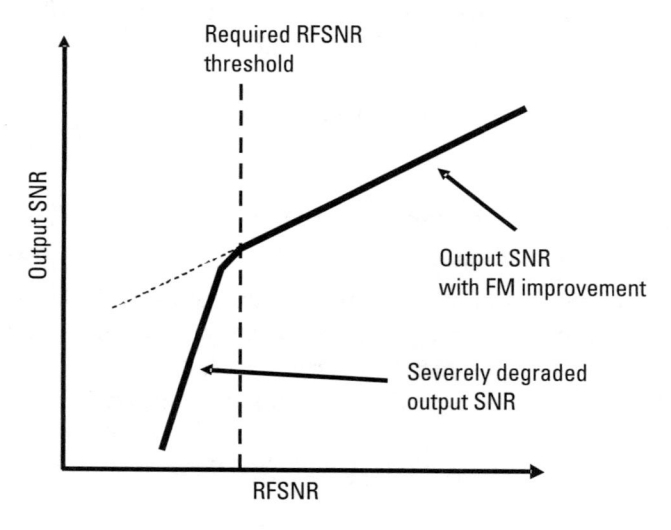

Figure 4.27 Output SNR versus RFSNR with required threshold.

curves in Figure 4.28. This shows the bit error rate as a function of E_b/N_o. E_b/N_o is the energy per bit divided by the noise per Hz of bandwidth and is determined by the formula:

$$E_b/N_o (\text{dB}) = RFSNR(\text{dB}) - 10\log(\text{bit rate}/\text{bandwidth})$$

If the ratio of bit rate to bandwidth is 1 bps/Hz, $E_b/N_o = RFSNR$.

For example, if the digital data is carried with noncoherent frequency shift keying, the bandwidth to bit rate ratio is one, and the RFSNR is 11 dB, the bit error rate in the recovered digital data would be approximately 7×10^{-4}.

4.5 Receiver System Dynamic Range

Dynamic range defines the difference between the strongest and weakest signals a receiver system can receive. The instantaneous dynamic range defines difference (in dB) between the weakest signal that can be received in the presence of the strongest.

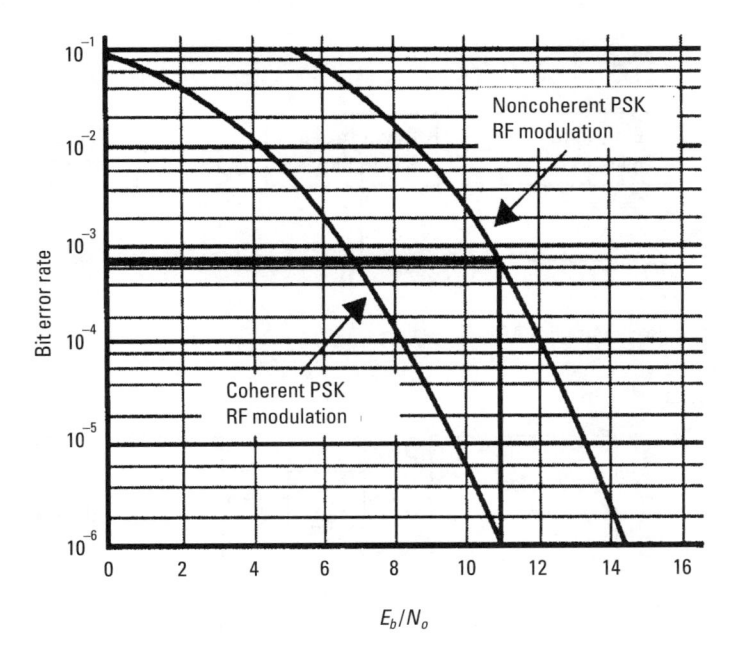

Figure 4.28 Bit error rate versus E_b/N_o.

In communication systems and radars, automatic gain control (AGC) allows receivers to accept a very wide range of signal strengths because the receiver is turned down to provide optimum reception of the strongest signal in the receiver's bandwidth as shown in Figure 4.29. However, AGC is seldom if ever appropriate to EW and reconnaissance receivers. It may be a matter of life or death to receive a weak threat signal while a very strong nonthreat signal is in-band.

Reconnaissance receiver systems must sometimes have switched attenuators in their front ends to allow a strong range and a weaker range of signals to be sequentially sampled. Many earlier systems were specified at 60-dB instantaneous dynamic range, but modern systems typically require 90 dB or more.

4.5.1 Analog Versus Digital Dynamic Range

Since many digital receivers are used in EW applications, it is important to remember that digital receivers comprise an analog front end followed by an analog-to-digital converter and then a computer in which the digital receiver functions are performed. (See Figure 4.30.) The analog and digital sections each have defined dynamic ranges. First, consider the dynamic range of the

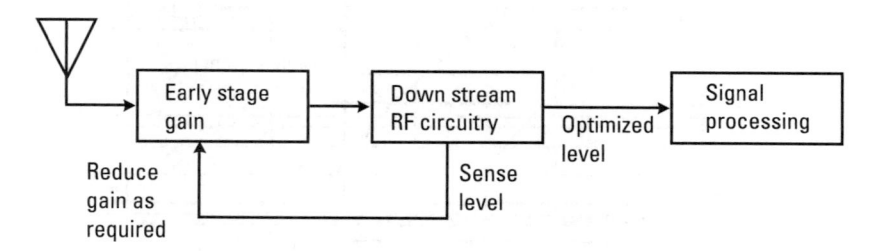

Figure 4.29 Receiver with AGC.

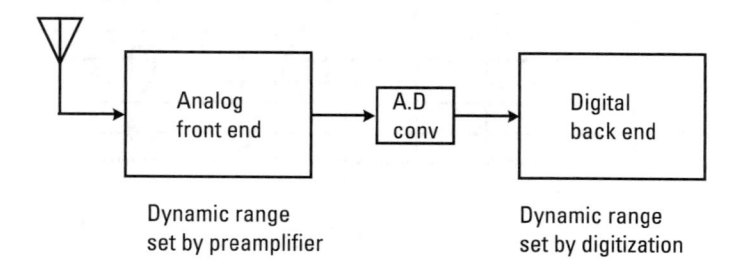

Figure 4.30 Digital receiver.

analog section, then that of the digital section. The two receiver sections should have the same dynamic range.

4.5.2 Analog Receiver Dynamic Range

The dynamic range is typically determined by a preamplifier, which is specified in terms of its gain, noise figure, and "intercept points." The impact of preamplifier gain and noise figure on system sensitivity is discussed earlier. The intercept points of the preamplifier impact the system dynamic range.

Figure 4.31 is used to determine the dynamic range from the intercept points. This diagram relates to the output of the preamplifier. The ordinate and abscissa of the graph are both logarithmic scales of preamplifier output power in dBm. The fundamental line represents the output power of a single amplified signal from the amplifier. Note that it has a 1:1 slope. The second-order response line represents the second-order spurious response at the output of the amplifier. This is the level of the spurious signal produced at the output of the amplifier at a frequency which is double the fundamental frequency or at the sum or difference frequencies of two input signals. The second-order response line has a 2:1 slope and intersects the fundamental line at what is called the second-order intercept point (also called "IP2"). Starting at the output power of the one or two signals causing the spurious response on the abscissa of the graph, move down to the second-order line, then left to the ordinate to read the level of the second-order spurious response output.

The third-order response line is the level of third-order spurs. A third-order spurious response is out put at two times the frequency of one input signal plus or minus the frequency of the second input signal (or one frequency plus or minus twice the other frequency). This line has a 3:1 slope and intersects the fundamental line at the third-order intercept point (IP3).

The intercept points are the signal levels at which the spurious response lines intercept the fundamental line in Figure 4.31. For the example shown, the second-order intercept point is +50 dBm and the third-order intercept is at +20 dBm. Two input signals at −27 dBm would produce third- and second-order spurious outputs at −100 dBm and −112 dBm, respectively, as shown in Figure 4.32.

A receiver design will normally eliminate the second-order spurs by selection of intermediate frequencies, and multiple conversions if necessary. However, the third-order spurs often cannot be avoided and limit the receiver's spurious free dynamic range.

Note that there is an intercept validity limit line shown on the graph. This is the level at which the amplifier is no longer "well behaved." It is near

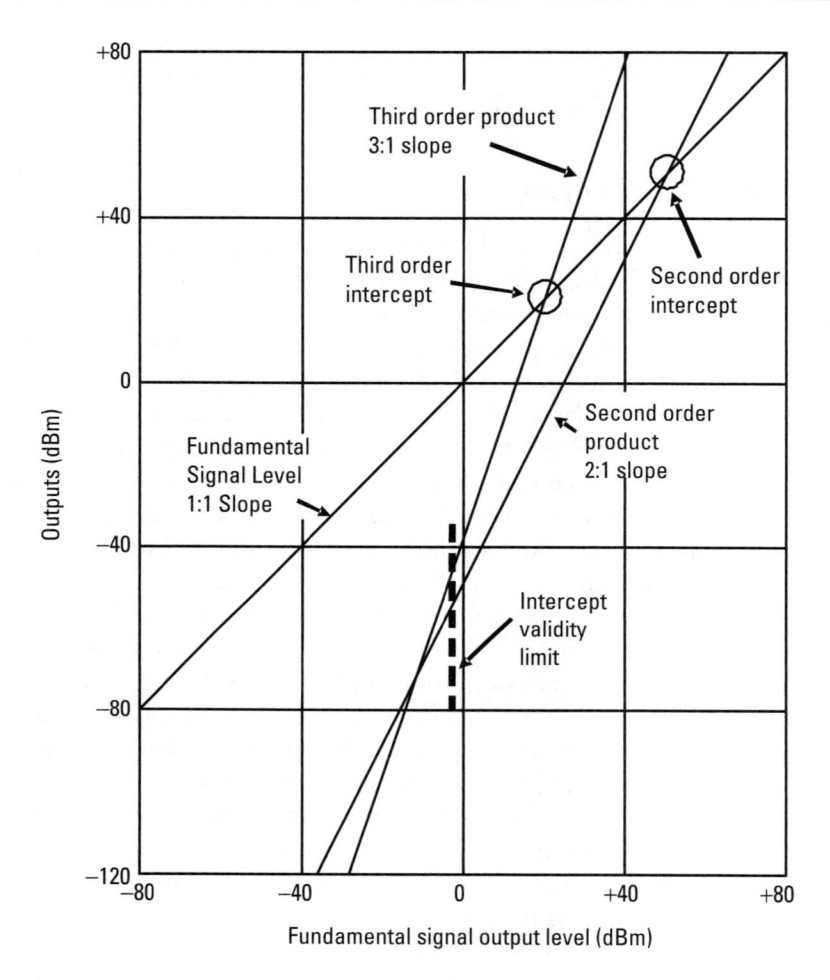

Figure 4.31 Intercept point chart.

the point at which compression in the amplifier output (caused by saturation) causes the actual levels of the spurious outputs to vary significantly from the second- and third-order line values. Thus, the dynamic range must be calculated to the left of this validity limit.

Determining Dynamic Range

Figure 4.33 adds some information to the graph in Figure 4.32. The sensitivity line is the level of a preamplifier output caused by a received signal (from the antenna) at the receiver system sensitivity level. In this example, the sensitivity level preamplifier output signal is −100 dBm. To calculate the dynamic

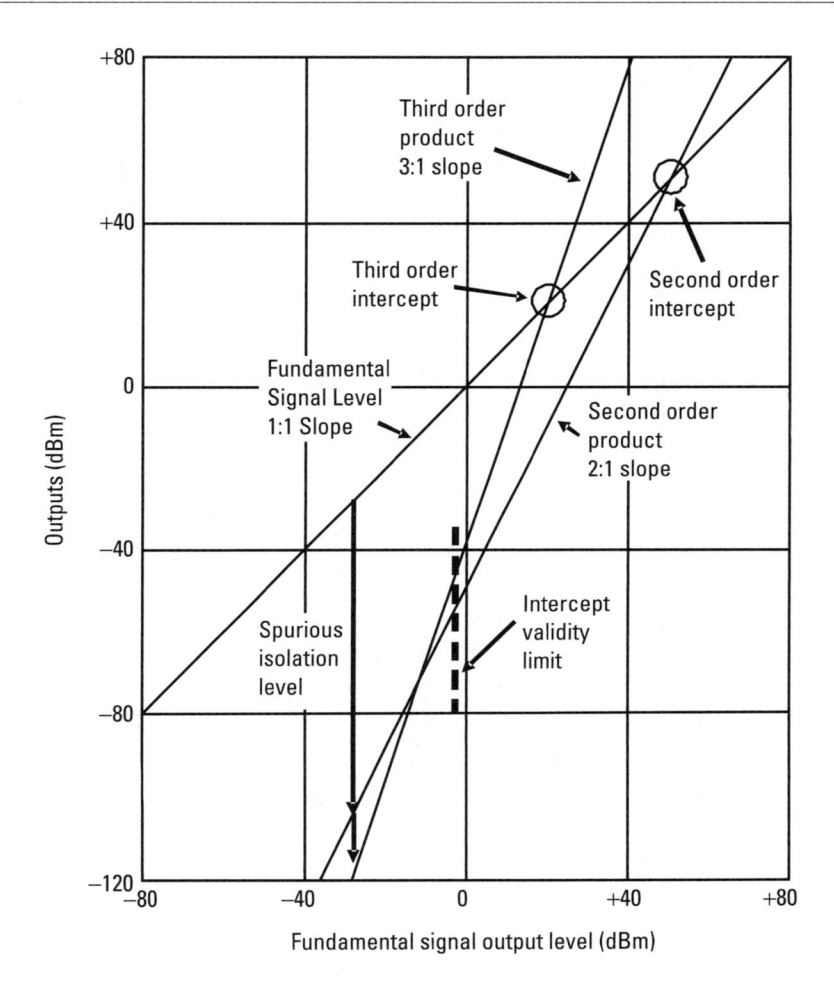

Figure 4.32 Spurious signal isolation.

range limited by third-order spurs, draw a vertical line from the intersection of the third-order intercept line and the sensitivity line up to the fundamental line. The vertical distance between these two lines (i.e., the difference in output signal strength in dB) defines the output signal strength of strong signals that will cause spurs at the sensitivity level. This means that if the signals causing spurs are at the fundamental line level (–32 dBm), the third-order spurious outputs will be at –100 dBm (which is the sensitivity level of the receiver at this point). So the dynamic range as shown in Figure 4.33 is 68 dB.

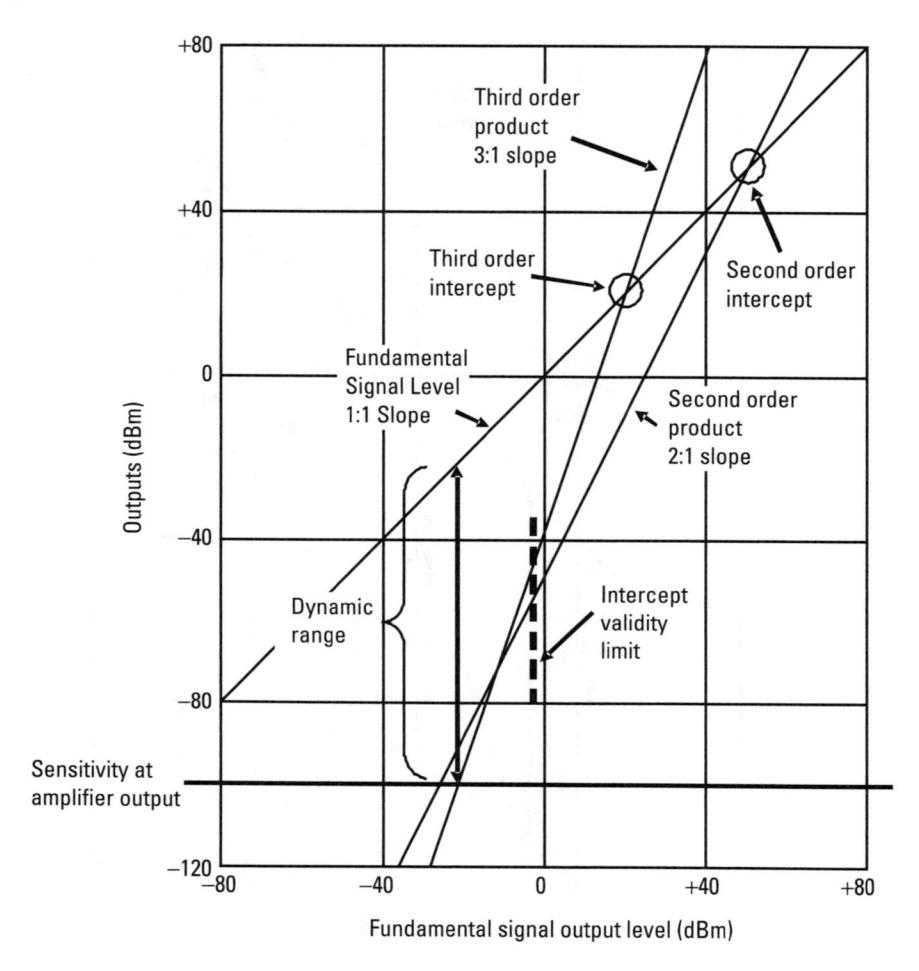

Figure 4.33 Spurious-free dynamic range.

4.5.3 Digital Dynamic Range

The dynamic range of the digital part of the receiver depends on the number of bits produced by the analog-to-digital converter. The weakest signal is digitized as one in the least significant bit (all other bits being zeros), while the strongest signal is digitized as all ones. Figure 4.34 shows the digitization of the maximum and minimum measurable levels in a 4-bit A/D converter. The dynamic range is then:

$$DR = 20 \log_{10}(2^n)$$

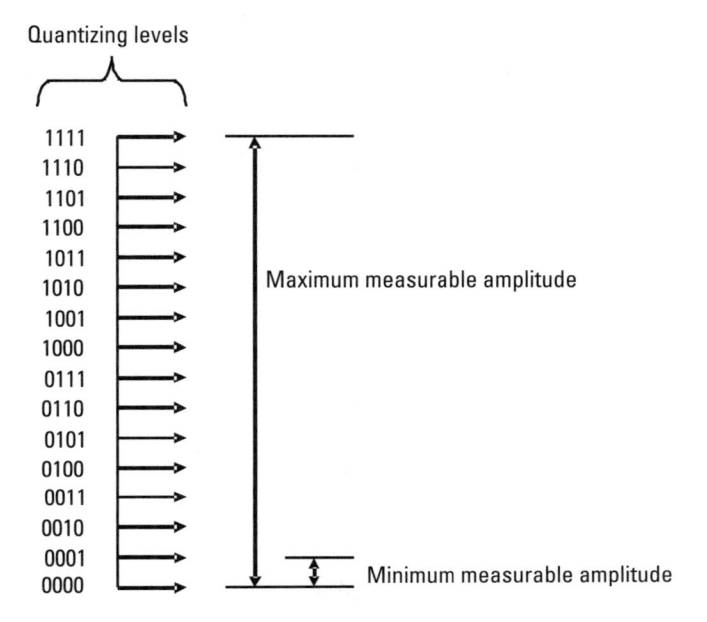

Figure 4.34 Digital receiver dynamic range.

where

DR is the dynamic range in dB

n is the number of bits to which the input signal is digitized

Note that the conversion to dB has a 20-multiplier rather than the 10-multiplier when signal-power ratios are converted. This is because the digitizer quantizing levels which determine the digital word produced are voltages. For example, 10 bits provides 60-dB dynamic range. Table 4.3 shows the digital dynamic range as a function of the number of digitizing bits.

4.6 Typical Receiver System Configurations

Almost all EW and reconnaissance receiving systems include multiple receivers. Since most systems include emitter location capability, they often require multiple identical receivers to simultaneously receive inputs from multiple antennas.

Another significant issue is the achievement of high probability of intercept during very short specified response times. Often, a receiver optimized for the search function finds new signals while other receivers perform

Table 4.3
Dynamic Range Versus Digitizing Bits

Number of Bits	Dynamic Range (dB)
4	24
5	30
6	36
7	42
8	48
9	54
10	60
11	66
12	72
13	78
14	84
15	90
16	96

prolonged analysis or intercept functions. Finally, sophisticated modern types of receivers can be required to handle specific problems with sophisticated threats or threat environment conditions.

This section considers reconnaissance receivers and multiple station–remote-controlled receiving systems.

4.6.1 Multiple Receiver Reconnaissance and Electronic Support Systems

The dominant difference between electronic support and reconnaissance receiver systems can be considered "attitude." Both receiver systems are designed to receive the same types of signals, but do so for different reasons. ES receivers are typically looking for threat signals of known types for immediate tactical reasons. Required response times are usually measured in low single-digit seconds, and data is collected only to determine which of the known types of signal is present—in what mode—at what location. Antennas tend to be wide (ideally 360 degrees total instantaneous coverage from the whole array) to provide extremely high probability of intercept. Receiver bandwidths tend to be wide, sacrificing sensitivity for fast response and high probability of intercept.

Reconnaissance receiver systems, on the other hand, typically have the luxury of more time to find enemy signals, but often must receive weaker signals and provide sufficient resolution and analysis to characterize new types of enemy signals. They may also deal with signal internals (that is the information being transmitted). Antennas can often be narrow, enhancing the ability to intercept signals from distant transmitters.

Both electronic support and reconnaissance receiver systems are used in all frequency ranges, and have a wide range of configurations. Often the same system performs both roles as required. A few configurations are selected here to facilitate discussion of important issues.

4.6.2 Multiple Receiver Systems

Figure 4.35 is a typical receiving system with multiple receivers, as is often the case in systems in all frequency ranges. Note that the antenna output can be distributed in a multiplexer if the receivers cover different frequency ranges. If all cover the same frequency range, a power divider is required. Search and acquisition is challenging, particularly if ranges to target emitters are near the detection limits. Acquisition receivers can be swept superheterodyne receivers with bandwidths and search rates optimized for the maximization of probability of intercept under the required geometry and timing constraints. Digital receivers, compressive receivers, and other broad coverage receiver types can also be employed for the acquisition function.

There are significant issues associated with the tasking of receivers. One common approach is to have many monitor receivers, each assigned to an

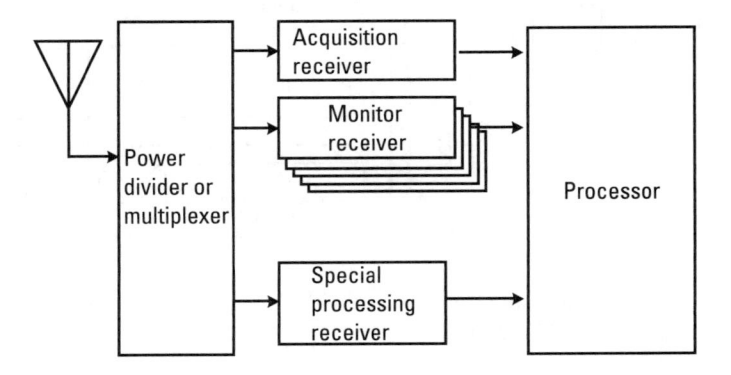

Figure 4.35 Receiving system with multiple receivers.

operator or to an automatic data recording or processing channel. When the acquisition receiver detects a signal, it may either stop its search for a short interval to perform an evaluation of the signal's interest, or it may task a special processing receiver to perform that task. Determination of interest and priority normally involves analysis of modulation externals—without consideration of the information carried by the signal (i.e., the modulation internals).

Once a signal of interest is identified and prioritized, a monitor receiver will be assigned to that signal for the duration of the signal or until a signal of higher priority must be monitored.

Local Oscillator Radiation

Because military receiving systems often use superheterodyne (or "superhet") receivers, local oscillator (LO) radiation is an important receiver specification. As discussed in Section 4.1.2, the superhet receiver tunes to signals by setting a local oscillator a fixed frequency offset above or below the frequency of interest. It also typically has a preselector bandpass filter at the frequency of interest.

Figure 4.36 shows the front ends of two superhet receivers in a receiver system. The LO signal is strong compared with received signals, and thus can be expected to propagate back through the preselector at some level. It is important to realize that the preselector is usually a relatively simple filter providing limited isolation. Therefore leaked LO energy will follow the two paths shown in the figure. If significant LO energy escapes from the antenna

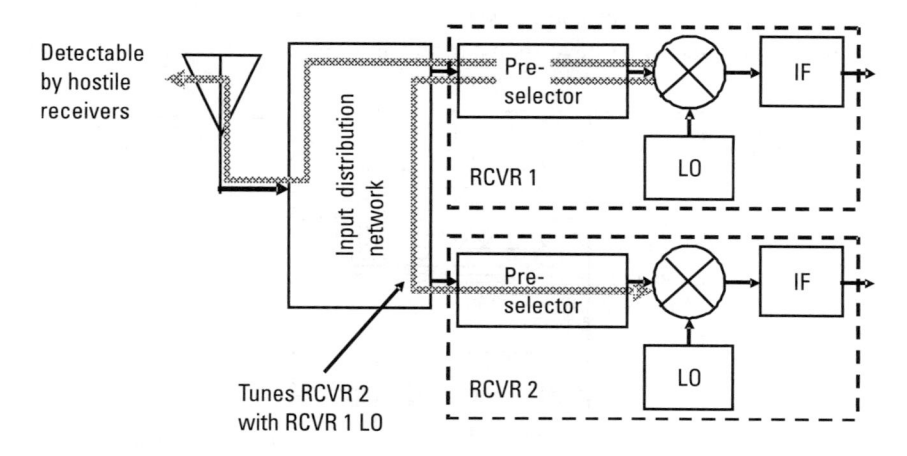

Figure 4.36 Local oscillator radiation.

it can be detected by a hostile receiver. It can also be received by other antennas in the same system or nearby friendly assets—degrading their performance. It is interesting to note that in countries requiring licenses for television receivers, LO radiation is used to detect nonlicensed TV sets.

Receiver Performance

Each signal path in a multiple receiver system is independently analyzed to determine its performance. Figure 4.37 shows part of a multiple receiver system. Each signal path is from the antenna through the preamplifier, one leg of the four-way power divider, and into one of the receivers. Consider examples for these two receivers. Individual component specifications for part of this system are shown in Figure 4.38. The required RFSNR for each receiver is 15 dB, and bandwidths are the effective bandwidth for the respective channels. To simplify the examples, we will ignore cable losses.

Using the techniques explained in Section 4.4, the sensitivity for each receiver channel is the sum of kTB, the noise figure, and the required RFSNR.

For channel 1, kTB is $-114 + 10 \log (50 \text{ kHz}/1 \text{ MHz}) = -127$ dBm. The system noise figure is 7 dB (including 3-dB degradation), and the required RFSNR = 15 dB. Therefore, the sensitivity is -105 dBm.

For channel 2, kTB is $-114 + 10 \log (250 \text{ kHz}/1 \text{ MHz}) = -120$ dBm. The system noise figure is 5 dB (including 1-dB degradation), and the required RFSNR = 15 dB. Therefore, the sensitivity is -100 dBm.

Using the technique explained in Section 4.5, the spurious free dynamic range for receiver 1 (considering only third-order spurs) is 68 dB [-17 (-85)]. The spurious free dynamic range for receiver 2 is 66 dB [-14 (-80)]. (Remember that the fundamental level is at the output of the preamplifier.)

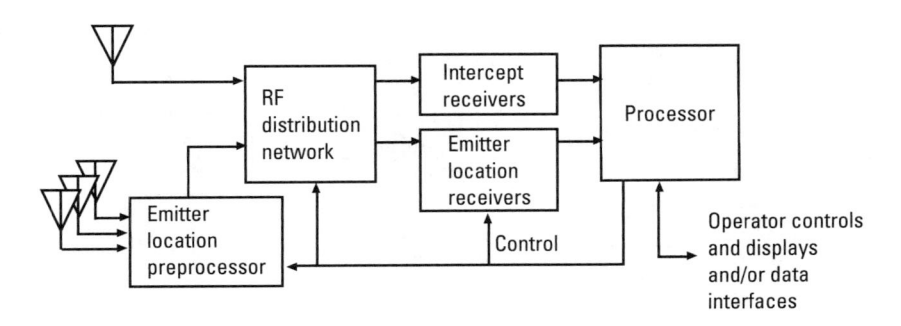

Figure 4.37 Front end of multiple receiver system.

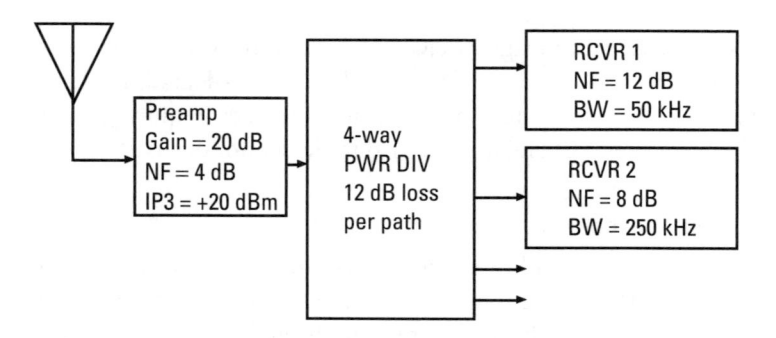

```
         Preamp                          RCVR 1
         Gain = 20 dB    4-way           NF = 12 dB
         NF = 4 dB       PWR DIV         BW = 50 kHz
         IP3 = +20 dBm   12 dB loss
                         per path        RCVR 2
                                         NF = 8 dB
                                         BW = 250 kHz
```

Figure 4.38 Front-end component specifications.

4.6.3 Remote Receiving Systems

There are many ground-based, fixed-wing, and helicopter-receiving systems that include multiple remote, cooperative receiving platforms or installations. Not only do these systems provide improved intercept geometry, but they also allow accurate location of hostile emitters through triangulation. Some systems have all remote receiving systems connected to a signal control station by control and data links as shown in Figure 4.39. In this type of system, several operators are typically located at the control station along with recording and analysis equipment. Other systems operate with each receiving system capable of being the master station—controlling the other receiver locations as subordinate stations. In these systems there are typically one or more operators at each receiver location. The master location can change as required during a single mission.

In general, control links are relatively narrowband, because the signals from the master (or control) station pass digital tuning and configuration

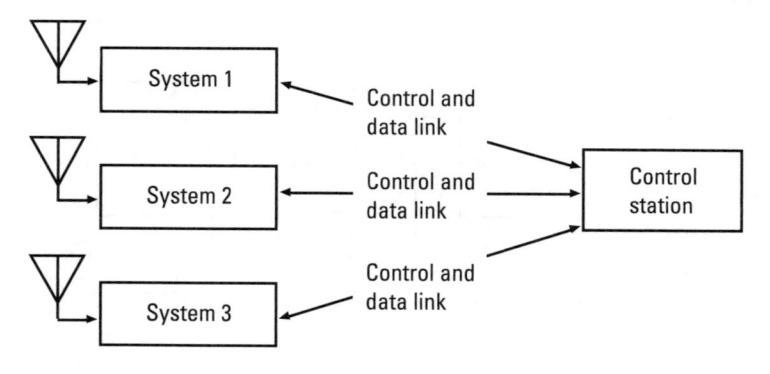

```
        System 1  ←—— Control and
                      data link
                                         ——→  Control
        System 2  ←—— Control and             station
                      data link          ——→

        System 3  ←—— Control and
                      data link
```

Figure 4.39 Multiple-station remote-controlled EW system.

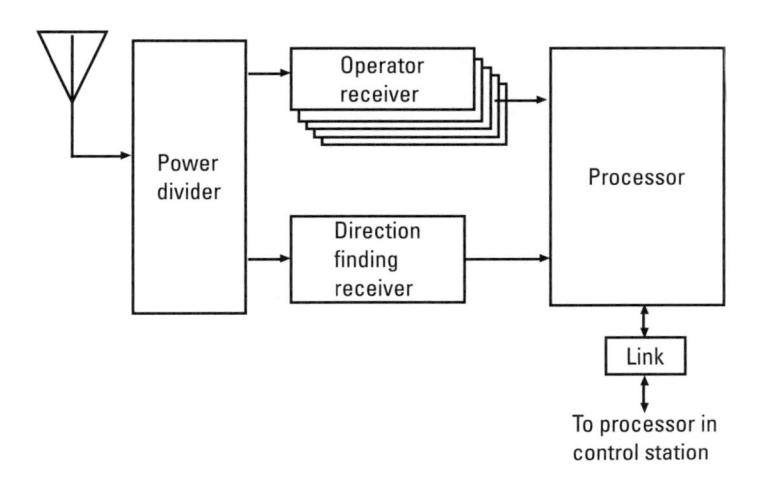

Figure 4.40 Remote receiving system with direction finding.

commands a few times per millisecond. However, the data links carry intercepted signal data from the receivers to the control station. Depending on the nature and number of signals received, data links can be very wide.

Figure 4.40 shows a single remote receiving station, which is one of several receiving stations linked to a single control station. There are several operator receivers, each linked to an operator in the control station. The operator monitors received signals and when he or she wants to determine the location of the transmitter being monitored, a location command from the operator station is sent to a central control station computer. The computer sends simultaneous commands to the direction finding (DF) receivers in all of the remote receiving systems. This causes them to measure the direction of arrival of the signal from the target transmitter at the same instant. These direction of arrival measurements are sent to the control station computer which adds the instantaneous location of the corresponding receiving platforms and calculates the location of the hostile emitter. If two operators request emitter locations at once, one request is delayed (on the order of a second) until the DF receivers in all receiving systems are available.

5

Communications Propagation

5.1 One-Way Link

In communication, the transmitter and receiver are in different locations. The purpose of communication systems of all types is to take information from one location to another. Thus, communication uses the "one way" communication link as shown in Figure 5.1. The one-way link includes a transmitter, a receiver, transmit and receive antennas, and everything that happens to the signal between those two antennas.

Figures 5.2 and 5.3 show important cases of the use of one-way links in electronic warfare. Figure 5.3 shows a communication link and a second link from the transmitter to an intercept receiver. Note that the transmit antenna gain to the desired receiver and to the intercept receiver may be different. Figure 5.4 shows a communication link and a second link from a jammer to the receiver. In this case, the receiving antenna may have different gain toward the desired transmitter and the jammer. Each of the links (in both figures) have the elements shown in the diagram of Figure 5.1.

5.2 The One-Way Link Equation

The one-way link equation gives the power into the receiver as a function of the other link values. Figure 5.4 is a diagram that represents this equation. The horizontal axis of this diagram is not to scale, it merely shows what

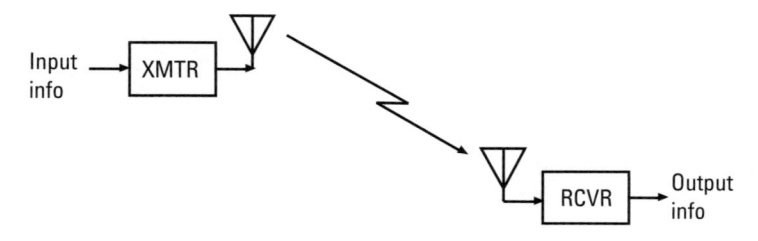

Figure 5.1 One-way communication link.

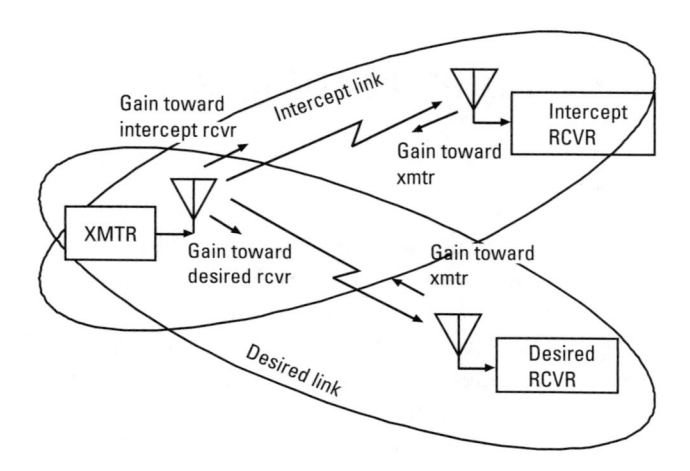

Figure 5.2 Intercepted communication link.

happens to the level of a signal as it passes through the link. The vertical scale is the signal strength (in dBm) at each point in the link. The transmitted power is the input to the transmit antenna. The antenna gain is shown as positive, although in practice any antenna can have positive or negative gain (in dB). It is important to add that the gain shown here is the antenna gain in the direction of the receiving antenna. The output of the transmit antenna is called the effective radiated power (ERP) in dBm. Note that the use of dBm units is not technically correct; in fact, the signal at this point is a power density, properly stated in microvolts per meter ($\mu v/m$). However, if we were to place a theoretical ideal isotropic antenna next to the transmit antenna (ignoring the near field issue) the output of that antenna would be the signal strength in dBm. Using the artifice of this assumed ideal antenna allows us to talk about signal strength through is whole link in dBm without converting units, and is thus commonly accepted practice. The formulas to convert back

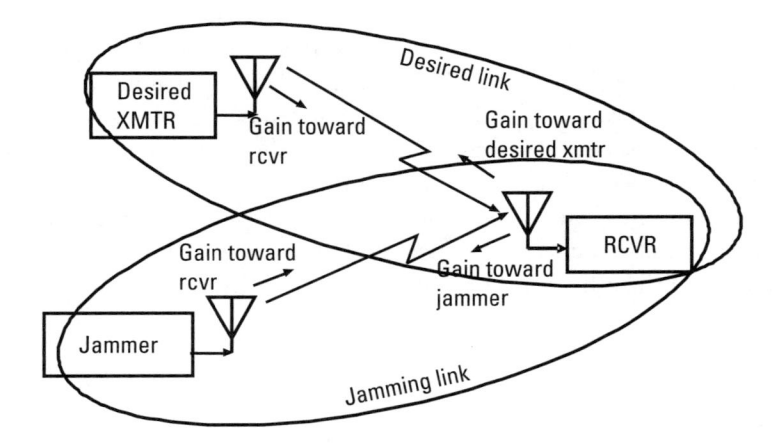

Figure 5.3 Jammed communication link.

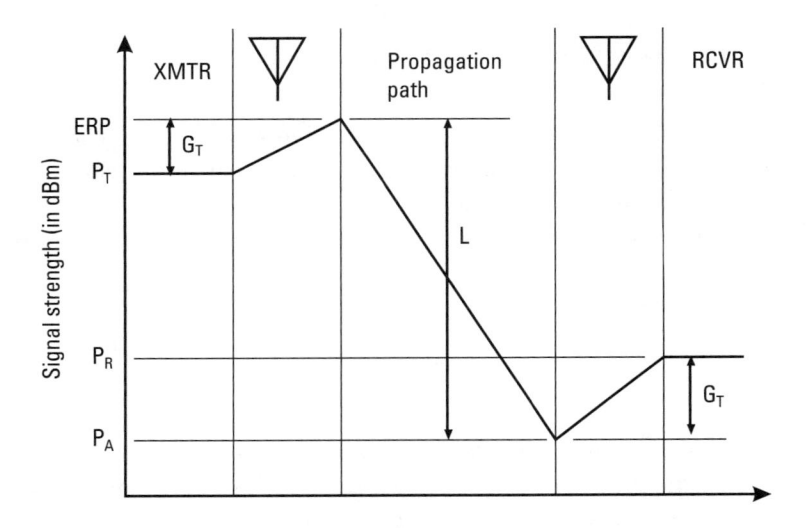

Figure 5.4 Graphic representation of the one-way communication link.

and forth between signal strength in dBm and field density in μv/m were given in Section 4.4.

Between the transmit and receive antennas, the signal is attenuated by the propagation loss. We will talk about the various types of propagation loss in detail later in this chapter.

The signal arriving at the receiving antenna does not have a commonly used symbol, but we will call it P_A for convenience in some of our later discussions. Since P_A is outside the antenna, it should really be in μv/m, but using the same ideal antenna artifice, we use the units dBm. The receiving antenna gain is shown as positive, although it can be either positive or negative (in dB) in real-world systems. The gain of the receiving antenna shown here is the gain in the direction of the transmitter.

The output of the receiving antenna is the input to the receiver system in dBm. We call it the received power (P_R). The one-way link equation gives P_R in terms of the other link components. In dB units, it is:

$$P_R = P_T + G_T - L + G_R$$

where

P_R = the received signal power in dBm

P_T = the transmitter output power in dBm

G_T = the transmit antenna gain in dB

L = the link loss from all causes in dB

P_R = the transmitter output power in dBm

In some literature, the link loss is dealt with as a "gain," which is, of course, negative (in dB). When this notation is used, the propagation gain is added in the formula rather than subtracted. In this book, we will consistently refer to loss as a positive number in dB, and therefore *subtract* loss in link equations.

In linear (i.e., non-dB) units, this formula is:

$$P_R = (P_T \; G_T \; G_R)/L$$

The power terms are in watts, kilowatts, and so forth, and must be in the same units. The gains and losses are pure (unitless) ratios. Since the link loss is in the denominator, it is a ratio greater than one. In subsequent discussions, the loss formulas both in dB and in linear form will consider loss to be a positive number.

Link Margin

If the received power is greater than absolutely required, the difference between received power and required power is the link margin. Link margin provides for signal attenuation effects which may or may not be present. Often, there are several effects which could possibly cause losses that are cumulatively very great, but not all of them are expected to occur at once.

The requirement of link margin is usually a trade-off of risks of inadequate performance against the cost, size, weight, and prime power required to avoid those risks.

One general approach to margin requirement is to ask how much of the time you can afford to be without the link. If you can do without it 10% of the time, you would like to have a 10-dB margin. If only 1% link outage is acceptable, you would like a 20-dB margin. If only 0.1% link outage is tolerable, you would like to have a 30-dB margin. However, the margin in practical communication and electronic warfare situations is usually significantly lower than these levels.

5.3 Propagation Losses

In the description of the link, we have clearly separated the transmitting and receiving antenna gains from the link losses. This implies that the link loss is calculated as though it were between two unity gain antennas. By definition, an isotropic antenna has unity gain—or 0-dB gain. All of the following discussion of link propagation losses will be for transmission between isotropic antennas.

There are a number of widely used propagation models, including: the Okumura and Hata models for outdoor propagation and the Saleh and SIRCIM models for indoor propagation. There is also small-scale fading, which is short-term fluctuation caused by multipath. The nature and uses of these more complex propagation models are discussed in detail in an excellent chapter (Chapter 84) in the *Communications Handbook* (ISBN 0-8493-8349-8) published cooperatively by CRC Press and IEEE Press. These propagation models all require computer models of the environment to support analysis of each reflection path in the propagation environment.

Because electronic warfare is dynamic by nature, it is common practice not to use these detailed computer analyses, but rather to use three important approximations to determine the appropriate propagation loss in practical applications. These three models are:

- Line of sight;
- Two ray;
- Knife-edge diffraction.

The above reference also discusses these three propagation models to some extent. Table 5.1 summarizes the conditions under which these three modes are used, and they are described in detail later, and multiple ways to calculate each type of loss are provided.

5.4 Line-of-Sight Propagation

Line-of-sight (LOS) propagation loss is also called free-space loss, spreading loss, or range-squared loss. It applies in space and between transmitters and receivers in any other environment in which there are no significant reflectors and the ground is far away in comparison with the signal wavelength. See Figure 5.5. This section will describe three ways to calculate LOS loss:

- A formula (in linear or dB form);
- A nomograph;
- The slide rule provided with this book.

Formula

The formula for LOS loss comes from optics, in which propagation loss is calculated by projecting the transmitting and receiving apertures on a unit sphere with its origin at the transmitter. This is converted to radio frequency propagation by considering the geometry of two isotropic antennas. As shown in Figure 5.6, the isotropic transmitting antenna propagates its signal spherically, with its total energy spread over the surface of the sphere. The sphere expands at the speed of light until its surface touches the receiving antenna. The area of the surface of a sphere is:

Table 5.1
Selection of Appropriate Propagation Loss

Clear Propagation Path	Low frequency, wide beams, near ground	Link longer than Fresnel zone distance	Use two-ray model
		Link shorter than Fresnel zone distance	Use line-of-sight model
	High frequency, narrow beams, far from ground		
Propagation path obstructed by terrain	Calculate additional loss from knife-edge diffraction		

Communications Propagation 125

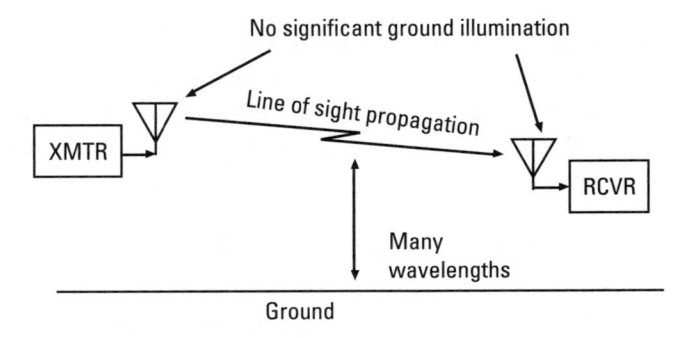

Figure 5.5 Transmission far from the ground.

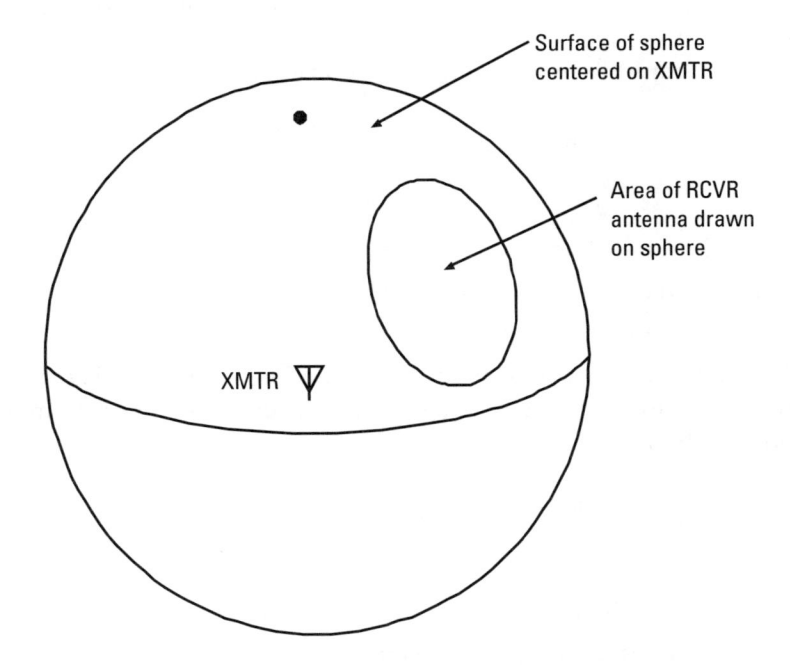

Figure 5.6 Geometry for calculation of line-of-sight loss.

$$4\pi d^2$$

where d (the radius of the sphere) is the distance from transmitter to receiver.

The effective area of the isotropic (i.e., unity gain) receiving antenna is:

$$\lambda^2/4\pi$$

where λ is the wavelength of the transmitted signal.

We want the loss to be a number larger than one, so we can divide the transmitted power by the loss to get the receive power. Thus, we determine the loss ratio by dividing the surface area of the sphere by the area of the receiving antenna:

$$\text{Line of sight loss} = (4\pi)^2 \, d^2/\lambda^2$$

where both the distance and the wavelength are in the same units (typically meters).

Note that some authors describe a propagation gain by which the transmitted signal is multiplied. This would invert the right side of the above formula.

If we convert from wavelength to frequency, the loss formula becomes:

$$\text{Line of sight loss} = (4\pi)^2 \, d^2 \, F^2/c^2$$

where

d is the transmission path distance in meters

F is the transmitted frequency in Hz

c is the speed of light (3×10^8 m/s)

Allowing distance to be input in km and frequency in MHz requires a conversion factor term. Combining terms and converting to dB form gives the loss in dB as:

$$L(\text{dB}) = 32.44 + 20 \log_{10} d + 20 \log_{10} F$$

where

d is the link distance in km

F is the transmit frequency in MHz

The 32.44-term combines the conversion factors and the c^2 and $(4\pi)^2$ terms converted to dB.

Alternate forms of this equation change the constant to:

- 36.52 if the distance is in statute miles;
- 37.74 if the distance is in nautical miles.

The formula is often used in applications to 1-dB accuracy. In this case the constants are simplified to 32, 37, and 38, respectively.

Nomograph

There is a widely used nomograph which gives the line-of-sight loss in dB as a function of the distance and the frequency. Like all nomographs, this one is just the graphical implementation of a formula. In this case, it is the formula given above for the line-of-sight loss. The nomograph is shown in Figure 5.7. To use this nomograph, draw a line between the frequency in MHz and the link distance in km. Your line crosses the center axis at the LOS loss in dB. In this figure, the loss at 1 GHz (i.e., 1,000 MHz) and 10 km is shown as just under 113 dB. Note that the above formula calculates the value at 112.44 dB.

Slide Rule

The slide rule provided with this book will allow you to quickly calculate line-of-sight loss. (It is called free-space loss on the slide rule.) Figure 5.8

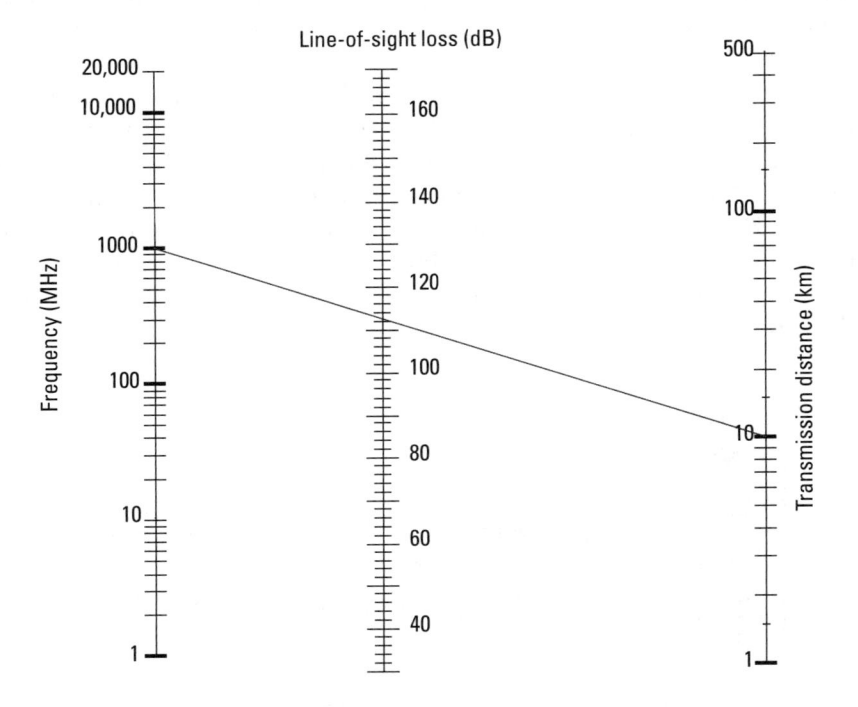

Figure 5.7 Line-of-sight loss nomograph.

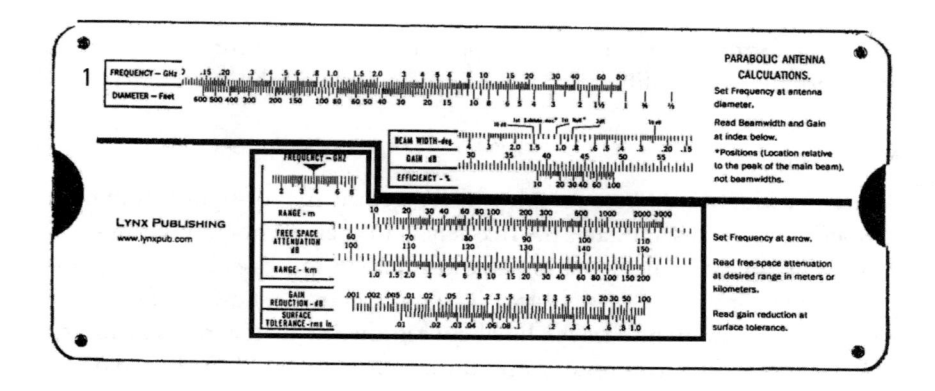

Figure 5.8 Front of antenna/propagation slide rule with line-of-sight attenuation scales highlighted.

shows the front (#1) side of the slide rule with the line-of-sight loss calculation area of the rule outlined. Figure 5.9 shows just the LOS calculation portion of the rule. To use the slide rule for LOS calculation, move the slide so that the transmit frequency is by the arrow at point A on Figure 5.9. Then, read the line-of-sight loss in dB against the link distance in km at point B in Figure 5.9.

In the example shown in this figure, the frequency is 300 MHz (0.3 GHz), the link distance is 25 km, and the LOS attenuation is 110 dB.

A shorter range scale for LOS loss is also provided on the slide rule. As shown in Figure 5.10, set the frequency at point A and read the free space against the link range in meters on the upper free-space attenuation scale. For this example, the frequency is still 300 MHz. Take the range as 25 meters (point C on the slide rule) and read the free-space attenuation at just less than 50 dB.

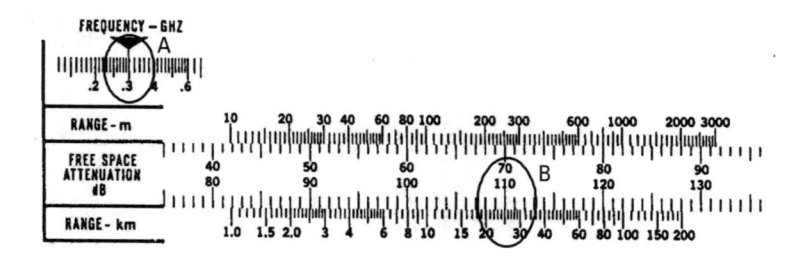

Figure 5.9 Close-up of free-space attenuation scales.

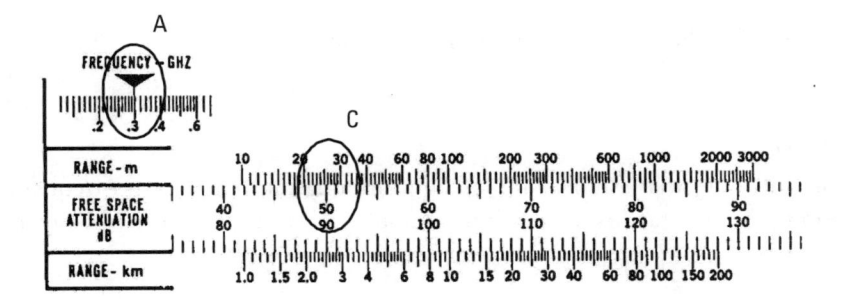

Figure 5.10 Close-up of free-space attenuation scales for short range.

5.5 Two-Ray Propagation

When the transmitting and receiving antennas are close to a single dominant reflecting surface (i.e., the ground or water) and the antenna patterns are wide enough to allow significant illumination of that surface, the two-ray propagation model must be considered. As we will see, the transmitted frequency and the actual antenna heights determine whether the two-ray or line-of-sight propagation model applies.

Two-ray propagation is also called "40 log d" or "d^4 attenuation" because the loss varies with the fourth power of the link distance. The dominant loss in two-ray propagation is the phase cancellation of the direct wave by the signal reflected from the ground or water as shown in Figure 5.11. The amount of attenuation depends on the link distance and the height of the transmitting and receiving antennas above the ground or water. This section provides three ways to calculate two-ray loss:

- A formula (in linear or dB form);
- A nomograph;

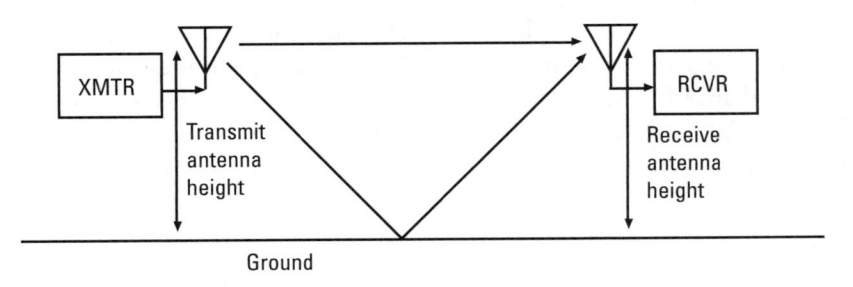

Figure 5.11 Direct and reflected rays close to the ground.

- The slide rule provided with this book.

Formula

You will note that (unlike line-of-sight attenuation) there is no frequency term in the two-ray loss expression. In nonlogarithmic form, the two-ray loss is:

$$L = d^4 / (h_T^2 \times h_R^2)$$

where
d = the link distance
h_T = the transmitting antenna height
h_R = the receiving antenna height

The link distance and antenna heights are all in the same units. The dB formula for the two-ray propagation loss is:

$$L = 120 + 40 \log(d) - 20 \log(h_T) - 20 \log(h_R)$$

where
d = the link distance in km
h_T = the transmitting antenna height in meters
h_R = the receiving antenna height in meters

Nomograph

Figure 5.12 gives a nomograph for the calculation of two-ray loss. To use this nomograph, first draw a line between the transmitting and receiving antenna heights. Then draw a line from the point at which the first line crosses the index line through the path length to the propagation loss line. In the example, two 10-meter high antennas are 30 km apart, and the attenuation is a little less than 140 dB. If you calculate the loss from either of the above formulas, you will find that the actual value is 139 dB. Note that the two antenna heights need not be equal for this nomograph to work.

Slide Rule

Figure 5.13 shows the back (#2) side of the slide rule with the scales for calculating two-ray loss highlighted. Figure 5.14 is a close-up of those scales.

Communications Propagation 131

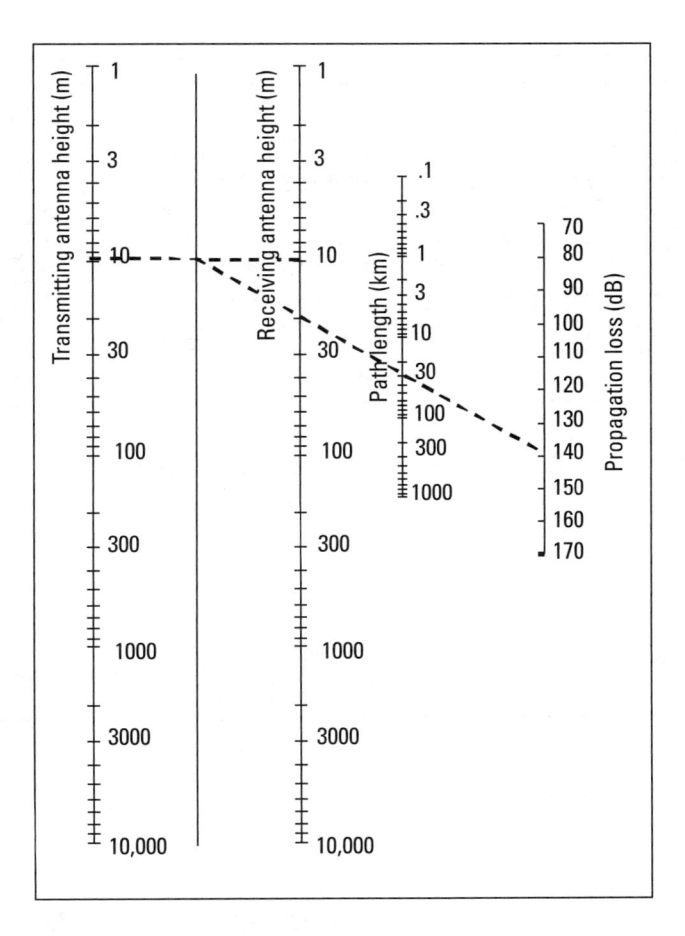

Figure 5.12 Two-ray propagation loss nomograph.

First move the slide so that the transmit antenna height (in meters) is aligned with the link distance in km. Then read the attenuation in dB at the receiving antenna height in meters. For the example shown in the figure, the link distance is 20 km. The transmitting antenna height (2 meters) is set next to 20 km at point A. Then the link loss is read at the receiving antenna height (30 meters) at point B. Note that the link attenuation is shown to be about 136.5 dB. You need to be a little careful when interpolating between numbered points on the attenuation scale; the attenuation increases as you move left on the scale, so the attenuation is 136.5 dB—not 143.5 dB.

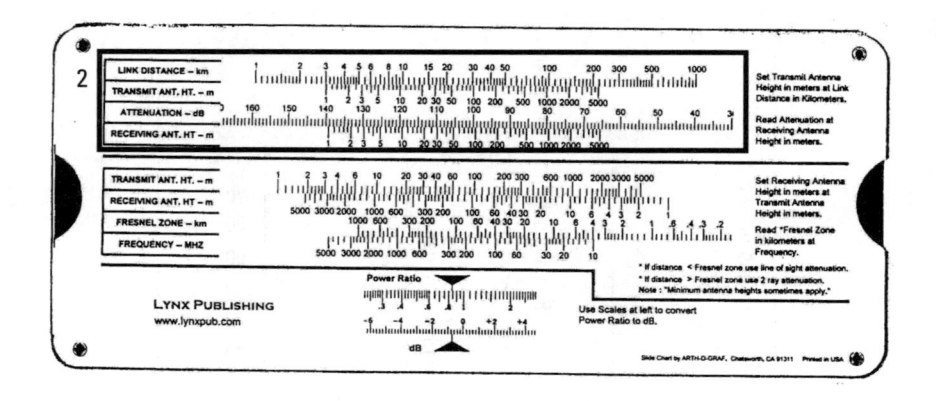

Figure 5.13 Back of slide rule with two-ray calculation scales highlighted.

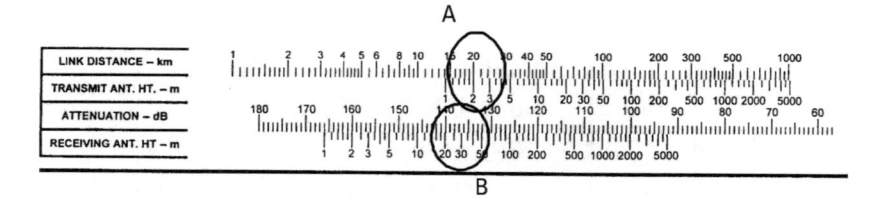

Figure 5.14 Close-up of two-ray propagation scales.

Minimum Antenna Height

Figure 5.15 shows minimum antenna height for two-ray propagation calculations versus transmission frequency. There are five lines on the graph, they are for:

- Transmission over sea water;
- Vertically polarized transmission over good soil;
- Vertically polarized transmission over poor soil;
- Horizontally polarized transmission over poor soil;
- Horizontally polarized transmission over good soil.

Good soil provides a good ground plane. If either antenna height is less than the minimum shown by the appropriate line in this graph, the minimum antenna height should be substituted for the actual antenna height

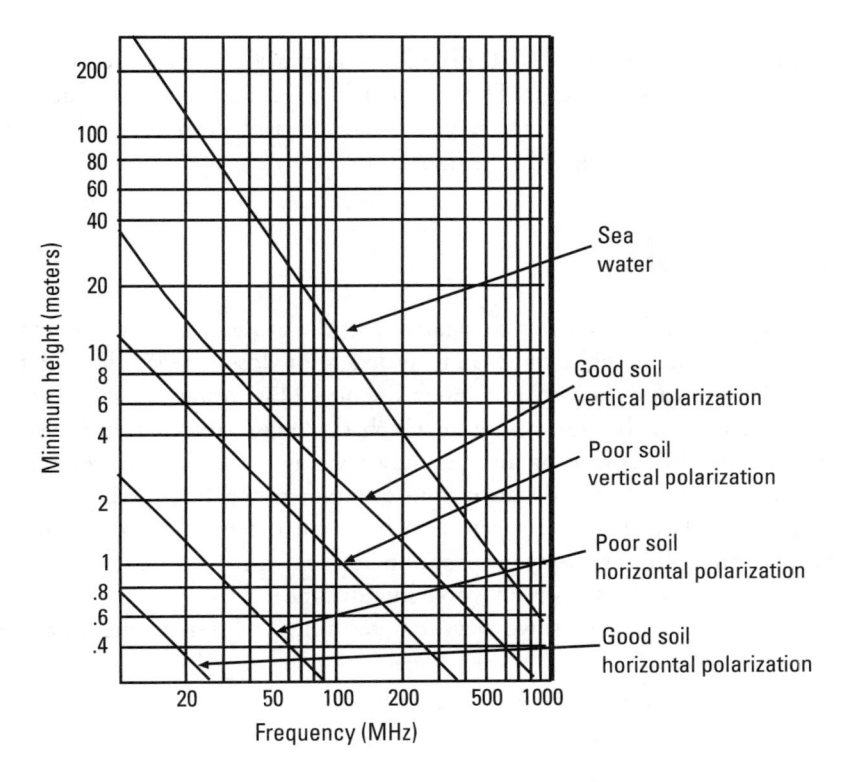

Figure 5.15 Minimum antenna heights for two-ray propagation.

before completing the two-ray attenuation calculation. Please note that if one antenna is actually at ground level, this chart is highly suspect.

A Note About Very Low Antennas

In the communication theory literature, discussions of very low antennas all seem to be constrained to antenna heights at least a half wavelength above the ground. A recent, far from complete, test (performed by the author) gives some insight into the performance of antennas lower than that. A 400-MHz, vertically polarized, 1-meter-high transmitter was moved various distances from a matched receiver while the receiver was lowered from 1-meter high to the ground. Over level, dry ground, the received power reduced by 24 dB when the receiving antenna was at the ground. With a 1-meter-deep ditch across the transmission path (near the receiver) this loss was reduced to 9 dB. Other recent (incomplete) testing seems to indicate that the loss between an antenna above the ground and a second antenna at the ground has a 20-dB cyclic variation as the range changes. The current practice (as this book goes to print) seems to be to add a 20-dB additional margin to predicted two-ray losses for EW applications. There is currently some significant activity among technical folks who are very interested in this problem, so look for some articles on this subject in the literature.

5.6 Fresnel Zone

As mentioned earlier, signals propagated near the ground or water can experience either line-of-sight or two-ray propagation loss—depending on the antenna heights and the transmission frequency. The Fresnel zone distance is the distance from the transmitter at which the phase cancellation becomes dominant over the spreading loss. As shown in Figure 5.16, if the receiver is less than the Fresnel zone distance from the transmitter, line-of-sight propagation takes place. If the receiver is farther than the Fresnel zone distance from the transmitter, two-ray propagation applies. In either case, the applicable propagation applies over the whole link distance.

This section provides two ways to determine the Fresnel zone distance:

- A formula (in two forms, neither in dB);
- The slide rule provided with this book.

Formula

The Fresnel zone distance is calculated from the following formula:

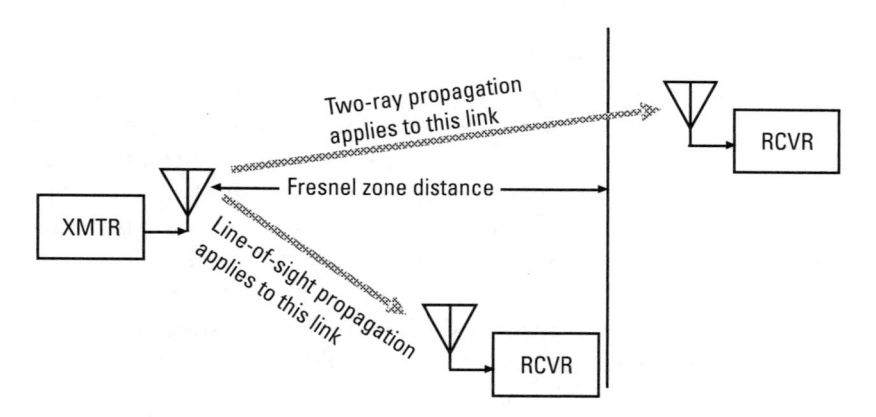

Figure 5.16 Selection of propagation mode based on Fresnel zone distance.

$$FZ = 4\pi h_T h_R / \lambda$$

where

FZ = the Fresnel zone distance in meters

h_T = the transmitting antenna height in meters

h_R = the receiving antenna height in meters

λ = the transmission wavelength in meters

Note that several different formulas for Fresnel zone are found in literature. This one is chosen because it yields the distance at which line-of-sight and two-ray attenuation are equal. A more convenient form of this equation is:

$$FZ = [h_T \times h_R \times F]/24{,}000$$

where

FZ = the Fresnel zone distance in km

h_T = the transmitting antenna height in meters

h_R = the receiving antenna height in meters

F = the transmission frequency in MHz

Slide Rule

Figure 5.17 shows the back (#2) side of the slide rule with the Fresnel zone calculation scales highlighted. Figure 5.18 shows a close-up of those scales.

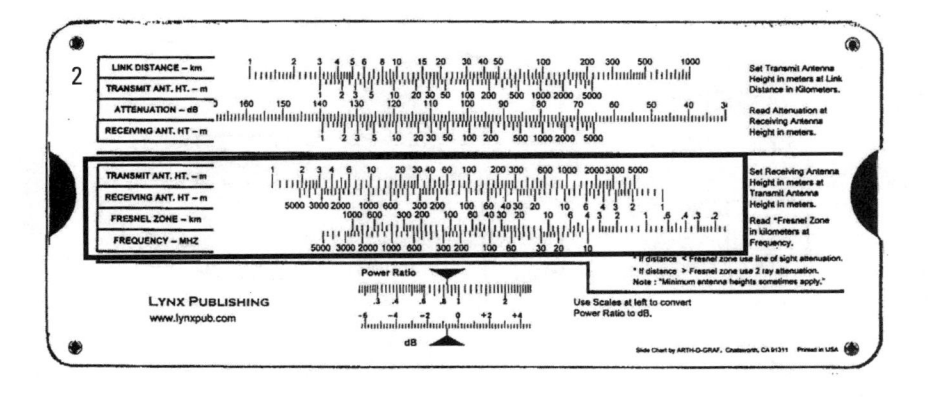

Figure 5.17 Back of slide rule with Fresnel zone scales highlighted.

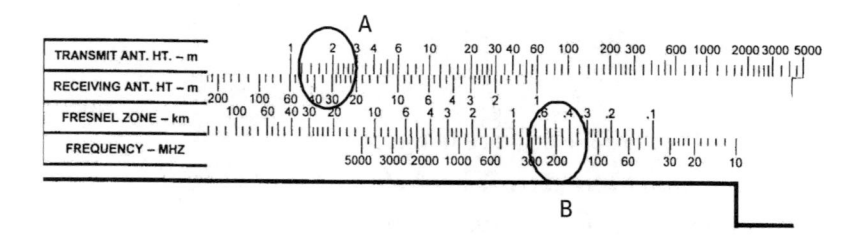

Figure 5.18 Close-up of Fresnel zone calculation scales.

To calculate the Fresnel zone distance, first move the slide so that the transmit antenna height in meters is aligned with the receiving antenna height in meters on the upper scale. Then read the Fresnel zone distance at the frequency in MHz on the lower scale. For the example shown in the figure, move the slide so that the transmit antenna height (2 meters) is lined up with the receiving antenna height (30 meters) at point A. Then read the Fresnel zone distance in km at the frequency (200 MHz) at point B. The Fresnel zone distance is found to be 0.5 km, so if the link is shorter than 500 meters, line-of-sight attenuation applies; if the link is longer than 500 meters, two-ray attenuation applies.

Complex Reflection Environment

In locations with very complex reflections—for example, when transmitting down a valley as shown in Figure 5.19, it is suggested in the literature that the line-of-sight propagation loss model will give a more accurate answer than the two-ray propagation model.

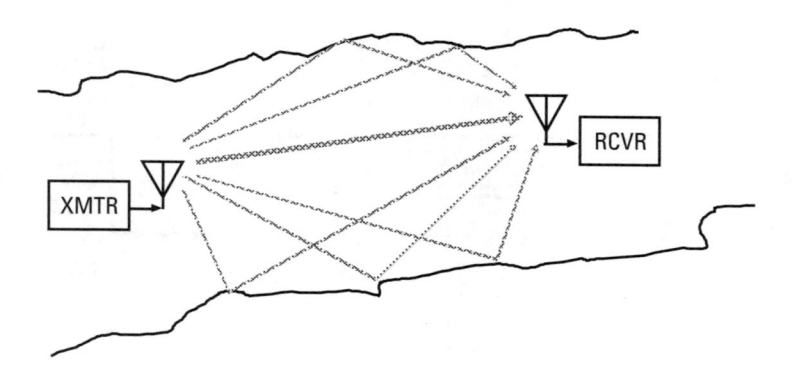

Figure 5.19 Transmission in a very complex reflection environment.

5.7 Knife-Edge Diffraction

Nonline-of sight propagation over a mountain or ridgeline is usually estimated as though it were propagation over a knife-edge. This is a very common practice and many EW professionals report that the actual losses experienced in terrain closely approximate those estimated by equivalent knife-edge diffraction (KED) estimation. In this section only a nomograph for the calculation of KED loss is provided.

The KED attenuation is added to the line-of-sight loss as it would be if the knife-edge were not present. Note that the line-of-sight loss rather than the two-ray loss applies when a knife-edge (or equivalent) is present. See Figure 5.20.

The geometry of the link over a knife-edge is shown in Figure 5.21. H is the distance from the top of the knife-edge to the line-of-sight as though the knife-edge were not present. The distance from the transmitter to the knife-edge is called d_1 and the distance from the knife-edge to the receiver is called d_2. For KED to take place, d_2 must be at least equal to d_1. If the receiver is closer to the knife-edge than the transmitter, it is in a blind zone in which only tropospheric scattering (with significant losses) provides link connection.

As shown in Figure 5.22, the knife-edge causes loss even if the line-of-sight passes above the peak—unless the line-of-sight path passes

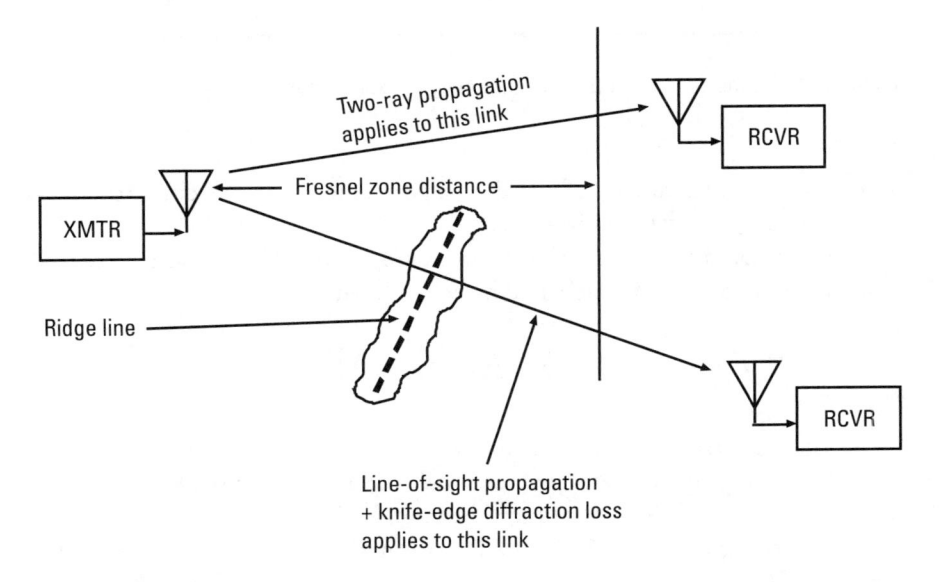

Figure 5.20 Propagation over a ridgeline versus two-ray propagation.

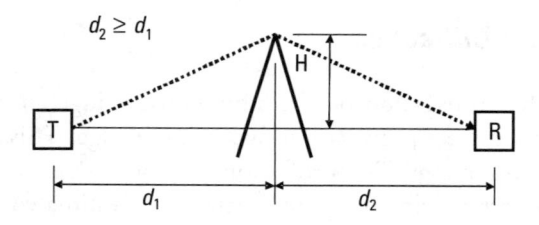

Figure 5.21 Knife-edge diffraction geometry.

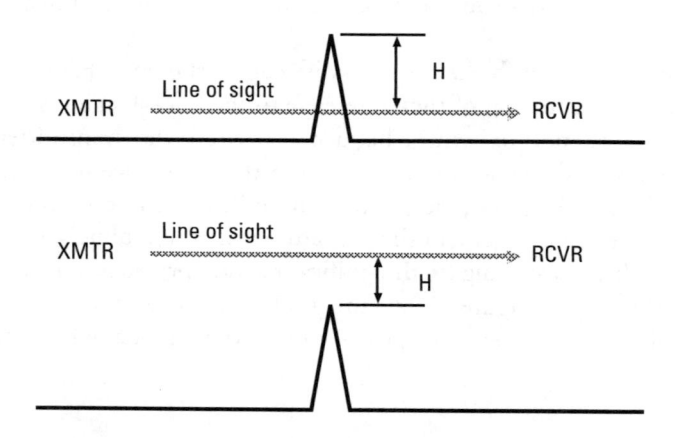

Figure 5.22 Line-of-sight path above or below the knife-edge.

several wavelengths above. Thus, the height value H can be either the distance above or below the knife-edge.

Figure 5.23 is a KED calculation nomograph. The left-hand scale is a distance value d which is calculated by the following formula:

$$d = \left[\sqrt{(2)}/(1 + d_1/d_2)\right] d_1$$

Table 5.2 shows some calculated values of d.

If you skip this step and just set $d = d_1$, the accuracy of the KED attenuation estimation will only be reduced by about 1.5 dB.

Returning to Figure 5.23, the line from d (in km) passes through the value of H (in meters). At this point, we don't care whether H is the distance above or below the knife-edge. Extend this line to the center index line.

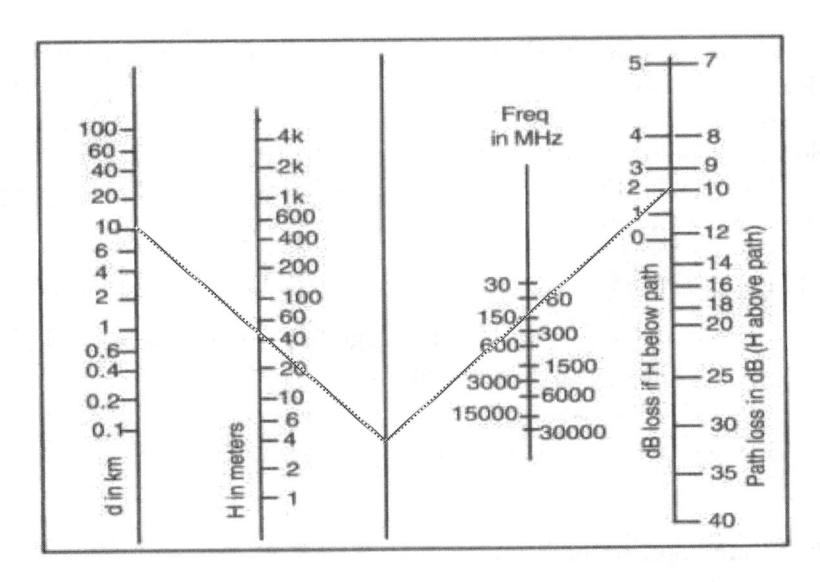

Figure 5.23 Knife-edge diffraction loss as a function of *d*, *H*, and frequency.

Table 5.2
Values of *d*

	d
$d_2 = d_1$	$0.707\, d_1$
$d_2 = 2\, d_1$	$0.943\, d_1$
$d_2 = 2.41\, d_1$	d_1
$d_2 = 5\, d_1$	$1.178\, d_1$
$d_2 \gg d_1$	$1.414\, d_1$

Another line passes from the intersection of the first line with the center index through the transmission frequency (in MHz) to the right-hand scale—which gives the KED attenuation. At this point, we identify whether H was above or below the knife-edge. If H is the distance above the knife-edge, the KED attenuation is read on the left-hand scale. If H is the distance below the knife-edge, the KED attenuation is read on the right-hand scale.

Consider an example which is drawn onto the nomograph. d_1 is 10 km, d_2 is 24.1 km, and the line-of-sight path passes 45 meters below the knife-edge.

d is 10 km (from Table 5.2) and H is 45 meters. The frequency is 150 MHz. If the line-of-sight path were 45 meters above the knife-edge, the KED attenuation would have been 2 dB. However, since the line-of-sight path is below the knife-edge, the KED attenuation is 10 dB.

The total link loss is then the line-of-sight loss without the knife-edge plus the KED attenuation:

$$\begin{aligned} \text{LOS loss} &= 32.44 + 20\log(d_1 + d_2) + 20\log(\text{frequency in MHz}) \\ &= 32.44 + 20\log(34.1) + 20\log(150) = 32.44 + 30.66 + 43.52 \\ &= \text{approximately } 106.6 \text{ dB} \end{aligned}$$

So the total link loss is $106.6 + 10 = 116.6$ dB.

5.8 Atmospheric and Rain Losses

5.8.1 Atmospheric Loss

This section covers atmospheric, rain and fog losses for links operating within the Earth's atmosphere. These losses are in addition to the propagation losses discussed earlier.

Figure 5.24 is a curve of atmospheric loss per kilometer of link distance versus frequency. To use this chart, start at frequency on the abscissa, go up to the curve and then left to the attenuation per km. The two peaks are caused by the two major contributors to atmospheric attenuation in the RF frequency range. The effect of water vapor peaks at 22 GHz and the effect of oxygen peaks at 60 GHz. You will note that the atmospheric loss is very low below 10 GHz. This loss is often ignored in HF, VHF, UHF, and low microwave frequency ranges.

The example on the figure is for a 50-GHz link. Go up from 50 GHz on the abscissa to the curve, then left to the ordinate, where the atmospheric loss is given as 0.4 dB per km.

Rain and Fog Loss

Rain loss is difficult to relate to real-world communication situations since rain is so dynamic. In general, these losses are calculated for some specified situation and probability. These calculated losses are then part of the determination of how much link margin must be provided.

Figure 5.25 is a diagram that might be used to trade-off various approaches to establishing the amount of rain loss which can be expected.

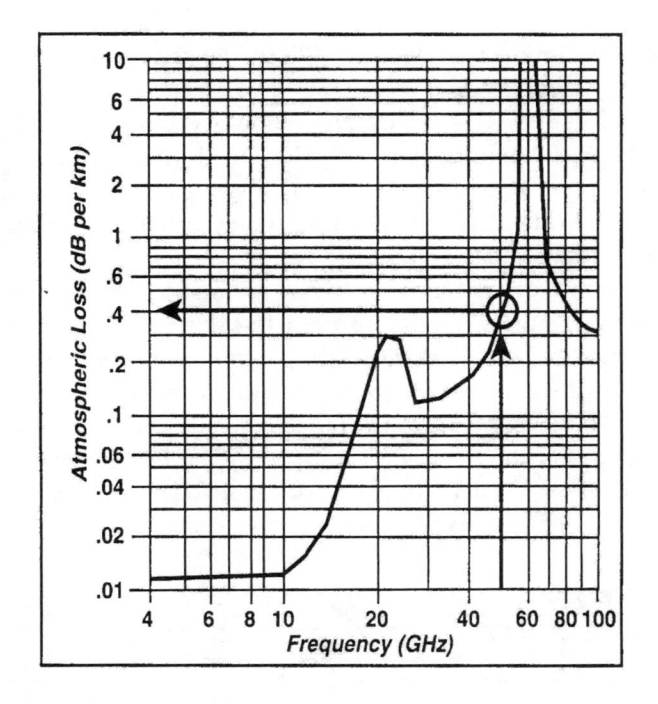

Figure 5.24 Atmospheric attenuation.

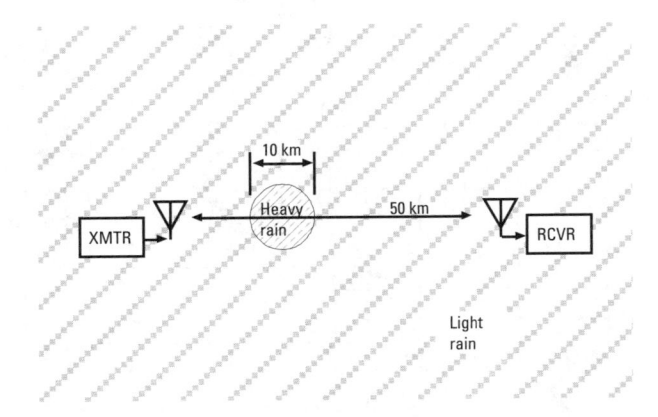

Figure 5.25 Link model for rain attenuation.

This comes from an evaluation of trade-offs for a particular unmanned aerial vehicle (UAV) link. The assumption is that there will be light rain over the 50-km maximum link range and that there will be 10 km of heavy rain cells in the path of the link. Thus, the link will experience 40 km of light rain and 10 km of heavy rain.

Table 5.3 defines what is meant by various levels of rain and fog. Figure 5.26 gives the attenuation per km for various levels of rain or fog (per Table 5.3) as a function of frequency. To use this chart, go up from the frequency on the abscissa to the curve representing the expected level of rain or fog. Then go left to the ordinate to read the loss per km.

The example drawn onto the figure is for 15 GHz, using the rain model shown in Figure 5.25. Go up from 15 GHz on the abscissa to the B curve (light rain) and go left to 0.033 dB per km on the ordinate. (Since the ordinate is a logarithmic curve, a value halfway between the 0.03 and 0.04 line is 0.033—0.5 being close to the log of 3.) Since the rain model calls for 40 km of light rain, the light rain causes 1.3 dB of loss. Then continue up from 15 GHz to the D curve (heavy rain). Then go left to the abscissa at 7.3 dB per km. Since we expect 10 km of heavy rain cells, the heavy rain is

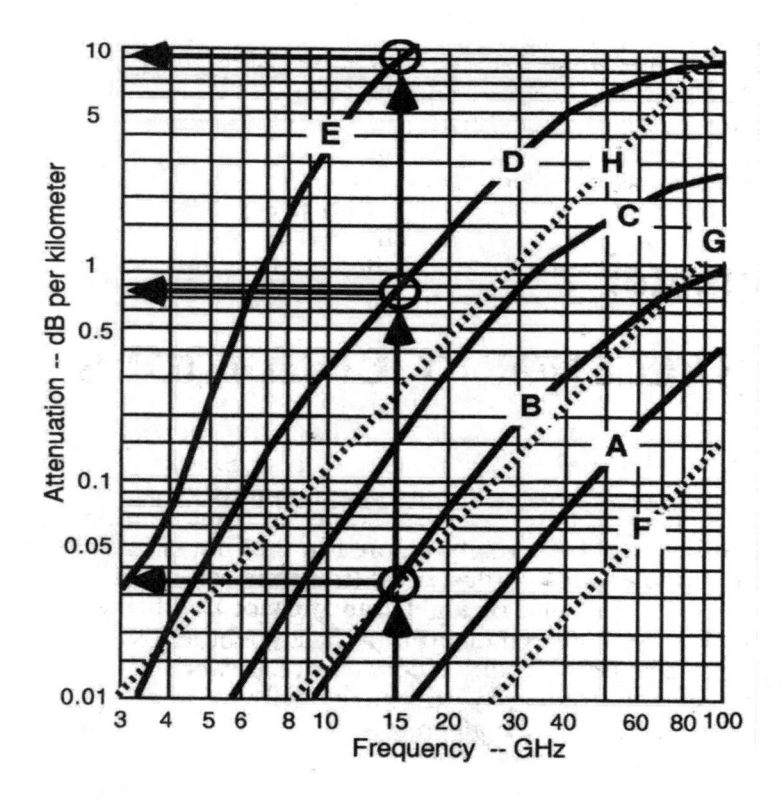

Figure 5.26 Rain and fog attenuation per km.

Table 5.3
Definitions of Rain and Fog Levels

Rain	A	0.25 mm/hr	0.01 in/hr	Drizzle
	B	1.0 mm/hr	0.04 in/hr	Light rain
	C	4.0 mm/hr	0.16 in/hr	Moderate rain
	D	16 mm/hr	0.64 in/hr	Heavy rain
	E	100 mm/hr	4.0 in/hr	Very heavy rain
Fog	F	0.032 gm/m^3	Visibility greater than 600 meters	
	G	0.32 gm/m^3	Visibility about 120 meters	
	H	2.3 gm/m^3	Visibility about 30 meters	

expected to cause 7.3 dB of loss. Thus the total rain loss is 8.6 dB above all other losses.

5.9 HF Propagation

This section is intended only to give general understanding of HF propagation. Note that we will discuss single-site locators along with other communications emitter location issues in Chapter 7. HF propagation is very complex, depending on time of day, time of year, location, and conditions (such as sun spot activity) that impact the ionosphere. An excellent article by Richard Groller in the June 1990 *JED* ("Single Station Location HF Direction Finding") is suggested as a starting point for further study. The next suggestion is a handbook such as *Reference Data for Radio Engineers (RDRE)* which includes probabilistic curves for HF propagation versus range and latitude. Finally, for specific ionospheric conditions, propagation parameters, and others, the Federal Communication Commission has a Web site with loads of data (www.fcc.gov).

In this coverage, we will discuss the ionosphere, ionospheric reflection, HF propagation paths, and single-site locator operation. The primary references for this section are Mr. Groller's article and the *RDRE*.

HF propagation can be line-of-sight, ground wave, or sky wave. Ground wave, which follows the Earth, is a strong function of the quality of the surface along the path. The FCC Web site has some curves for this propagation mode. Beyond about 160 km, HF propagation depends on sky waves reflected from the ionosphere.

The Ionosphere

The ionosphere is a region of ionized gasses from about 50 to 500 km above the Earth's surface. Its primary interest here is that it reflects radio transmissions in the medium- and high-frequency ranges. As shown in Figure 5.27, the ionosphere is divided into several layers:

- The D layer is from about 50 to 90 km above the Earth. It is an absorptive layer, with absorption decreasing with frequency. Its absorption peaks at noon and is minimal after sunset.

- The E layer is from about 90 to 130 km above the Earth. It reflects radio signals for short- and medium-range HF propagation during the daytime. Its intensity is a function of solar radiation and varies with the seasons and sunspot activity.

- Sporadic-E is a condition causing a short-term transient layer of ionization that is present in local summer, primarily in Southeast Asia and the South China Sea. It causes short-term changes in HF propagation.

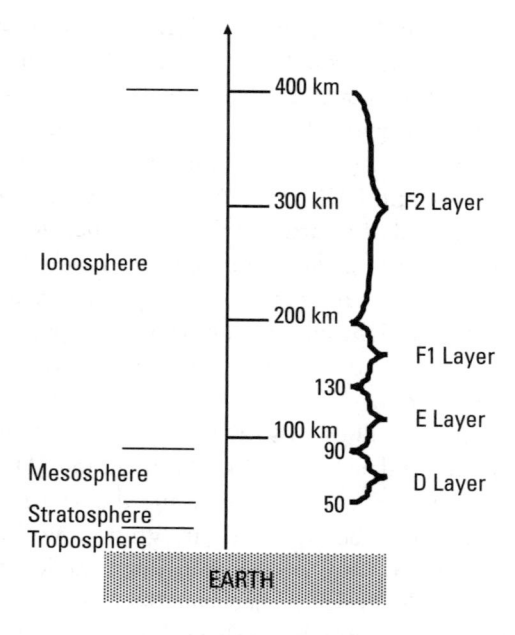

Figure 5.27 The ionosphere is characterized with D, E, F1, and F2 layers.

- The F1 layer extends from about 175 to 250 km above the Earth. It exists only during daytime and is strongest during summer and periods of high sun spot activity. It is most prominent at the middle latitudes.

- The F2 layer extends from about 250 to 400 km above the Earth. It is permanent, but extremely variable. It allows long range and nighttime HF propagation.

Ionospheric Reflection

Reflection from the ionosphere is characterized by virtual height and critical frequency. The virtual height as shown in Figure 5.28 is the apparent point of reflection of a signal from an ionspheric layer. This is the height measured by sounders which transmit vertically and measure the round-trip propagation time. As frequency is increased, the virtual height increases until the critical frequency is reached. At this frequency, the transmission passes through the ionospheric layer. If there is a higher layer, the virtual height increases to the higher layer.

The maximum frequency at which reflection can occur is also a function of the elevation angle (θ in Figure 5.28) and the critical frequency (F_{CR}). The maximum usable frequency (MUF) is determined by the formula:

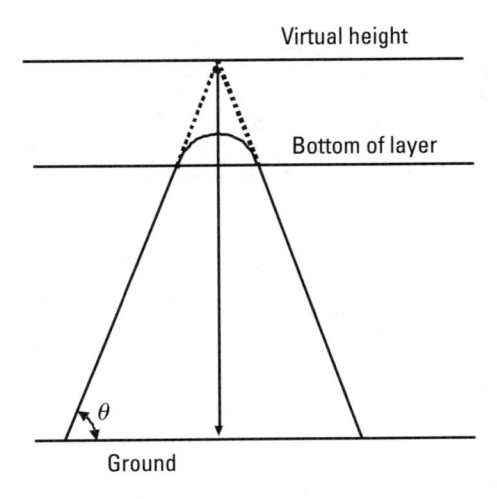

Figure 5.28 The virtual height of the ionosphere is the apparent reflection point for HF transmissions.

$$MUF = F_{CR} + Sec(\theta)$$

HF Propagation Paths

As shown in Figure 5.29, there can be several different transmission paths between a transmitter and a receiver depending on ionospheric conditions. If the sky wave passes through one layer, it may be reflected from a higher layer. There can be one or more hops from the E layer, depending on the transmission distance. If the E layer is penetrated, one or more hops from the F layer can occur. At night, this would be from the F2 layer and in the daytime from the F1 layer. Depending on the local density of various layers, there can also be hops from the F layer to the E layer, back to the F layer and finally to the Earth.

The received power from sky wave propagation is predicted by the following formula:

$$P_R = P_T + G_T + G_R - (L_B + L_i + L_G + Y_P + L_F)$$

where

P_T is the transmitter power
G_T is the transmit antenna gain
G_R is the receiving antenna gain
L_B is the spreading loss

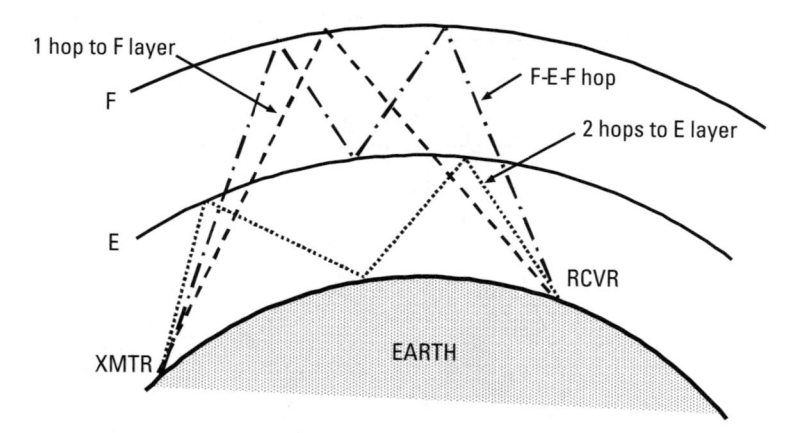

Figure 5.29 There are several possible propagation paths from a transmitter to a distant receiver, depending on the signal frequency, the ionospheric conditions, and the transmitter/receiver location geometry.

L_I is the ionospheric absorption loss

L_G is the ground reflection loss (for multiple hops)

Y_P is the miscellaneous loss (focusing, multipath, polarization, and so forth)

L_F is the fading loss

5.10 Satellite Links

Communication satellites have their bandwidth rented out by the Hz, and there is a whole different approach to calculating satellite link performance. Our purpose here is to discuss the way a terrestrial system handles transmission to or from a satellite.

First, consider how far it is from the satellite to a ground station which is either transmitting or receiving. If we assume a low-Earth satellite which orbits the Earth each two hours, it will have an average altitude of 1,698 km (see Figure 5.30). If the satellite is 5° above the local horizon at the Earth station, the link distance will be 4,424 km. If the satellite is in a synchronous

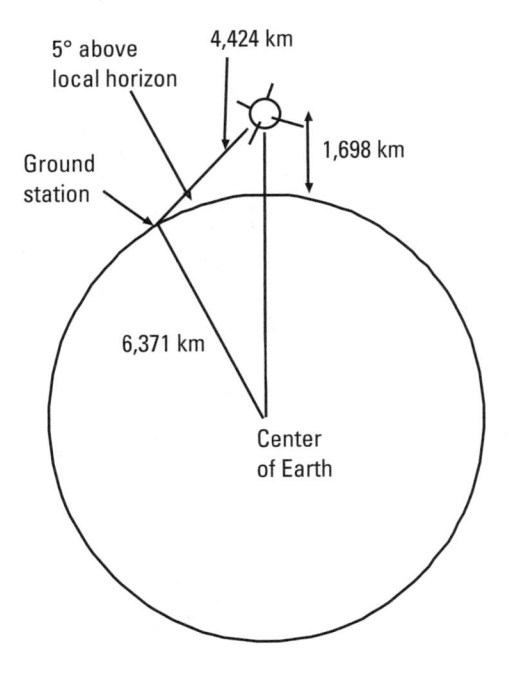

Figure 5.30 Range to low-Earth satellite.

orbit (i.e., stays over one point on the equator), it has an altitude of 38,000 km. If it is 5° above the horizon from the ground station, the path length is 41,346 km (see Figure 5.31). If the satellite is in synchronous orbit and has

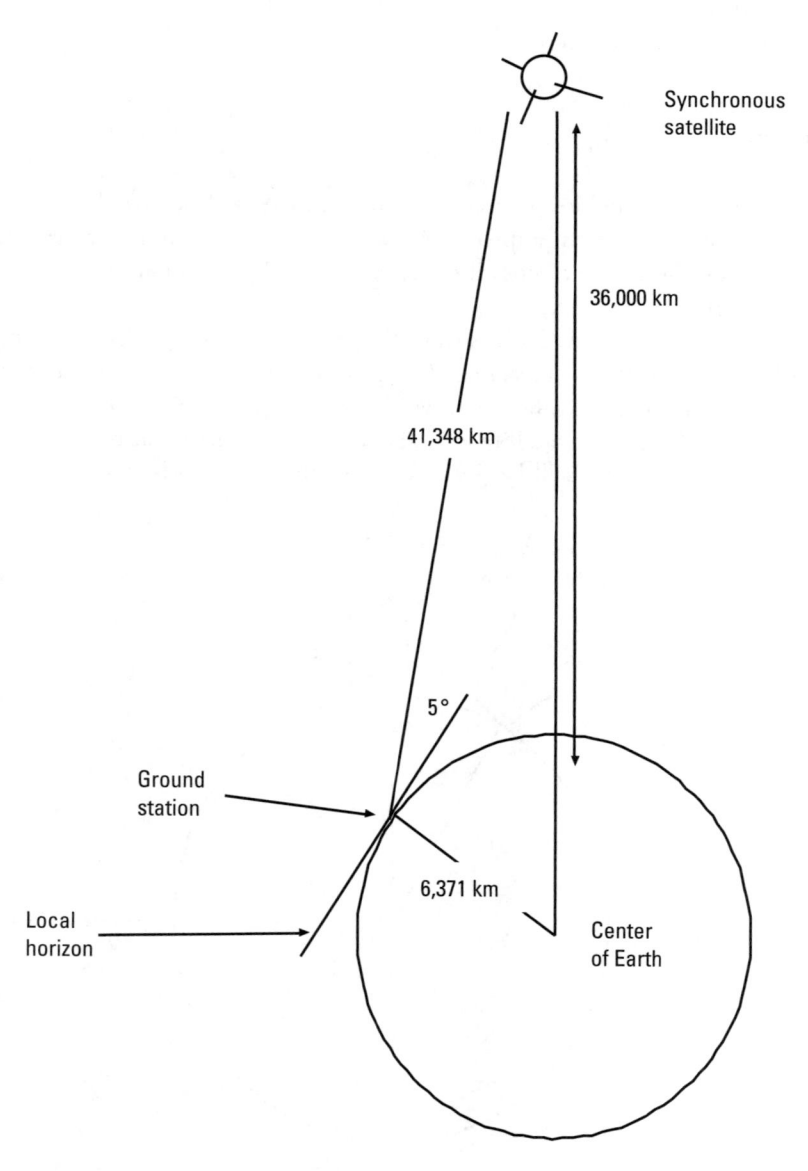

Figure 5.31 Range to a synchronous satellite.

an Earth-coverage antenna as shown in Figure 5.32, the beamwidth of that antenna is 17.3°.

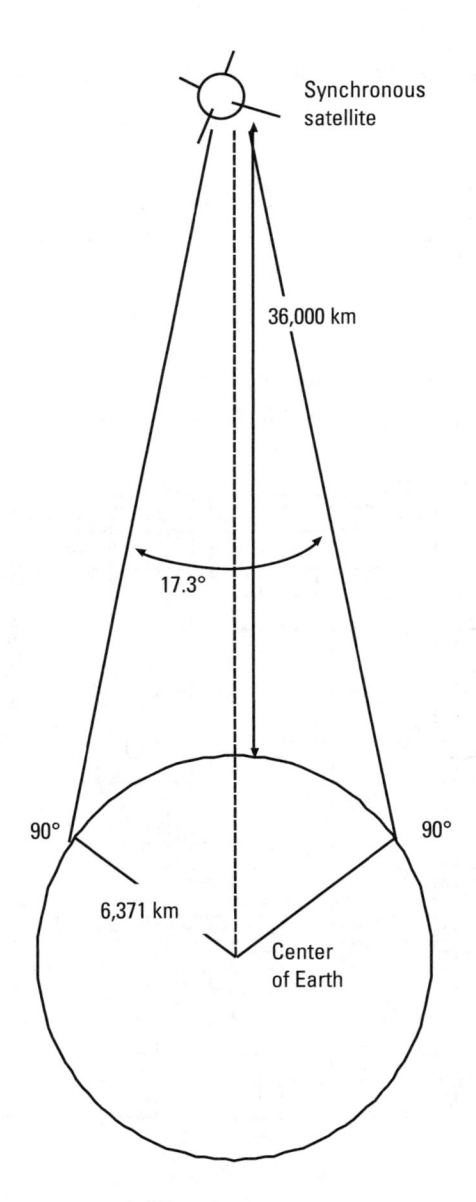

Figure 5.32 Earth-coverage satellite antenna.

Line-of-Sight Loss

The satellite link clearly meets the criteria for line-of-sight propagation, so the LOS loss can be calculated from the formula:

$$\text{LOS loss} = 32.44 + 20\log(\text{distance}) + 20\log(\text{frequency})$$

Atmospheric Loss

The atmospheric loss can be found from Figure 5.33, which gives the attenuation through the whole atmosphere as a function of elevation angle from the ground station and the frequency of transmission.

Rain Loss

Rain loss is a little more complicated, since rain falls only from the 0° isotherm. That is the elevation at which the atmosphere is at the freezing point.

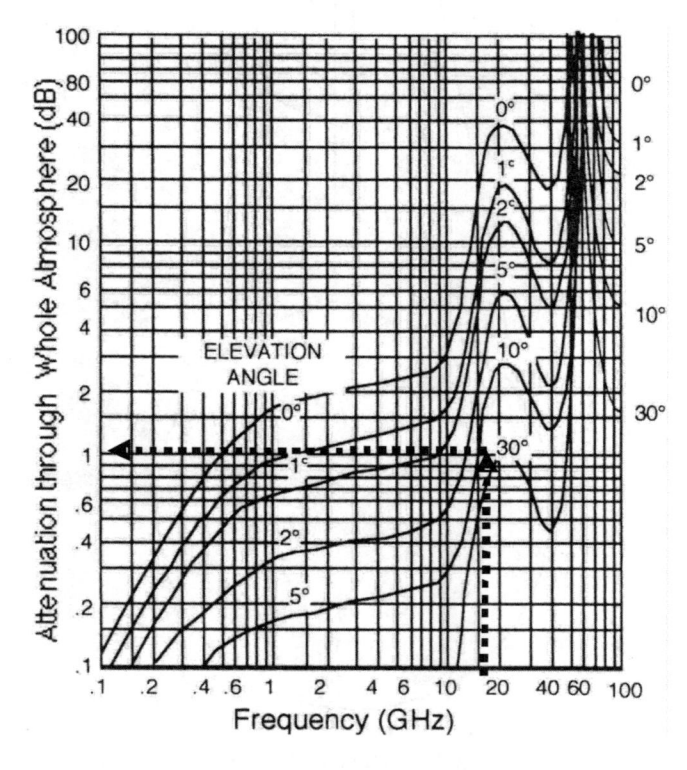

Figure 5.33 Attenuation through full atmosphere.

The 0° isotherm is found in many communication satellite books as curves of probability that it is at, or below, a given altitude versus latitude. A sample value that we will be using in an example is: At 40° latitude (north or south), there is a 0.01% probability that the 0° isotherm is at or below 3.3-km altitude. Now we must calculate the path length that falls between the ground station and the 0° isotherm. The path through the rain is calculated by the geometry shown in Figure 5.34 by taking the sine of the elevation angle of the satellite from the ground station and dividing the 0° isotherm altitude by that sine value. Once we know the path length through the rain, we multiply that distance by the loss per km from Figure 5.26.

An Example

Let's consider the link for the case in which a ground station at 40° latitude is intercepting a signal from a synchronous satellite which is 5° above the local horizon at 5 GHz and we want it to operate in heavy rain with a 0.01% probability 0° isotherm.

The link loss is:

$$\text{LOS loss} = 32.44 + 20 \log(5{,}000) + 20 \log(38{,}000) = 198.0 \text{ dB}$$

Atmospheric loss at 5° elevation at 5 GHz is 0.4 dB.

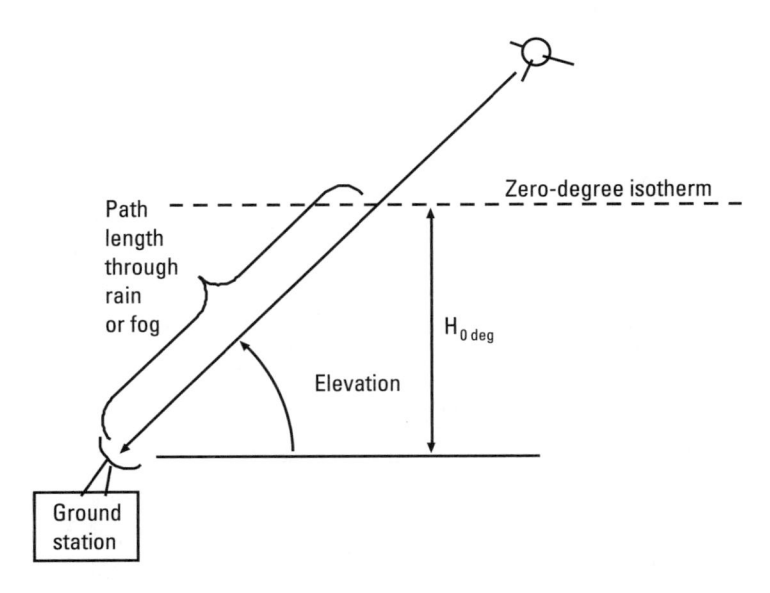

Figure 5.34 Path length through rain and fog.

The distance through the rain is 3.3 km/sin(5°) = 37.9 km.
Heavy rain causes 0.43 dB loss per km at 5 GHz.
Rain loss is 37.9 × 0.43 dB = 16.3 dB.
So the total link loss is: 198.0 + 0.4 + 16.3 = 214.7 dB.

6

Search for Communication Emitters

Military organizations go to great trouble to keep their operating frequencies from being known by enemies. However, in general, it is necessary to know the frequency at which an enemy is operating in order to perform the various EW operations. Thus, frequency search is an important EW function. When an EW receiving system employs directional antennas, angular search is also an issue. This chapter discusses both angular and frequency search for communications emitters. However, because so many communications receiving systems use 360°-coverage emitter-location systems, the main emphasis is on frequency search. It starts with general search considerations, then covers search techniques for conventional communication signals, and finally covers search techniques for low probability of intercept (LPI) signals.

Ideally, the receiving part of an EW system would be able to see in all directions at once—at all frequencies—for all modulations—with extremely high sensitivity. While such a receiving system could be designed, its size, complexity, and cost would make it impractical for most applications. Therefore, the practical EW receiving subsystem represents a trade-off of all of the above mentioned factors to achieve the best probability of intercept within the imposed size, weight, power, and cost constraints. This problem has been made significantly more difficult by the presence and increasing use of LPI communication.

6.1 Probability of Intercept (POI)

Other definitions of probability of intercept are used, but in EW, the following definition is generally accepted:

The probability that the EW system will detect the presence and some parameters of a particular threat signal between the time it first reaches the EW system's location and the time at which it is too late for the EW system to do its job.

Most EW receivers are specified to achieve a probability of intercept of 90% or 100% for each of the signals in its threat list, when a specified set of signals is present in a specified scenario, within a specified time.

6.2 Search Strategies

In general, the approach used in search for threat signals is one of the following:

- General search;
- Directed search;
- Sequential qualification.

6.2.1 General Search

General search assumes no prior knowledge of the presence of specific signals of interest. It is often called "the first day of the world" approach. It is also called the less dignified, "garbage collection." Every possible direction of arrival and frequency is considered without preference or priority. The product of a general search is a "map" of the environment that allows more sophisticated subsequent search or direct action against important enemy assets discovered.

6.2.2 Directed Search

Directed search takes advantage of some knowledge of the environment. It is practical to store the frequency, modulation, and priority of many signals, even in small-scale receiving systems. When specific signals of interest are assigned high priority, those frequencies (and directions of arrival if appropriate) are recalled from memory and checked first. Then, the rest of the environment is covered in priority order. The most important (or perhaps the most dangerous) frequency bands or locations are checked next, and then the rest of the environment is searched. Frequencies or locations known not to be of interest are skipped to save time. Frequencies and locations of high-interest signals are revisited often, while lower priority signals are revisited less often.

6.2.3 Sequentially Qualified Search

Sequential qualification involves the quick measurement of some parameter of any signal found, so that priorities can be applied to determine if it is worth more time to look for more emitter parameters. In general, the parameters which can be measured in the least time are the first sorting parameters.

The most common approach is to first search a prioritized frequency range for any signal energy. This may take only microseconds per channel. When energy is found, the time is taken to determine the next parameter.

The second step may be to determine the modulation of the signal or its general location. The modulation can be determined by spectral analysis, which is quite fast when fast Fourier transforms (FFT) can be employed. The general location (i.e., is the emitter in friendly or enemy territory) is generally available in minimum time when the EW system has emitter-location capability. As will be discussed in Chapter 7, emitter-location systems generally make multiple data collections and calculations which are averaged before a full-accuracy emitter-location report is generated. The first calculation is generally complete in a very small part of the report generation time. If this first calculation result is available, it will generally be of much lower precision than the final report, but good enough to determine if milliseconds should be invested in the determination of additional qualification parameters in the search.

Several such levels of qualification may be used before long-duration collection or analysis of a signal is initiated.

6.2.4 A Useful Search Tool

Figure 6.1 shows a commonly used tool to assist in the development or evaluation of a frequency search approach. It is a graph of frequency versus time on which the characteristics of target signals and the time versus frequency coverage of one or more receivers can be plotted. The frequency scale should cover the whole frequency range of interest (or some part of that range) and the time scale should be long enough to show the search strategy. The signal depictions show each signal bandwidth versus expected duration of the

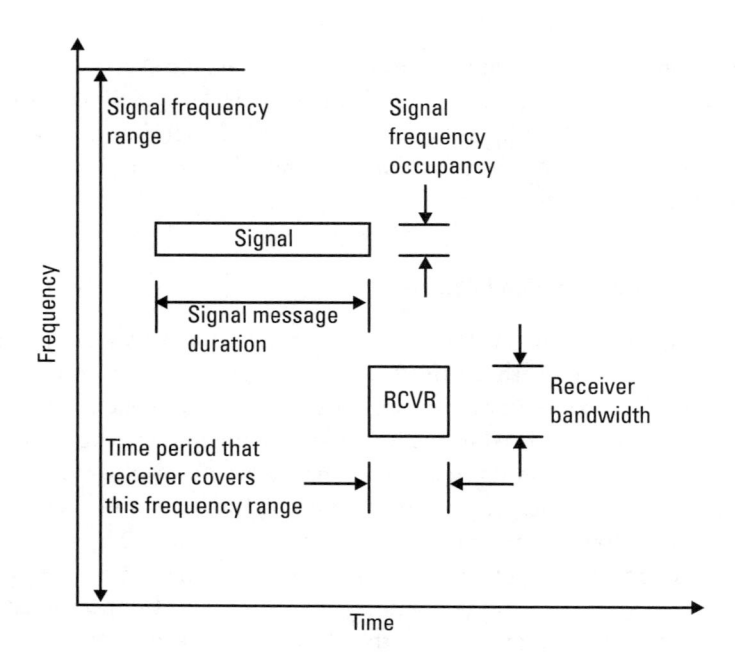

Figure 6.1 Search planning tool.

signal. If the signals are periodic or change frequency in some predictable way, these characteristics can be shown on the graph. The receiver is shown tuned to a particular frequency with its bandwidth and the time during which it covers that specific frequency increment.

A typical sweeping receiver strategy is shown in Figure 6.2. The parallelograms show the frequency versus time coverage of the swept receiver. The receiver bandwidth is the height of the parallelogram at any frequency and the slope is the receiver's tuning rate. Note that signal A is optimally received (its whole bandwidth over its whole duration). Signal B is received if you do not need to see its whole bandwidth, and Signal C is received if you do not need to see its whole duration. You can set the rules to fit the nature of the signals and the purpose of your search.

6.2.5 Practical Considerations Affecting Search

Theoretically, a receiver can sweep at a rate which allows a signal to remain within the receiver bandwidth for a time equal to the inverse of the bandwidth (e.g., 1 μs in a 1-MHz bandwidth). However, system software requires

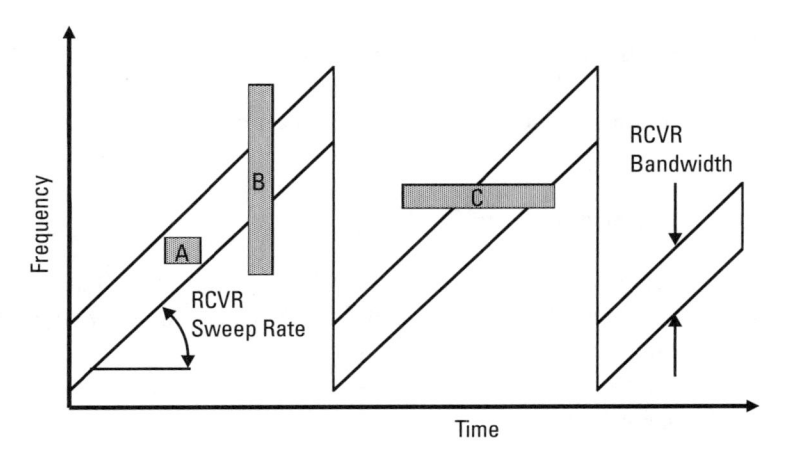

Figure 6.2 Receiver sweep for signals of interest.

time to determine if a signal is present. This might require as much as 100 to 200 μs—which can be significantly more than a time equal to 1/bandwidth.

Performing processing on each signal present (such as modulation analysis or emitter location) will typically take longer to identify the signal as a signal of interest. This level of processing may take as much as one or more milliseconds for each signal found.

Technology Issues

Years ago, military intercept receivers were mechanically tuned, so it was necessary to search by manually tuning or by automatically tuning across the whole covered band in a single sweep that was more or less linear. This approach was commonly called garbage collection because you needed to look at every signal in the environment and then pick the few signals of interest from a typically large collection of signals that were not of interest. Identification of signals of interest required rather complex analysis by a trained operator. Remember that 50 years ago, computers were rooms full of vacuum tubes requiring large quantities of forced air cooling and that their capabilities were miniscule compared to modern computers.

With the availability of digitally tuned receivers and one-chassis (and eventually one-chip) computers with massive memory and blinding speed, much more civilized search approaches became practical. Now you can store the frequencies of known signals of interest and automatically check each one before looking for new signals of interest. An FFT can be performed on each potential signal of interest to allow computerized spectral analysis. An interest/no-interest determination can be made from the results of the spectral analysis; perhaps enhanced by a quick look at the general emitter location (if the system has direction finding or emitter location capability).

6.3 System Configurations

The configuration of an EW system has infinite possibilities. The purpose here is only to discuss the configuration features which impact search. The three general configurations considered are:

- Single receiver;
- Search and monitor receivers;
- Search monitor and special receivers;
- Systems with emitter location capability.

Single Receiver

When a system has only one receiver as shown in Figure 6.3, that receiver must search for a signal of interest and then monitor that signal as long as required before search can resume. While this approach simplifies the system, it has great disadvantages in the probability of intercept in a rapidly changing environment or when various signals of interest have different priority.

Search and Monitor

It is very common to use one receiver to search and have one or more receivers than can be assigned to signals of interest for long-term monitoring or data collection. The search receiver can be either the same type as the

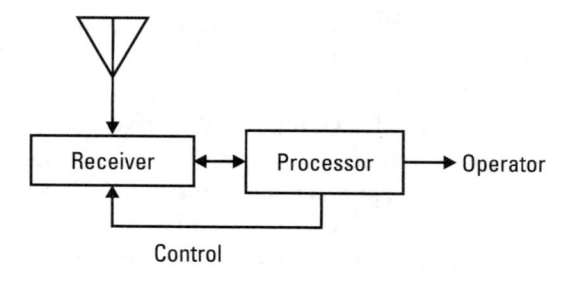

Figure 6.3 Single receiver system.

monitor receivers, or it can be a wideband receiver that determines only frequency see Figure 6.4.

Special Receivers

Special receivers have special attributes that are used as required, particularly in systems which employ the sequential qualification search strategy. Figure 6.5 shows a receiver system with search and monitor receivers supplemented with a special receiver. Examples of special receivers are:

- Digital modulation analysis receiver;
- Direction-finding receiver;
- Special receiver for LPI signals.

When the search receiver encounters a signal, it typically only determines the frequency. This may not be enough information to determine whether or not a monitor receiver should be assigned to that signal. A digital receiver can perform an FFT to produce the frequency spectrum of the signal. Analysis of this spectrum can determine:

- The modulation of the signal;
- The modulation parameters;
- If it is encrypted, what type of encrypter.

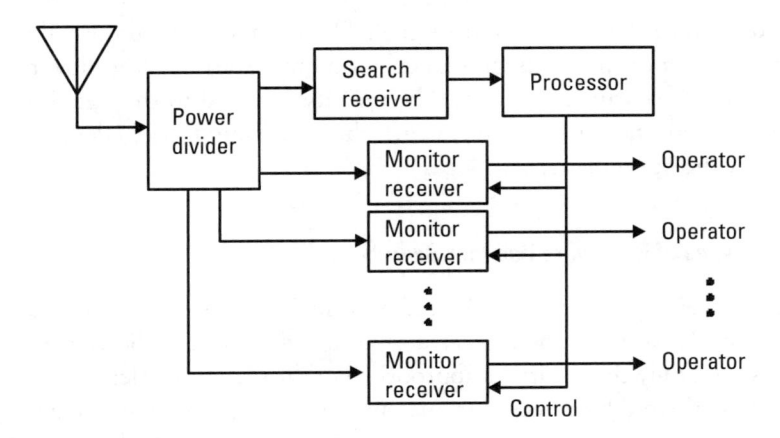

Figure 6.4 System with search and monitor receivers.

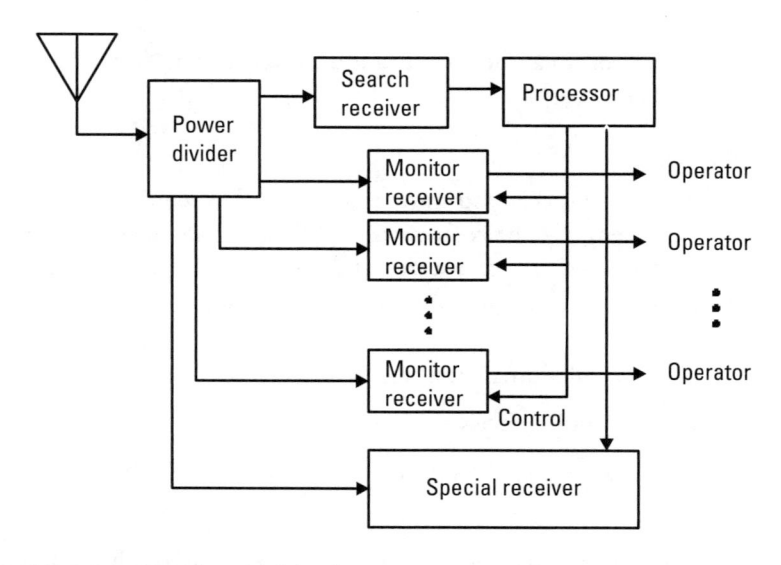

Figure 6.5 System with special receiver.

All of this information may be enough to evaluate the priority of the signal to determine if one of the system's monitor receivers should be assigned to it.

A direction-finding receiver is part of a direction-finding (DF) system, which will typically have its own antenna array. Most direction-finding systems can make a quick, but not too accurate estimate of the angle of arrival of the signal. This can be reported out to a processor as a parameter which will help determine the priority of the signal. The DF system then normally averages several measurements and calculations to determine the direction of arrival to full specified accuracy. The priority assigned to the signal may be used to determine if the signal is worth the processing time to develop a full accuracy direction of arrival.

6.3.1 Types of Receivers Used for Search

The receiver types used in EW and reconnaissance systems include those shown in Table 6.1. These receive types are described in Section 4.1: the table shows only their features appropriate to the search problem.

Superheterodyne receivers measure frequency and recover any type of signal modulation. They typically receive only one signal at a time, so they are not affected by multiple simultaneous signals. They can have good

Table 6.1
Search Receivers

Receiver Type	Sensitivity	Information Determined
Instantaneous frequency measurement	Low	Provides only frequency, can handle only one signal at a time
Superheterodyne	High	Provides frequency, recovers modulation, receives one of multiple simultaneous signals
Channelized	High	Provides frequency, recovers modulation, can receive multiple simultaneous signals
Compressive	High	Provides only frequency
Digital	High	Provides frequency, recovers modulation, can receive multiple simultaneous signals, performs spectral and other analysis

sensitivity, depending on the bandwidth. One important feature of superheterodyne receivers is that they can be designed with almost any bandwidth, providing a trade-off between frequency coverage and sensitivity

Compressive (or micro-scan) receivers sweep a wide frequency range very quickly—often within a single pulse width. They measure the frequency and received signal strength of multiple simultaneous signals, and have good sensitivity; however, they cannot recover signal modulation.

Channelized receivers simultaneously measure frequency and recover the full modulation for multiple signals as long as they are in different channels. They can also provide good sensitivity, depending on the channel bandwidth. However, the narrower the bandwidth the more channels are required to cover a given frequency range.

Digital receivers digitize a large frequency segment which is then filtered and demodulated in software. They can measure frequency and recover the full modulation for multiple simultaneous signals. They can provide good sensitivity.

6.3.2 Digitally Tuned Receiver

Figure 6.6 shows a digitally tuned superheterodyne receiver. A digitally tuned receiver has a synthesizer local oscillator and an electronically tuned preselector to allow very fast selection of any signal frequency within the tuning range. Tuning can be either by operator or computer control.

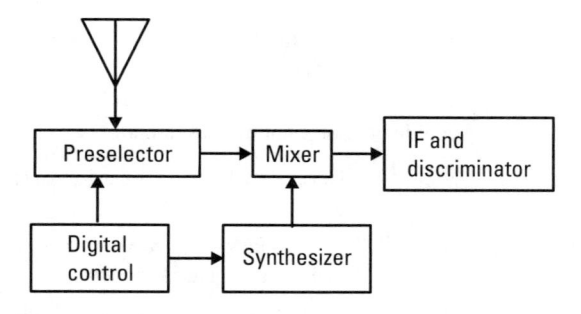

Figure 6.6 Digital receiver system.

Figure 6.7 shows a phase-lock-loop synthesizer block diagram. Note that the voltage-tuned oscillator is phase locked to a multiple of the frequency of an accurate and stable crystal oscillator. This means that the tuning of a digitally tuned receiver is accurate and repeatable—making the search approach described above practical. Note that the bandwidth of the feedback loop in the synthesizer is set at the optimum compromise between low-noise signal output (i.e., narrow loop bandwidth) and high-tuning speed (i.e., wide loop bandwidth). In the search mode, time must be allowed for the synthesizer to settle before beginning the analysis of any signals in the selected instantaneous receiver bandwidth coverage.

When a digitally tuned receiver is used in the search mode, it is tuned to discrete frequency assignments as shown in Figure 6.8. The search need not move linearly through the whole band of interest, but can check specific frequencies or scan frequency subbands of high interest in any order desired. It is often desirable to provide 50% overlap of receiver tuning steps. This prevents a band-edge intercept of a signal of interest. On the other hand, a 50% overlap will require twice as much time to cover the signal range of interest.

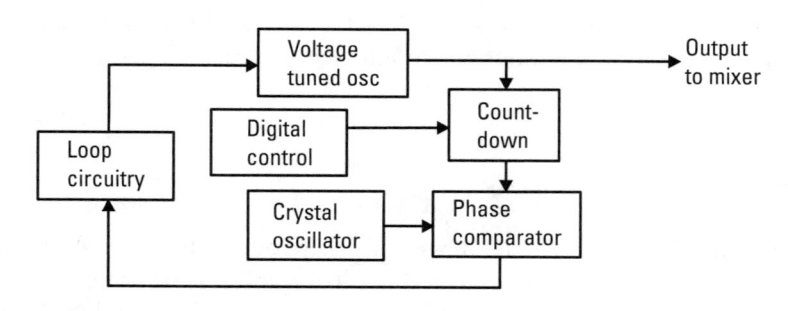

Figure 6.7 Phase-locked loop synthesizer.

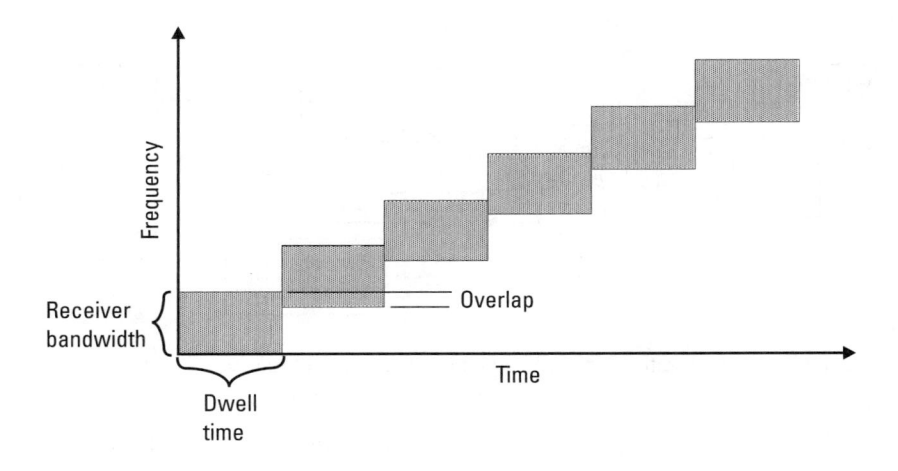

Figure 6.8 Search with digital receiver.

The amount of overlap is a trade-off that must be made to optimize the search in a specific situation.

6.3.3 Digital Receivers

Since digital receivers have a great deal of flexibility, they may one day handle the whole search and monitor job. They are restricted by the state of the art in digitization and computer processing (versus size and power requirements)—but the state of the art in these areas is changing almost daily. Watch digital receivers for the near future. The use of a digital frequency in the search mode is discussed in Section 6.3.2.

6.3.4 Frequency Measuring Receivers

Figure 6.9 shows another approach. Again, the antenna output must be power divided and one receiver provides set-on information for multiple narrowband receivers. However, now a wideband frequency measurement receiver is used. The frequency measuring receiver can be an IFM receiver, a compressive receiver, or (if practical) a Bragg cell receiver. Since this receiver can only measure the frequency of the signals present, the processor must assign the set-on receivers based only on frequency. The processor will keep a record of all signals which have been recently found. Typically, it will assign monitor receivers only to new or high-priority signals.

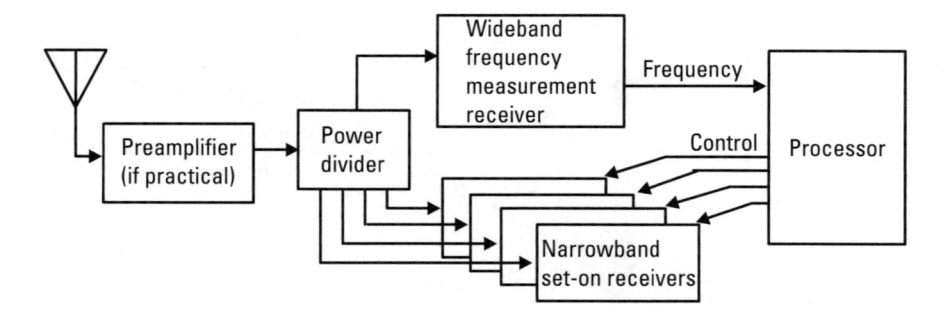

Figure 6.9 Search with wideband frequency measurement receiver.

Because some types of frequency measurement receivers have poorer sensitivity than the narrowband set-on receivers, they may not be able to receive some of the signals that could be monitored. A mitigating factor is that it often requires less received signal strength to detect the presence of a signal and measure its RF frequency than is required to get the full signal modulation.

6.3.5 Energy Detection Receivers

Noise-like signals such as DSSS spread spectrum signals do not have the obvious modulation parameters used by conventional receivers to handle conventional signals. A widely spread signal presents such a low signal-to-noise ratio values that it appear to be just a slight increase in the noise level. To detect the presence of such a signal at long range requires an energy detection receiver. There are several types of receivers which can be useful in detecting DSSS signals. The following three types are described here as typical:

- Integrate and dump;
- Correlative detector;
- Chip detector.

Integrate and Dump

When attempting to detect noise like signal, such as DSSS spread spectrum signals, it is not normally possible to observe the modulation parameters observed by receivers in handling conventional signals.

Figure 6.10 shows a basic integrate-and-dump receiver. The power law detector produces an output voltage proportional to the input power. The receiver output is integrated over some period that makes sense. If for example, we know the approximate chip rate of a DSSS signal, the integration period would be the chip period. At the end of the integration period, the receiver samples the signal and holds the integrated value in a register. Then, the integrator is dumped to zero so that the integration can be repeated.

A variation of the integrate-and-dump approach uses two channels, one delayed by half an integration period as shown in Figure 6.11. Assuming that we know the chip period but cannot phase lock to the signal, we do not know when the chip transitions occur. This two-channel approach means that we will always have a measurement in the last half of the chip—which will have greater integrated value than the first half.

Correlative Detector

A system like that shown in Figure 6.12 has two independent receivers, each attached to an antenna. Two antennas are required to avoid cross-coupling of the two receivers. The noise from each receiver is the kTB noise generated inside the receiver itself. A correlator following the two receivers will show a

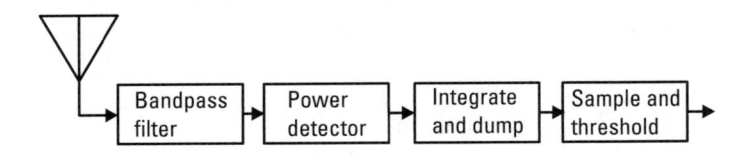

Figure 6.10 Integrate-and-dump receiver.

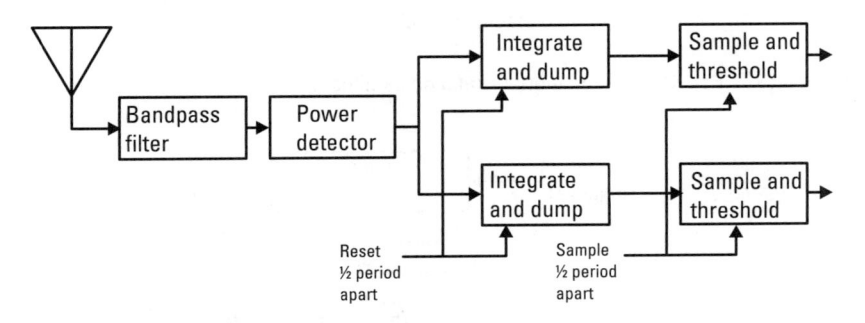

Figure 6.11 Parallel integrate-and-dump receivers.

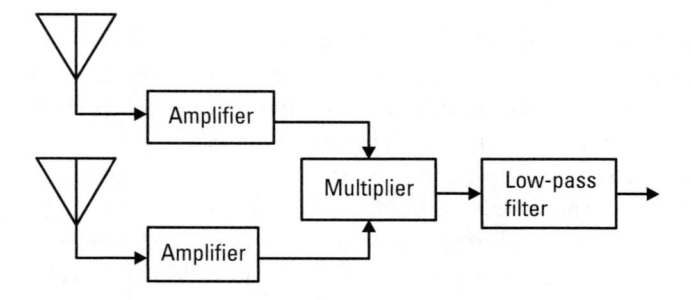

Figure 6.12 Correlative radiometer.

relatively low correlation value because the two noise signals are not correlated. However, if a noise-like signal enters the two antennas, its noise will cause an increase in correlation at the output of the two receivers.

Chip Detector

If a random digital signal is observed on an oscilloscope with positive going synchronization, it will look like Figure 6.13. Note that the chip rate of a DSSS signal must be very stable to allow synchronization and despreading. This causes the chip transitions to fall on top of each other in the ladder pattern shown. If a receiver (as shown in Figure 6.14) has a tapped delay line with taps spaced at the chip period, the energy from the chips will be folded

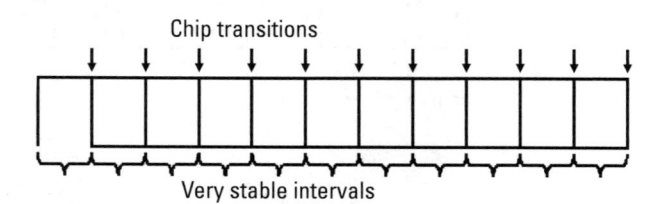

Figure 6.13 Pseudo-random digital signal on oscilloscope.

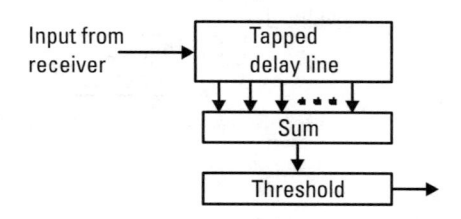

Figure 6.14 Time collapsing threshold.

onto each other creating a significant signal that can be detected. This will allow the chip detector receiver to detect the presence of the spread spectrum signal. Note that the tapped delay line can be implemented in either hardware or software.

Binary Moving Window

A double thresholding technique called "binary moving window" can be useful in detecting indistinct signals. This concept is shown in Figure 6.15. In the case of a DSSS signal search, we would use an integrate-and-dump approach to measure the energy in each chip. These would be compared to a threshold to determine if a signal is present. Since this threshold is not cleanly above the nonsignal-present level, it cannot detect a signal with high confidence. However, if the number of hits over some longer period is used as the detection criterion, confidence in the detection is significantly increased. The longer period needs to make some sense in the intercept situation. A good example is the duration of a typical transmission—perhaps 5 seconds.

6.4 The Signal Environment

The signal environment is defined as all of the signals that reach the antenna of a receiver within the frequency range covered by that receiver. The environment includes not only the threat signals which the receiver intends to receive, but also signals generated by friendly forces and those generated by neutral forces and noncombatants. There may more friendly and neutral signals in the environment that threat signals, but the receiving system must deal with all of the signals reaching its antenna in order to eliminate the signals which are not of interest and identify threats.

To repeat the often repeated generality, the signal environment is extremely dense and its density is increasing. Like most generalities, this one is usually true, but does not tell the whole story. The signal environment in

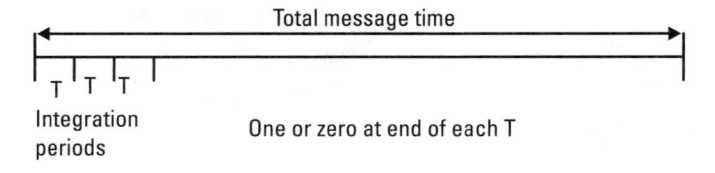

Figure 6.15 Binary moving window.

which an EW or reconnaissance system must do its job is a function of the location of the system, its altitude, its sensitivity, and the specific frequency range it covers. Further, the impact of the environment is strongly affected by the nature of the signals the receiver must find and what information it must extract from those signals to identify the signals of interest.

6.4.1 Angular Coverage

For ship and ground-mounted EW systems, angular search coverage is typically 360° in azimuth and from the horizon to an elevation of 10° to 30° depending on mission. Although these systems may be required to provide protection from airborne threats which can be at any elevation, the relatively small volume at higher elevations means that threat emitters will spend little time at these elevation angles as shown in Figure 6.16. Another factor is that by the time a platform carrying a threat emitter is observed at a high-elevation angle, it is so close that the received signals are at a high enough power level to make their detection hard to avoid.

Airborne receiving systems, because they are elevated, can "see" much farther than surface-located systems. As a practical matter, the platforms used for this kind of operation perform their mission primarily with the yaw plane level (e.g., with wings level for fixed wing aircraft). Emitter location is normally a part of the search process, and it is common to report only the latitude and longitude of emitters of interest. Thus, when the aircraft is in other than a straight and level orientation, there are location errors. These errors could be avoided by measuring the observed elevation of the emitter, but this would significantly increase the system complexity. One common solution to this dilemma is to stop reporting location data when the aircraft orientation is other than level. The criteria for valid data will be given in terms of roll angle—for example, 10° in either direction might be the maximum for valid

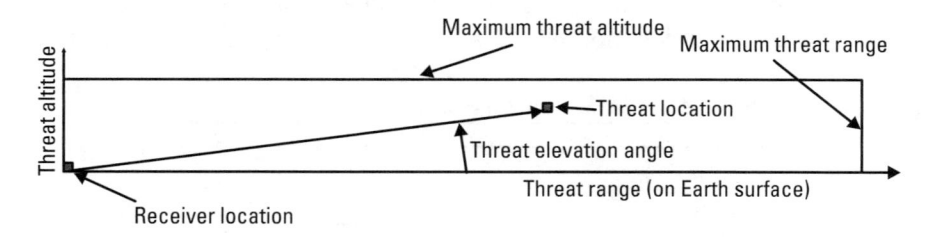

Figure 6.16 Range and elevation to threat signal from surface receiver.

location data. This means that the total search area would be as shown in Figure 6.17.

6.4.2 Channel Occupancy

Modern warfare, which requires a great deal of mobility for almost all assets, is highly dependant on radio communication. This includes large numbers of both voice and data links. The tactical communication environment is often described as having 10% channel occupancy. This is a little misleading because it refers the likelihood that at any microsecond you would expect 10% of all available RF channels to be active. If you stay on each channel for a few seconds, the occupancy rate is much higher—closer to 100%. This means that any search for a specific emitter must find it among a thick forest of nontargeted emitters.

The main value of this characterization is the understanding that any search scheme that depends on seeing signals in a fairly low-density environment is useful only in rare sparse signal environments. The presence of many signals has significant impact on any approach that uses a rapid search with stops for further analysis on signals found.

6.4.3 Sensitivity

Another element determining signal density is the receiver sensitivity (plus any associated antenna gain). As discussed in detail in Chapter 5, received signal strength decreases as the square or fourth power of distance between the transmitter and the receiver, depending on the propagation mode. Receiver sensitivity is defined as the weakest signal from which a receiver can recover the required information—and most EW receivers include some kind of thresholding mechanism so that signals below their sensitivity level need not be considered. Thus, receivers with low sensitivity and those using

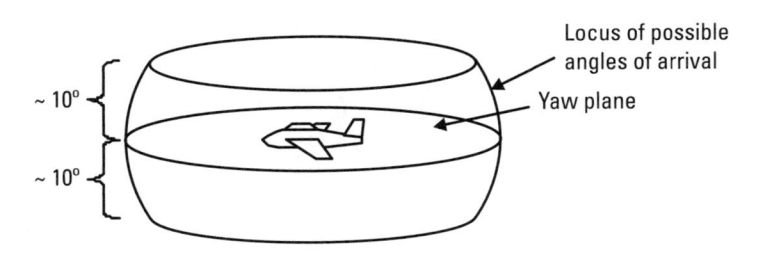

Figure 6.17 Area search from reconnaissance aircraft.

low-gain antennas deal with far fewer signals than highly sensitivity receivers or those which benefit from high antenna gain. This simplifies the search problem by reducing the number of signals which must be considered by the system in identifying threat emitters.

6.5 Radio Horizon

For signals at VHF and higher frequencies, which can be considered to be restricted to line-of-sight transmission, only those signals above the radio horizon will be in the signal environment. The radio horizon is the Earth surface distance from the receiver to the most distant transmitter for which line-of-sight radio propagation can occur. This is primarily a function of the curvature of the Earth, and is extended beyond the optical horizon (an average of about 15%) by atmospheric refraction. The usual way to determine radio horizon is to solve the triangles shown in Figure 6.18. The radius of the Earth in this diagram is 1.33 times the true Earth radius to account for the refraction factor (called the "4/3 Earth" factor). The line-of-sight distance between a transmitter and a receiver can be found from the formula:

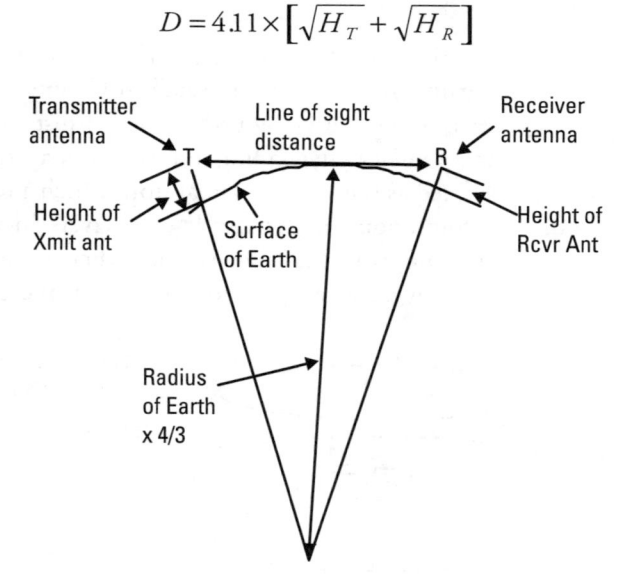

$$D = 4.11 \times \left[\sqrt{H_T} + \sqrt{H_R} \right]$$

Figure 6.18 Radio horizon geometry.

where

D = the transmitter to receiver distance in km

H_T = the transmitter height in meters

H_R = the receiver height in meters

Thus the radio horizon has a relative definition—depending on the altitude of both the receiver and any transmitters present. All else being equal, you would expect the number of emitters seen by a receiver to be proportional to the Earth surface area which is within its radio horizon range—but of course the emitter density also depends on what is happening within that range.

Figure 6.19 is a 4/3 Earth radio horizon chart based on the above equation. Its usefulness is that it allows the introduction of terrain features into the horizon calculation. Note that this chart can be used not only to determine the horizon limited maximum range between a transmitter and a receiver, but can also be used for a radar and a target. (Substitute radar for receiver and target for transmitter in the following discussion.)

Figure 6.20 shows the data that is drawn onto the chart of Figure 6.19 to calculate the horizon over spherical, sea-level Earth. For this example, the

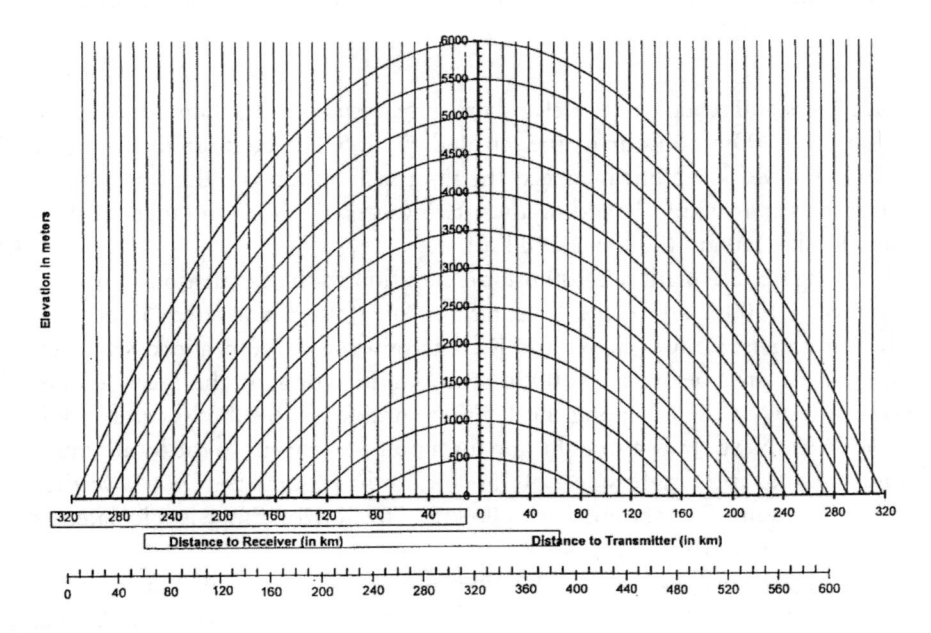

Figure 6.19 Horizon calculation nomograph.

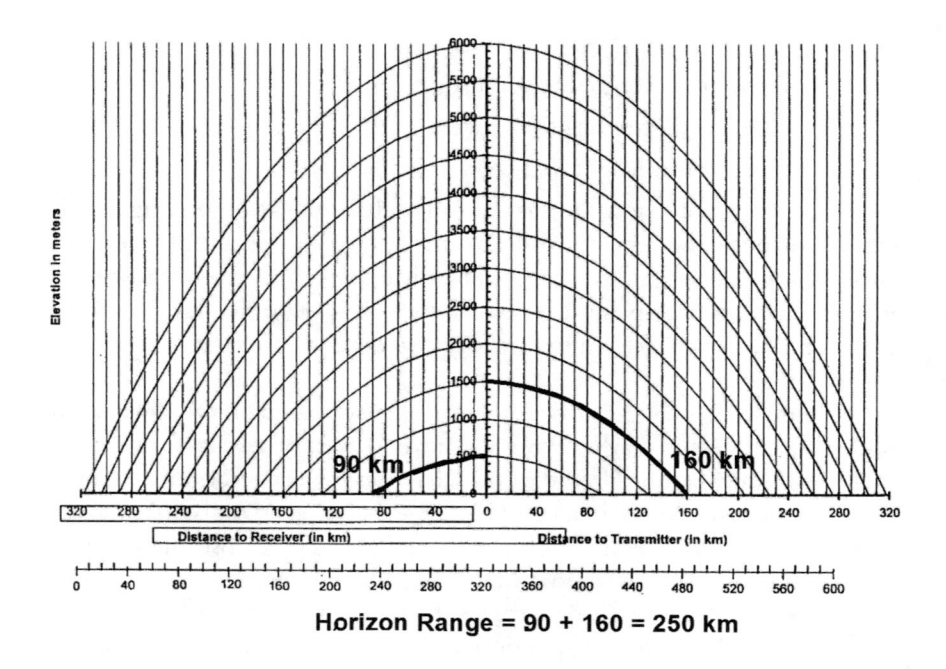

Figure 6.20 Horizon calculation for transmitter at 1,500m and receiver at 500m above sea level.

receiver is 500m above sea level and the transmitter is at 1,500m. Draw a line along the 500m contour to the left and the 1,500m contour to the right. Note that the 500m line intersects the abscissa at 90-km and the 1,500m line intersects at 160 km. By adding these two values, the horizon range is calculated to be 250 km (90 km + 160 km).

Figure 6.21 adds a 1,000m local elevation to the calculation. Now the receiver is 500m above the local terrain, or 1,500m above sea level. The transmitter is 1,500m above the local terrain, or 2,500m above sea level. Shade the area below the 1,000m contour to represent the local terrain. Then, draw to the left along the 1,500m contour to the abscissa at 160 km (point A). Next, draw a straight line from point A tangent to the local terrain area, ending at the right side 2,500m contour line (point B). Draw a vertical line from point B to the abscissa at 86 km. The horizon range is calculated as 246 km (160 km + 86 km).

Figure 6.22 adds a 2,000m ridgeline or mountain 180 km from the receiver. From the intersection of the left 1,500m contour with the abscissa (point A), draw in the ridgeline with its peak (at point B) on the 2,000m

Search for Communication Emitters 173

Horizon Range = 160 + 86 = 246 km

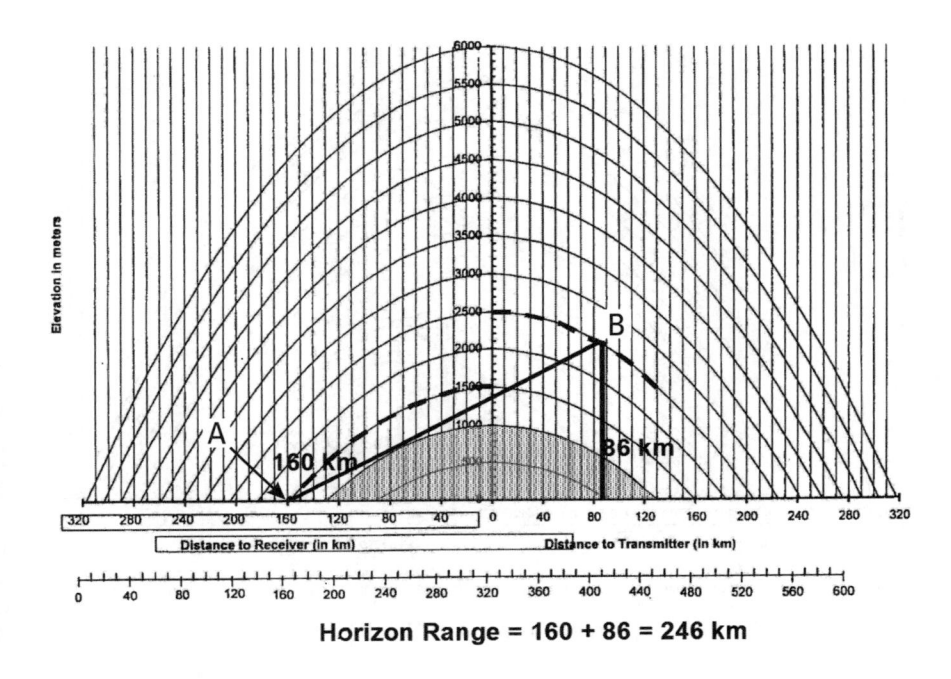

Figure 6.21 Horizon calculation for transmitter at 1,500m and receiver at 500m above 1,000m-high local terrain.

contour 180 km to the right. Note that point B is directly above the 20 km (to the right) point on the abscissa. Now, draw a straight line through point B to the right hand 2,500m contour (this is point C). Finally, draw straight down from point C to the abscissa at 52 km. The horizon range is calculated as 212 km (160 km + 52 km).

Lower Frequencies

When the receiver is operating below 30 MHz, the signals have significant "beyond the horizon" propagation modes, so the signal density is not so directly a function of altitude. VHF and UHF signals can also be received beyond line of sight, but the received signal strength is a function of frequency and the Earth geometry over which they are transmitted. The higher the frequency and the greater the nonline-of-sight angle, the greater the attenuation will be. For practical purposes, microwave signals can be considered to be limited to the radio horizon.

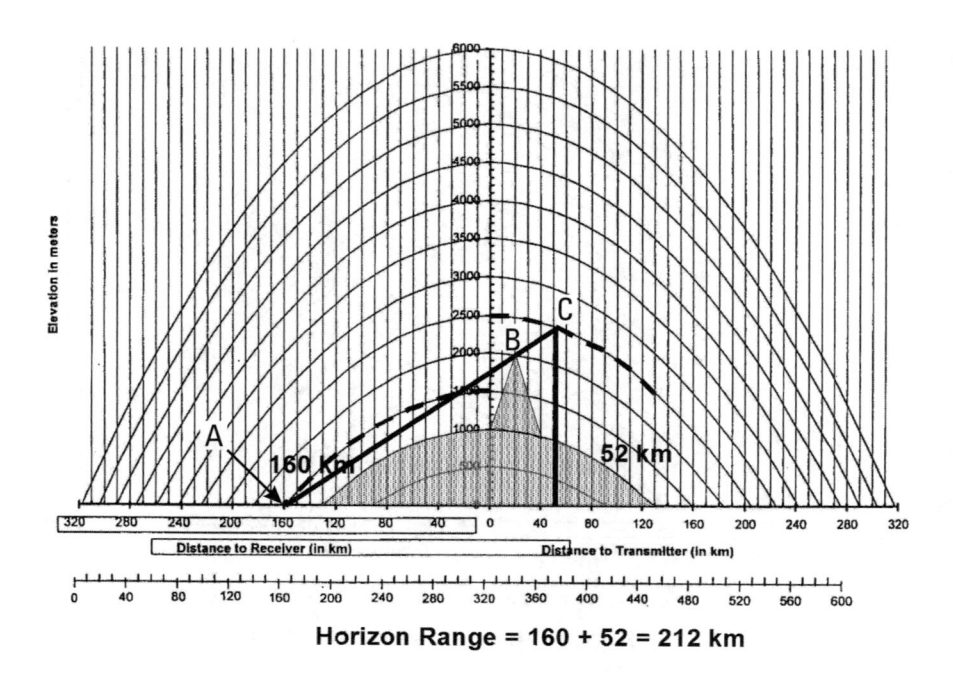

Figure 6.22 Horizon calculation with 2,000m ridgeline 180 km from the receiver.

6.6 Search for Low Probability of Intercept Signals

Low probability of intercept (LPI) signals are—by design—challenging to the receiving systems attempting to detect them. LPI signals are very broadly defined, including any feature which makes the signal harder to detect or the emitter harder to locate. The simplest LPI feature is emission control—reducing the transmitter power to the minimum level that will allow the threat signal (radar or communication) to provide adequate signal-to-noise ratio in the related receiver. The lower transmitter power reduces the range at which any particular hostile receiver can detect the transmitted signal. A similar LPI measure is the use of narrow-beam antennas or antennas with suppressed side lobes. Since these antennas emit less off-axis power, the signal is more difficult for a hostile receiver to detect. If the signal duration is reduced, the receiver has less time in which to search for the signal in frequency and/or angle of arrival—thus reducing its probability of intercept.

However, when we think of LPI signals, we most often think of signal modulations which reduce the signal's detectability as described in Section 2.4. LPI modulations spread the signal's energy in frequency, so that the frequency spectrum of the transmitted signal is orders of magnitude wider than required to carry the signal's information (the information bandwidth) as shown in Figure 6.23. Spreading the signal energy reduces the signal strength per information bandwidth. The noise in a receiver is a function of its bandwidth (as described in Chapter 4), so the signal-to-noise ratio in any receiver attempting to receive and process the signal in its full (spread) bandwidth will be greatly reduced by the signal spreading. As shown in the same figure, a receiver with bandwidth equal to the information bandwidth receives a significantly lower signal strength from the spread signal.

The challenge that all LPI modulations pose to the search function is that they force an unfavorable trade-off of sensitivity versus bandwidth. In some cases, the structure of the spreading technique allows some advantage to the receiver, but this requires some level of knowledge about the modulation characteristics, and can significantly increase the complexity of the receiver and/or its associated processor.

6.6.1 LPI Search Strategies

The basic LPI search techniques always involve optimization of intercept bandwidth and other considerations which are peculiar to the type of spread spectrum signal. Search for the following three types of spread spectrum signals is discussed in this section:

- Frequency hopping signals;
- Chirp signals;
- Direct sequence spread spectrum signals.

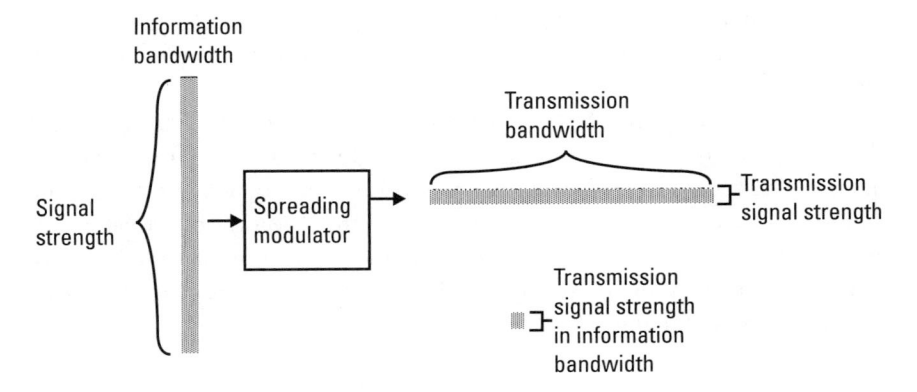

Figure 6.23 Spread spectrum signal strength.

6.6.2 Frequency Hoppers

Frequency hoppers are the easiest spread spectrum signals to detect because they move their entire transmitted power to one information bandwidth at a time. The complication is that the frequency is changed to a pseudo-randomly chosen value every few milliseconds for "slow hoppers," and every few microseconds for "fast hoppers." Slow hoppers have multiple information bits per hop and fast hoppers have multiple hops per information bit. In either case, the identification of a frequency hopper depends on determining the location of the emitter. A hopper will have multiple frequencies at a single location, whereas a nonhopping signal will have only a single frequency at the emitter location. Techniques for locating frequency hoppers will be discussed later in Chapter 7.

Slow Hoppers

There are two approaches for the detection of slow hoppers. One is to search across the hopping range with a narrowband receiver. Statistically, the sweeping receiver will receive a few intercepts as it randomly encounters the hop frequencies during a transmission. Such receivers can normally search empty hop slots in a few tens of microseconds, but must stop for the order of a millisecond to perform analysis on any signal found (usually including direction of arrival measurement). To maximize the frequency of these encounters, the sweeping receiver bandwidth covers multiple information bandwidths. The optimum search bandwidth is 4 to 6 times as wide as the hopper's channel bandwidth (the information bandwidth). When the search receiver bandwidth is greater than this level, the stops on single frequency transmissions within the hopping range delay the search too much. A binary moving window approach is used to qualify frequency hopper intercepts. For example, if 5 or 10 hits at different frequencies are encountered at a single angle of arrival, detection of a frequency hopper at that direction of arrival is reported.

The second approach is to use a digital receiver with FFT processing to determine the frequency of all signals present during a small part of the hop period. The timing of this frequency search is described along with the digital receiver description in Section 4.3.2. Since the direction of arrival is necessary to the identification of a frequency hopper, direction of arrival analysis must be performed on any signal found during the frequency search. This is discussed later in Chapter 7.

Fast Hoppers

Since fast hoppers have multiple hops per bit, they are much harder to detect than slow hoppers. Because fast hoppers must change frequency very quickly, they most likely use direct synthesizers which have multiple oscillators and are thus significantly more complex than the synthesizers used for slow hoppers. This makes it reasonable to assume that the fast hopper would not have too many hop frequencies. Thus, it may be practical to employ a channelized receiver with one channel at each hop frequency. Analysis of the outputs of the channels will detect the presence of the fast hopper.

Energy detection techniques may also be appropriate, depending on the number of hop channels and the received signal-to-noise ratio.

6.6.3 Chirp Signals

The vulnerability of chirped signals to detection is that their full signal power passes through every frequency within its chirp range. This means that a receiver designed to measure only the frequency of a received signal (without capturing the modulation) may be able to achieve a number of "hits" on a chirped signal. Analysis of this data will show that the signal is chirped and give some level of information about its frequency scanning characteristics. It is possible to design a "carrier frequency only" receiver with greater sensitivity versus instantaneous RF bandwidth than would be required to recover the signal's modulation.

6.6.4 Direct Sequence Spread Spectrum Signal

There are two basic ways to detect the presence of a direct sequence spread spectrum (DSSS) signal. One is through energy detection with various filtering options. In general, this requires that the received signal be very strong. The other approach is to take advantage of some characteristic of the transmitted signal. When observed at double its transmitted frequency, a binary phase shift keyed (BPSK) modulated signal will have all of its bits in phase. Figure 6.24 illustrates this concept by showing phase diagrams of the BPSK signal at the transmitted frequency and at twice that frequency. Likewise, the bits of a quadriphase signal will be at the same phase at four times the transmitted frequency. Thus, at the second or fourth harmonic of the transmit signal, the spreading modulation would disappear, despreading the signal. Unfortunately, the output amplifier and the transmitting antenna will greatly reduce these harmonics, so this technique will probably only be useful at very short ranges from the target transmitter.

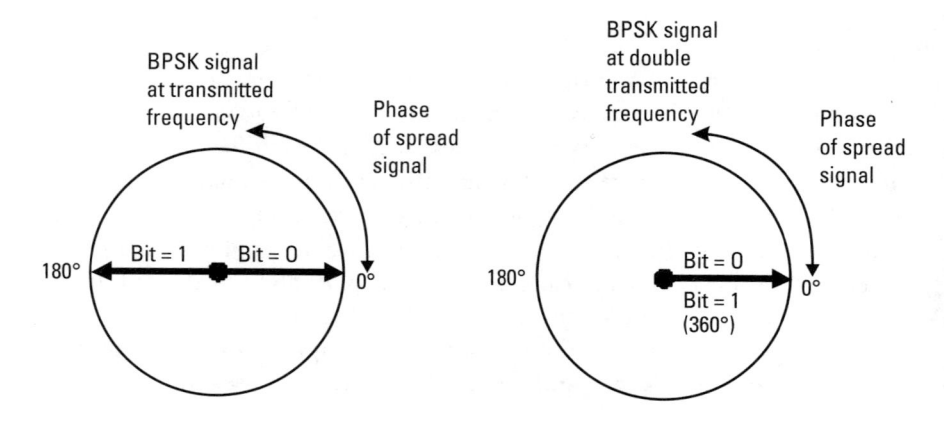

Figure 6.24 Phase diagram of BPSK signal.

Chip detectors and energy detectors as described in Section 6.3.4 are also practical ways to detect the presence of DSSS signals

6.7 Look Through

In general, any type of electronic warfare receiving system is challenged to detect all of the threat signals present in the brief time available for the search function. There is almost always a wide frequency range to cover and a there are a few signal types that can only be received using narrowband receiver assets. This process is made even more challenging when there is a jammer on the same platform with the receiver, or operating in close proximity—because the jammer has the potential to blind the receiver to incoming signals. Consider that EW receiver sensitivities are in the range of −65 to −120 dBm and that jammers typically output hundreds or thousands of watts. The ERP of a 100-watt jammer is +50 dBm plus antenna gain, so the jammer output can be expected to be 100 to 150 dB (or more) stronger than the signals for which the receiver is searching.

Whenever possible, the receiver and the associated jammer are isolated operationally, that is the receiver performs its search function in cooperation with the jammer so that it is searching bands or frequency ranges in which no jamming is momentarily taking place. Where spot jamming and some types of deceptive jamming are used, this operational isolation can allow the receiver to perform a fairly efficient search—if there is some level of isolation to keep the jammer from saturating the receiver's front-end components.

Unfortunately, this will seldom solve the whole problem, so other measures must be employed. When broadband jamming is used, a whole band will usually be denied to the receiver unless adequate isolation can be achieved.

The first choice look-through approach is to achieve as much isolation between the jammer and the receiver as possible. Antenna gain pattern isolation is important, but with the wide angle antennas common in communication band radios and jammers it may be difficult to achieve. Any difference between the gain of the jamming antenna toward the threat being jammed and toward its own receiver antenna reduces the interference. Likewise, any difference between the receiving antenna gain in the direction of the threats and toward the jammer helps. Wide-beam or full-azimuth coverage antennas can achieve isolation when they are physically blocked from each other (for example, one on the top of an aircraft and the other on the bottom).

The physical separation of the antennas also helps. A formula for spreading loss between two omni-directional antennas for short ranges is:

$$L = -27.6 + 20\log_{10}(F) + 20\log_{10}(D)$$

where
    L is the spreading loss in dB
    F is the frequency in MHz
    D is the distance in meters

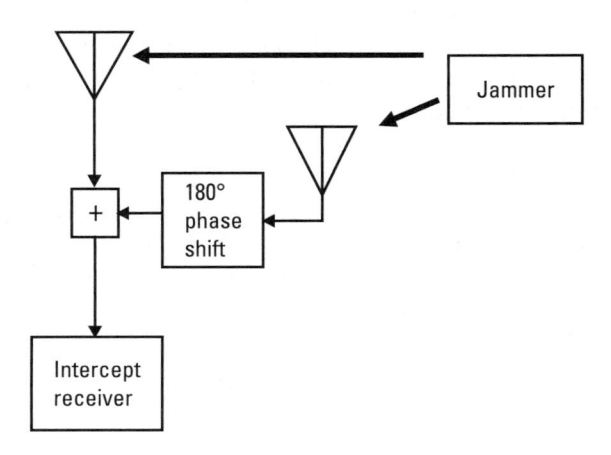

Figure 6.25 Coherent jamming canceller.

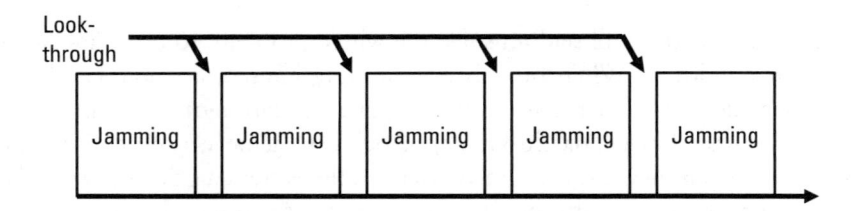

Figure 6.26 Time sharing look-through.

The space loss for short ranges can also be determined from the antenna and propagation calculator included with this book—as shown in Figure 6.10.

Thus a jammer operating at 4 GHz 10 meters from a receiver would have 64.4 dB of isolation just from the distance between the jamming and receiving antennas.

If the jamming and receiving antennas have different polarization, additional isolation is provided. For example, there is approximately 25 dB of isolation between right- and left-hand circularly polarized antennas. In general, polarization isolation is less than this in wide frequency band antennas and can be better than this in very narrowband antennas.

Finally, radar-absorptive materials can be used to provide additional isolation, particularly at high microwave frequencies.

If adequate isolation between jamming and receiving antennas cannot be achieved, it may still be practical to cancel the jamming signal in the receiver input. This requires that a second antenna be oriented toward the jammer. It will receive the jamming signal by the direct and many reflection paths. This secondary jammer signal will then be 180° phased shifted and added to the receiver input, as shown in Figure 6.25.

If jammer cancellation is not practical, it will be necessary to provide short look-through periods as shown in Figure 6.26, during which the receiver can perform its search functions. The timing and duration of look-through periods is a trade-off of jamming effectiveness against probability of intercept of threat signals by the receiver. The look-through periods must be short enough to prevent the jammed threat receiver from receiving adequate unjammed signal to recover the transmitted information. Remember that digital signals are defeated by 33% jamming unless the signals have significant error-correction coding. (See Chapter 9.)

6.8 Fratricide

Fratricide refers to the accidental jamming of friendly communications. It could just as well apply to the jamming of friendly receivers which are searching for enemy signals. This is a huge problem in any modern military operating area. With the presence of improvised explosive devices (IEDs) that are triggered by a large variety of commercial devices that transmit radio signals, jammers must be used to prevent detonation of IEDs. If broadband jammers are used, they will make it very difficult for essential friendly command and control communication and search for enemy signals to take place. This is less a problem when jammers are used only against active hostile transmissions.

In general, the most successful antifratricide approaches depend on limiting jamming power and duty cycle to the minimum that will provide protection and careful communication planning to take maximum advantage of parts of the frequency spectrum not expected to contain weapon-triggering transmissions.

When jamming enemy LPI signals, it is important to realize that friendly LPI transmissions are operated in the same frequency ranges. The use of jammers which are deployed as close as possible to the enemy receivers and the use of directional antennas aimed at enemy receivers (and away from friendly receivers) can help a great deal.

6.9 Search Strategy Examples

6.9.1 Narrowband Search

Narrowband search involves tuning a single receiver as quickly as possible over the frequency range which contains a signal of interest during the time the signal is expected to be present. In general, the tuning rate of the searching receiver (the amount of frequency spectrum searched per unit time) is limited by requiring a signal to remain in the bandwidth for a time equal to the inverse of the bandwidth as explained above. In modern digitally tuned receivers, this translates to dwelling on each tuning step for a period equal to the inverse of the bandwidth. This is often described as "searching at a one-over-the-bandwidth rate." Note that restrictions on control and processing speed in some receiving systems can further restrict the search rate.

There are two more restrictions on the search approach: one is that the receiver bandwidth must be wide enough to accept the signal being detected, and the second is that the receiver must have adequate sensitivity to receive the signal with adequate quality. The sensitivity is a function of bandwidth because of the kTB factor mentioned in Section 4.4.

We want to find a 25-kHz wide communication signal which is between 30 and 88 MHz. We will assume that the signal is up for 1/2 second. Note that a signal this short is probably a "key click," which at one time was the shortest signal an intercept system had to worry about. In this example, times will be rounded to the nearest millisecond.

Our receiving antenna covers 360 degrees of azimuth, and the search receiver bandwidth is 25 kHz. The receiver must dwell at each tuning step for a time equal to the inverse of the bandwidth. To avoid band edge intercepts, we will overlap our tuning steps by 50%.

$$\text{Dwell} = 1/\text{bandwidth} = 1/25\,\text{kHz} = 40\,\mu s$$

Figure 6.27 shows the search problem in the diagram format we discussed in Section 6.2.4. Note the overlap of the receiver coverage which causes us to change frequency only 12.5 kHz with each tuning step.

For 100% probability of finding the signal of interest, the receiver must cover the whole 58 MHz in one-half second. The number bandwidths required to cover the signal range is:

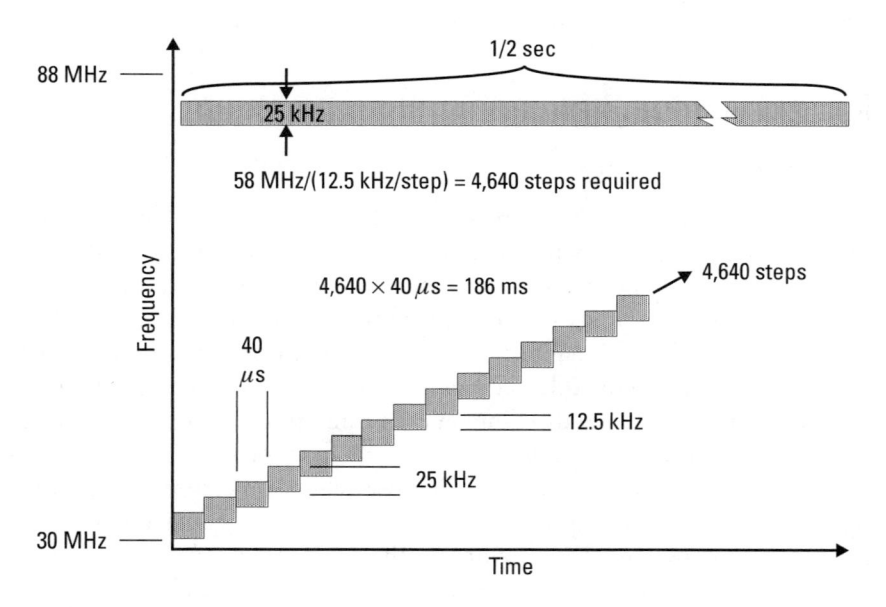

Figure 6.27 Search plan diagram.

$$58 \text{ MHz}/25 \text{ kHz} = 2{,}320$$

With 50% overlap, the 58-MHz frequency range requires 4,640 tuning steps.

At 40-μs dwell per step, 4,640 steps requires 186 ms.

This means that the receiver can find the signal of interest in less than one-half of the assumed minimum signal duration, so 100% probability of intercept is easily achieved.

Require Longer Dwell for Signal ID

The above analysis assumes that we have an optimum search and that the signal will be instantly recognized as our signal of interest. To make the problem more interesting, let's assume that we have a processor which can recognize the modulation of the signal in 200 μs. The search planning diagram is changed as shown in Figure 6.28. This means that we must hold at each frequency for that dwell time, so it takes 928 ms to cover the 58-MHz search range:

$$200 \mu\text{s} \times 4{,}640 = 928 \text{ ms}$$

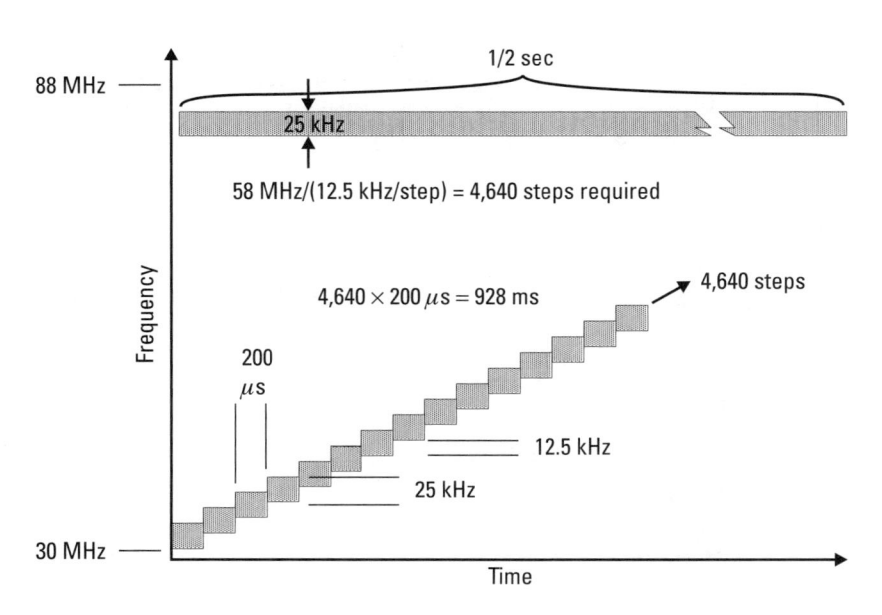

Figure 6.28 Search plan diagram with 200-μs dwell.

The search does not find the signal within the specified half-second.

Increase the Receiver Bandwidth

If the search receiver bandwidth is increased to 150 kHz (covering 6-target signal channels) and we assume that the 200-μs processing time also allows determination of the signal frequency within the bandwidth, the search is enhanced. See Figure 6.29. Now it takes only 773 steps to cover the frequency range of interest.

$$4{,}640/6 = 773$$

At 200 μs per receiver tuning step, it takes only 155 ms to cover the 2,320 channels (with 50% overlap).

$$773 \times 200\ \mu s = 155\ \text{ms}$$

Note that this increase in bandwidth would reduce the receiver sensitivity by almost 8 dB. If the receiver gets much wider than this, the increasing probability of multiple signals in the bandwidth becomes problematic.

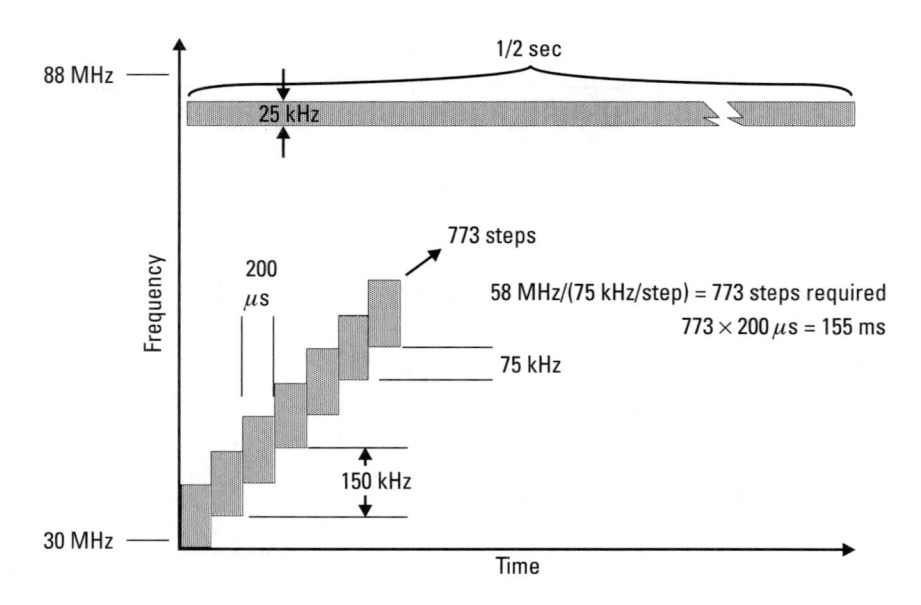

Figure 6.29 Search plan diagram with increased bandwidth.

Add DF Requirement

To make the problem even more interesting, let's assume that our receiver is part of a direction-finding system and that we must determine the direction of arrival (DOA) of the signal of interest. If the direction finder requires 1 ms to determine the DOA, it adds only 1 ms to our search time—if there are no other signals present.

6.9.2 Hand-Off from Wideband Receiver

There are two wideband receiver search strategies recommended for communications EW systems. One is to dedicate one of several receivers to the search function and set its bandwidth higher than the information bandwidth of targets of interest. An FM discriminator at the output of this receiver can determine the exact frequency of a discovered signal of interest to allow exact set-on of one of the remaining receivers.

The second approach is to determine the frequency of all signals present with a wideband frequency-measuring receiver such as a compressive receiver. The compressive receiver hands off the frequencies of all signals found to a processor. The processor identifies signals of interest from stored frequencies of signals of interest, signals previously analyzed (which need not be reconsidered), frequencies of known signals which are not of interest, and others. Then the processor, or a human supervisor, decides which encountered signals have high enough priority to rate assignment of a monitor receiver. Detailed analysis and monitoring or recording of signals of interest are performed using the monitor receiver.

6.9.3 Search with a Digital Receiver

This approach was discussed along with the digital receiver description in Section 4.3.2. The digital receiver as configured was able to provide the frequency of the signal of interest to 20 kHz over a 20-MHz bandwidth in 25 μs. Since three 20-MHz receiver tuning steps were required to cover the 58-MHz search range, the frequency was found in 75 μs. The frequencies of all signals present within the 58-MHz range are used to assign monitor receivers as discussed in Section 6.9.2.

7

Location of Communications Emitters

One of the most important requirements placed on EW systems is the location of threat emitters. Communication emitters pose particular challenges because of their relatively low frequencies. Lower frequency implies larger wavelength, hence larger antenna apertures. In general, communications electronic support (ES) systems are required to provide instantaneous 360° angular coverage, and adequate sensitivity to locate distant emitters. They must typically be able to accept all communications modulations, sometimes including those associated with low probability of intercept (LPI) transmissions. In all cases, communications ES systems deal with noncooperative (i.e., hostile) emitters. Thus, the techniques available for location of cooperative systems are by definition unavailable.

In this chapter, we will discuss the common approaches, and most important techniques. We will first discuss the location of normal (i.e., non-LPI) emitters and then the approaches to the location of LPI emitters. In the discussions of all system applications, the high-signal density expected in the modern military environment will be an important consideration.

7.1 Emitter Location Approaches

The approaches described here have to do with the intercept geometry. Not all apply to all location techniques.

7.1.1 Triangulation

Triangulation is the most common approach to the location of noncooperative communications emitters. As shown in Figure 7.1, this involves the use of two or more receiving systems at different locations. Each such system must be able to determine the direction of arrival (DOA) of the target signal. It must also have some way to establish an angular reference—typically true North. For convenience we will call these direction-finding (DF) systems in the following discussion.

Since terrain obstruction or some other condition might cause two DF systems to see different signals (in the typical dense signal environment), it is common practice to perform triangulation with three or more DF systems. As shown in Figure 7.2, the DOA vectors from three DF systems will form a triangle. Ideally, all three would cross at the emitter location, and if the triangle is small enough, the three-line intersections can be averaged to calculate the reported emitter location.

These DF sites are normally quite distant from each other, so the DOA information must be communicated to a single analysis location before the emitter location can be calculated. This also implies that the location of each DF site is known.

It is important that each of the DF sites be able to receive the target signal. If the DF systems are mounted on flying platforms, they will normally be expected to have line-of-sight to the target emitter. Ground-based systems can be expected to provide more accurate location if the terrain allows

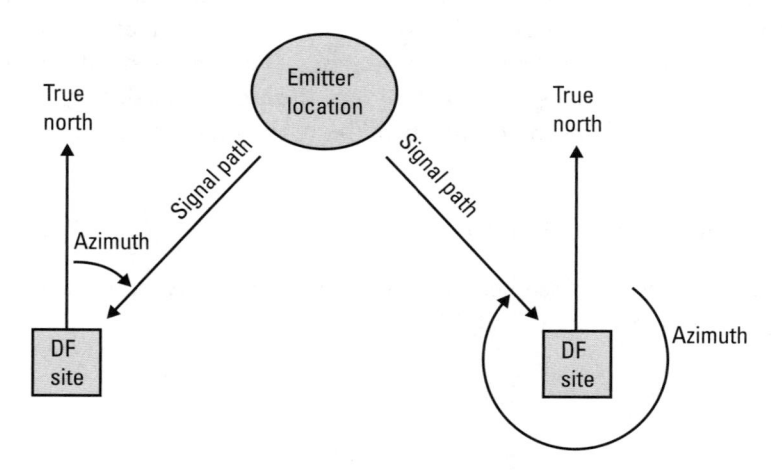

Figure 7.1 Triangulation geometry.

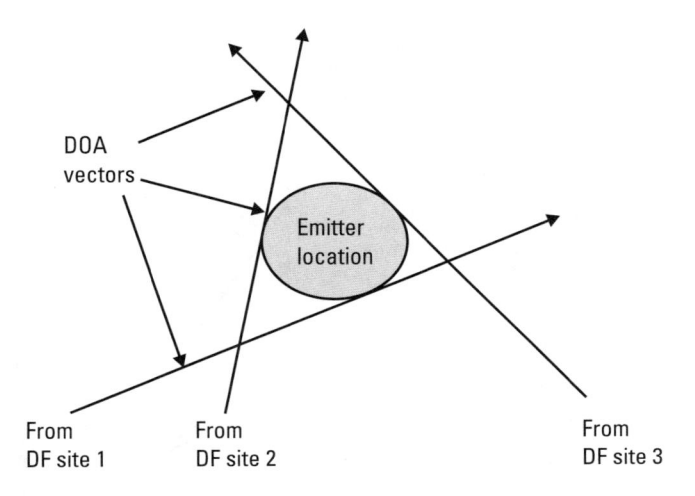

Figure 7.2 Detail of triangulation at target emitter.

line-of-sight, but should be able to determine the location of over-the-horizon emitters with some acceptable accuracy.

Note that the optimum geometry for triangulation provides 90° of angle between the two DF sites as seen from the emitter location.

Triangulation can also be performed from a single, moving DF system, as shown in Figure 7.3. This normally only applies to airborne platforms. The lines of bearing should still cross at 90° at the target. Therefore, the speed of the platform on which the DF system is mounted and the distance between the flight path and the target will dictate the time required for an accurate emitter location.

For example, if the DF platform is flying at 100 knots and passes about 30 km from the target emitter, it will take almost 10 minutes to achieve the optimum location geometry. This may be quite practical for stationary emitters, but may be too slow to track moving emitters. For this approach to yield acceptable accuracy, the movement of target emitters must not be greater than the required location accuracy over the time when data is being collected. Note that acceptance of less than optimum geometry (hence location accuracy) may provide the best operational performance.

7.1.2 Single-Site Location

There are two cases in which the location of a hostile transmitter can be determined from the azimuth and range from a single-emitter location site.

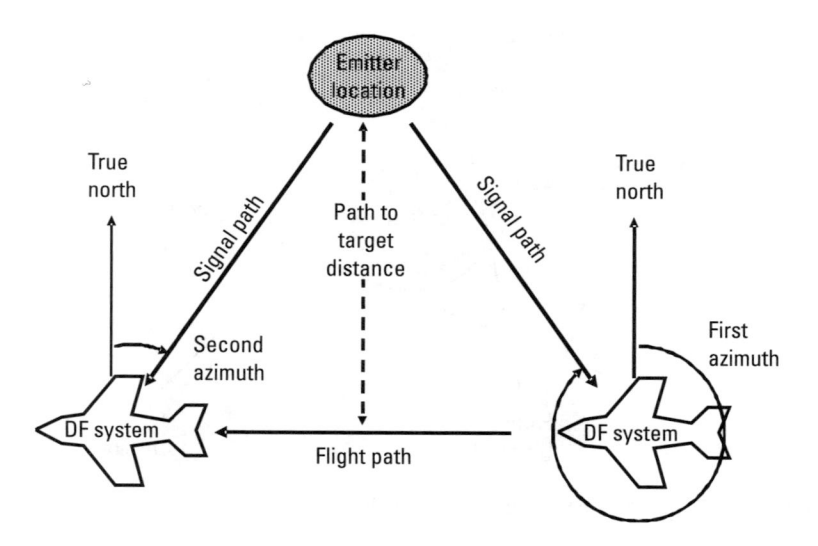

Figure 7.3 Triangulation from a moving DF system.

One applies to ground-based systems dealing with signals below about 30 MHz and the second applies to airborne systems.

Signals below approximately 30 MHz can be located by a single-site locator (SSL) as shown in Figure 7.4. These signals are refracted by the ionosphere. They are said to be "reflected" by the ionosphere because they return with the reciprocal angle as shown in Figure 7.5. If both the azimuth and

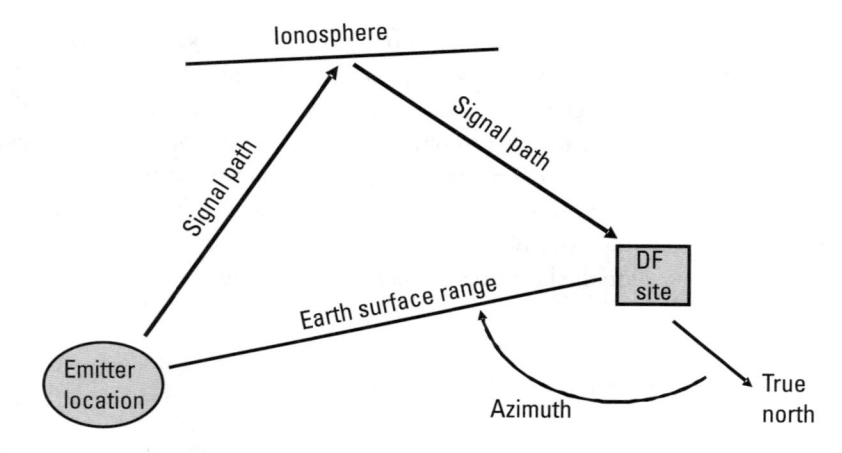

Figure 7.4 Determination of range from elevation below 30 MHz.

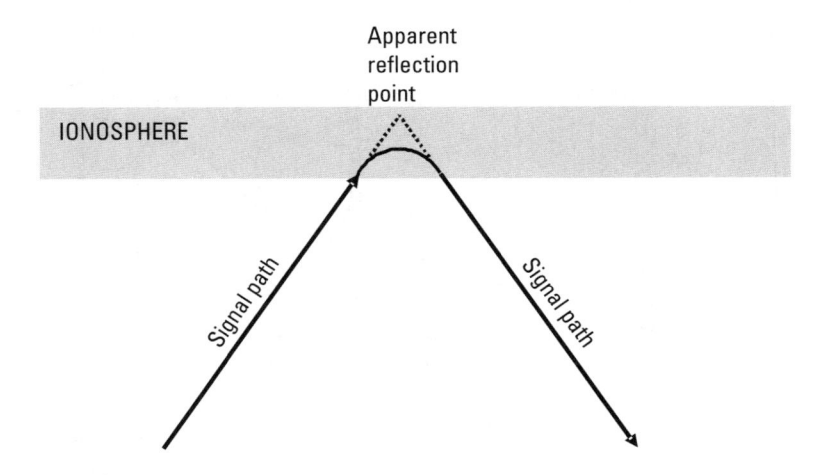

Figure 7.5 Ionospheric refraction.

elevation angle of the signal arriving at the emitter location site are measured, the transmitter can be located. The range is calculated from the elevation angle and the "height" of the ionosphere at the "reflection" point because the angle of reflection from the ionosphere is the same as the angle of incidence. The most difficult part of this process is the accurate characterization of the ionosphere at the point of reflection. Normally, the range calculation is significantly less accurate than the azimuth measurement, causing an elongated zone of location probability.

7.1.3 Azimuth and Elevation

If an airborne emitter location system measures both the azimuth and elevation to a noncooperative emitter on the ground, the emitter location can be calculated as shown in Figure 7.6. The range determination requires that the aircraft know its location over the ground and its elevation. It must also have a digital map of the local terrain. The Earth surface range to the emitter is the distance from the subvehicle point to the intersection of the signal path vector with the ground.

7.1.4 Other Location Approaches

Precision emitter location approaches, to be described later, use comparison of parameters of a target signal as received at two distant sites to calculate a mathematically derived locus of possible emitter locations as shown in Figure 7.7. The techniques used can place the emitter very close to this locus, but

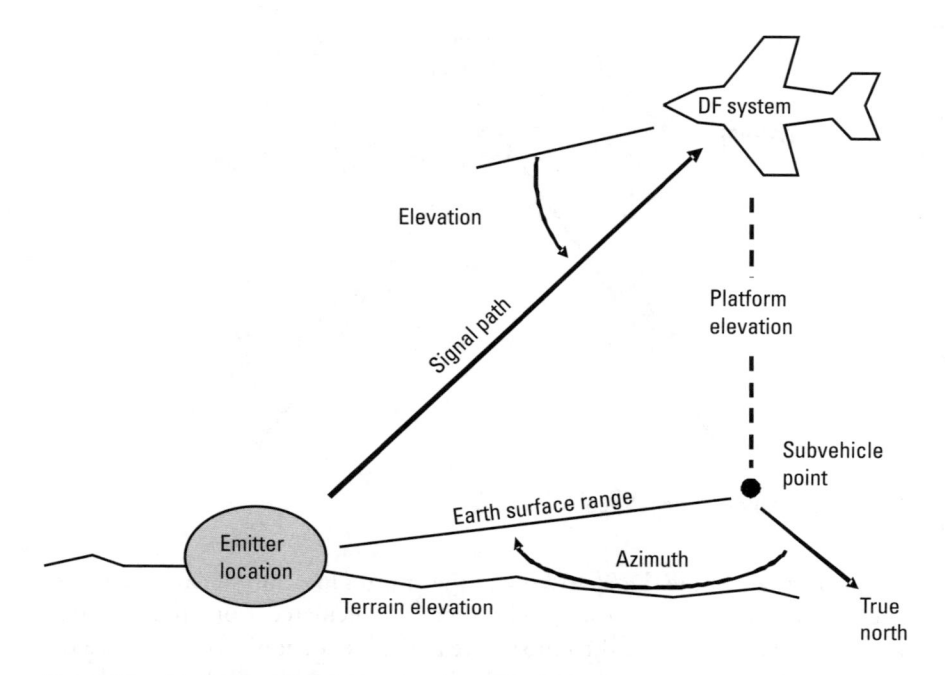

Figure 7.6 Emitter location from azimuth and elevation from an airborne DF system.

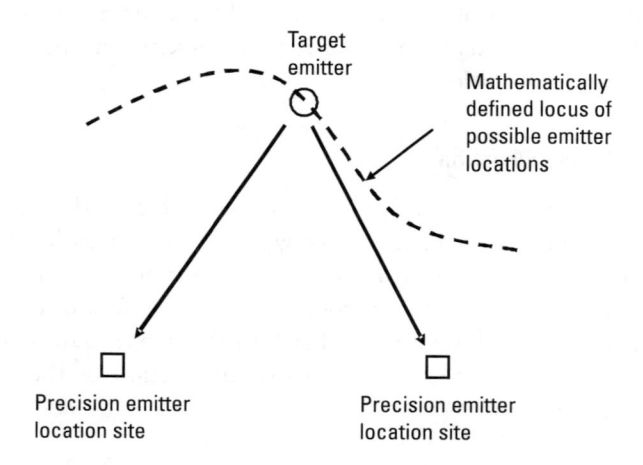

Figure 7.7 Location of an emitter on a calculated locus.

the locus is typically many kilometers long. By adding a third site, a second and a third locus curve can be calculated. These three locus curves cross at the emitter location.

7.2 Accuracy Definitions

The location accuracy provided by emitter location systems is significant to the application of the location information. For example, if the purpose of emitter location is to support electronic order of battle development, 1 or 2 km may be close enough. If the purpose is targeting, the location will need to be within the burst radius of the weapon being applied.

Accuracy definitions are the way that different systems and approaches are compared.

7.2.1 RMS Error

The accuracy of DOA measurement systems is typically stated in terms of the root mean square (RMS) error. This is considered the effective accuracy of a DF system. It does not define the peak errors that might be present. The system could conceivably have a relatively small RMS error even though there are a very few large peak errors. It is assumed, when defining the RMS error of a DF system that the errors are caused by randomly varying conditions, such as noise. There have been systems in which there were known large systematic errors caused by the way the system was implemented. When these few large errors were averaged with many lower errors, an acceptable RMS error was achieved. However, there were predictable conditions in which errors several times the RMS error values were experienced—reducing the operational dependability of emitter location. Where this kind of known peak errors are corrected in processing, a proper RMS error specification is achieved.

To determine the RMS error, a large number of DF measurements are made at fairly evenly distributed frequencies and angles of arrival. Data is usually collected over 360° of azimuth for a two-dimensional system or 4π steridians for an azimuth and elevation system. For each data collection point, the true angle of arrival must be known. In ground systems, this is accomplished by use of a calibrated turntable on which the DF system is mounted or by use of an independent tracker which measures the true angle to the test transmitter at a significantly higher accuracy that that specified for the DF system (ideally a full order of magnitude). In airborne DF systems, the true angle of arrival is calculated from the known location of the test transmitter and the location and orientation of the airborne platform from its inertial navigation system.

Each time a DOA is measured by the DF system (during this system testing), it is subtracted from the true angle of arrival. This error measurement is then squared. The squared errors are then averaged and the square root taken. This is the RMS error of the system. The RMS error can be broken into two components as follows:

$$(\text{RMS Error})^2 = (\text{Standard Deviation})^2 + (\text{Mean Error})^2$$

Thus, if the mean error is mathematically removed, the RMS error equals the standard deviation from the true angle of arrival. In practice, removal of the mean error means that all output readings are offset to compensate for the calculated mean error. Since the mean error may be largely caused by misalignment of the antenna array or reflections from the geometry immediately around the antenna array, this correction is normally quite reasonable.

If the causes of errors can be considered normally distributed, the standard deviation (σ) is 34.13%. Thus, as shown in Figure 7.8, the ± RMS error lines describe an area around the true line of bearing that have a 68.26% chance of containing any measured angle of arrival. Looking at this a different way, it means that if the system measures a specific angle, there is a 68.26% chance that the true emitter location is within the wedge-shaped

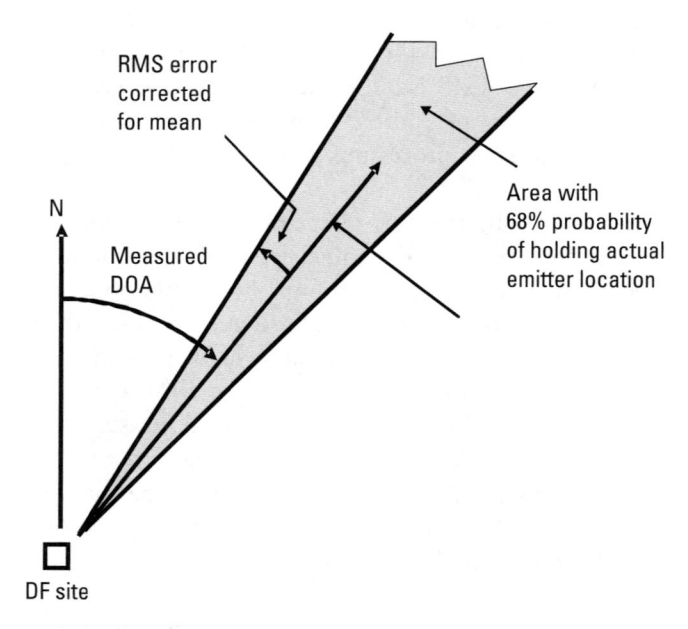

Figure 7.8 Location probability from RMS error.

area shown. If two DF sites are equal distance from the located emitter and are 90° apart as seen from the emitter, they are said to be ideally located, because they will provide he best location accuracy. As shown in Figure 7.9, two ideally located DF sites create an intersection of triangular areas that has a 46.6% (i.e., 68.26%^{2}) probability of containing the actual emitter location. This assumes, of course, that the measured mean error has been removed during data processing.

7.2.2 Circular Error Probable and Elliptical Error Probable

Circular error probable (CEP) is a bombing and artillery term that refers to the radius of a circle around an aiming stake in which half of a number of dropped bombs or fired artillery shells fall. We use this term in emitter location system evaluation to indicate the radius of a circle around a measured emitter location which has a 50% probability of containing the true emitter location as shown in Figure 7.10. The smaller the CEP, the more accurate the system. The term "90% CEP" is also used to describe the circle around the measure location with a 90% chance of containing the true emitter location.

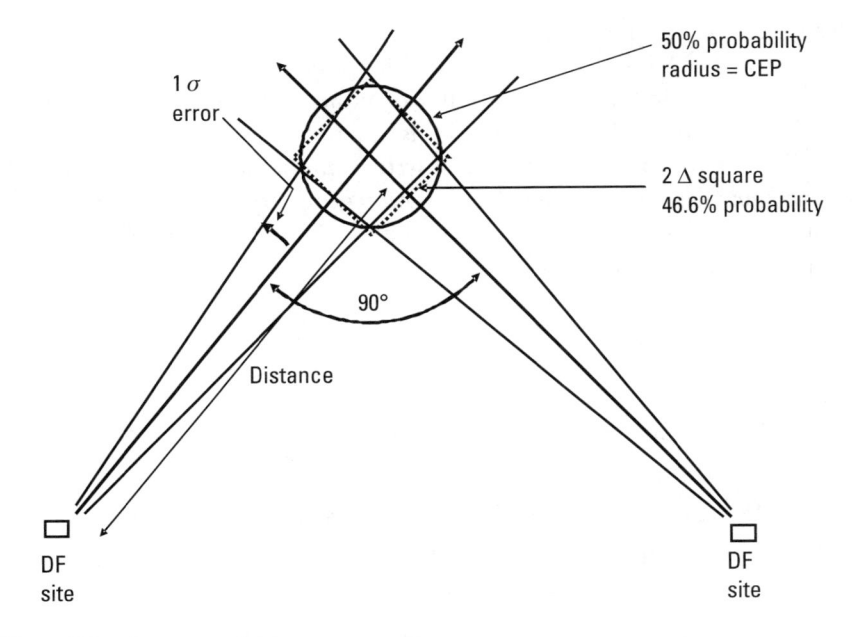

Figure 7.9 Area uncertainty versus angular accuracy.

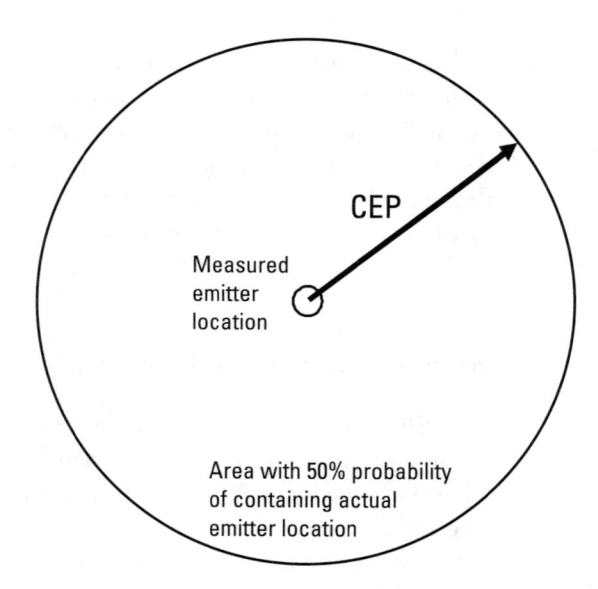

Figure 7.10 Circular error probable.

To approximate the CEP for this situation, we will first determine the area included in the area within the ± RMS error angle limits from the two ideally located DF sites as shown in Figure 7.9. This area is approximated as a square with 2Δ on a side, where Δ is the distance between the calculated emitter location and the RMS error line on either side. The CEP is the radius of a circle which has the same area as this square.

The approximate CEP for a DF system can be calculated from the following equation. This assumes ideal location geometry. Note that this is only a convenient approximation. The actual measured CEP is a strong function of the actual geometry.

$$CEP = 1.17d \tan(RMS)$$

where

CEP is the circular error probable in km

d is the distance from each DF site in km

RMS is the DF system RMS error in degrees

Remember that the CEP describes a circle with a 50% probability of containing the emitter location. For 90% CEP the formula (again just an approximation) is:

$$90\% \text{ CEP} = 1.57d \tan(\text{RMS})$$

For example, two ideally located DF sites 100 km from an emitter with 1° RMS and the mean errors removed will provide 2-km CEP and 2.7-km 90% CEP.

Elliptical Error Probable

The elliptical error probable (EEP) is the ellipse which has a 50% probability of containing the actual emitter when a location has been measured by two sites which do not have ideal geometry to the target. The 90% EEP is also often considered. The EEP can be calculated from the RMS error and the intercept geometry, and can be drawn on a map as shown in Figure 7.11 to indicate not only the measured location of the emitter, but also the confidence a commander can place in the location measurement.

The CEP can also be determined from the EEP by the following formula:

$$\text{CEP} = 0.75 \times \sqrt{(a^2 + b^2)}$$

where a and b are the semimajor and semiminor axes of the EEP ellipse.

The CEP and EEP are also defined for precision emitter location techniques, and these will be described later.

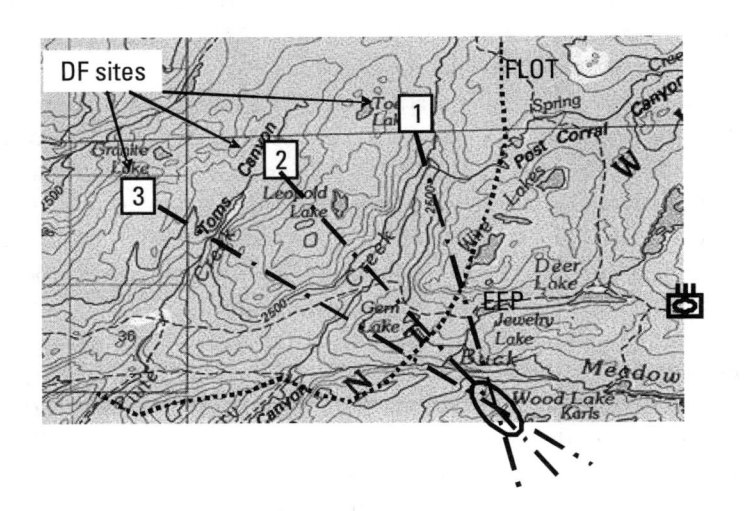

Figure 7.11 EEP shown on a map.

7.2.3 Calibration

Calibration involves the collection of error data as described earlier. However, this error data is used to generate calibration tables. These tables, in computer memory, hold the angular correction for many values of *measured* DOA and frequency. When a direction of arrival is measured at a particular frequency, it is adjusted by the calculated angular error and the corrected angle of arrival is reported out. If a measured DOA falls between two calibration points (in angle and or frequency) the correction factor is determined by interpolation between the two closest stored calibration points. Note that slightly different calibration schemes yield better results for some specific DF techniques. These will be discussed along with those techniques.

7.3 Site Location and North Reference

For triangulation or single-site emitter location to be performed, the location of each DF site must be known and input to the process. For angle-of-arrival (AOA) systems, there must also be a directional reference (often to true North). Site location is also required for the precision emitter location techniques mentioned earlier. As shown in Figure 7.12, errors in site location and reference direction will cause errors in the AOA determined for target emitters. This figure is deliberately exaggerated to show the effects of errors. Typically, site location and reference direction errors are of the order of magnitude of the measurement accuracy errors. As you will see, these errors are typically only a few degrees.

Figure 7.13 (also deliberately exaggerated) shows the location errors caused by measurement, site location, and directional reference errors. If an error contribution is fixed, it must be directly added to the location accuracy. Site location errors are typically considered fixed. On the other hand, when sources of errors are random and independent of each other, they are "RMSed" together. That is, the resulting RMS error is the square root of the sum of the squares of the various error contributions.

Before the mid-1980s, the location of DF sites was quite challenging. Ground-based DF systems required that the DF site location be determined by survey techniques and entered into the system manually. The North reference required either that the DF antenna array be oriented and stabilized to a specified orientation, or that the antenna array orientation be automatically measured and input. Automatic North sensing was particularly important for mobile sites.

Location of Communications Emitters 199

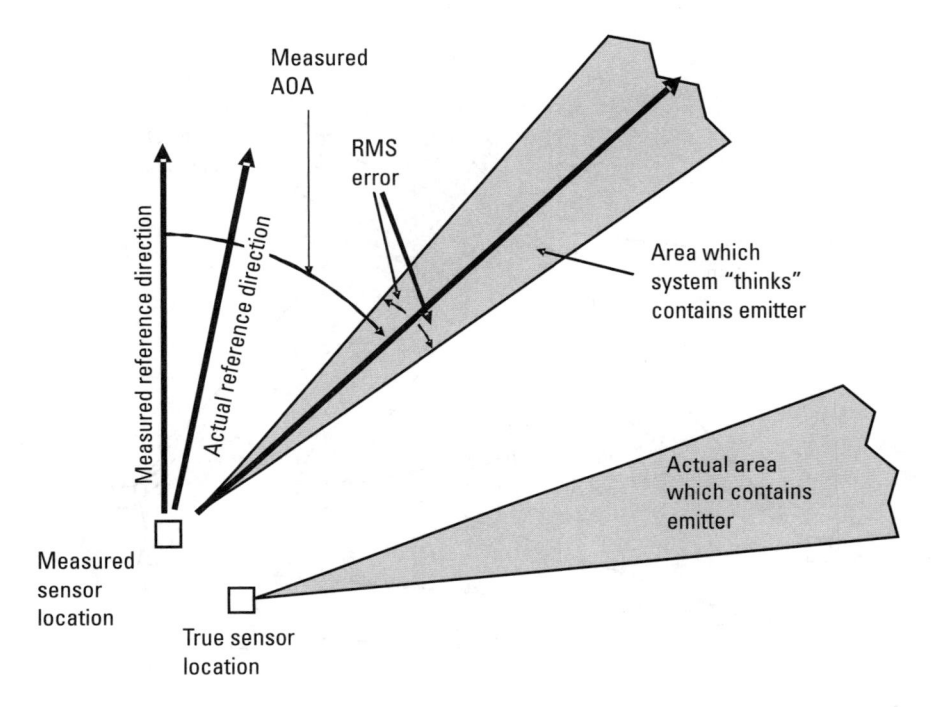

Figure 7.12 Site location and reference accuracy implications.

A magnetometer is an instrument which senses the local magnetic field and provides an electronic output. It is functionally a digital reading magnetic compass. When a magnetometer was integrated into the antenna array of a ground-based system, its (magnetic) North reference could be automatically entered into the computer in which the triangulation was being performed. The local declination (i.e., variation of magnetic North from true North) had to be manually input to the system to calculate the azimuth reference from each site. The magnetometer accuracy was typically about 1.5°. As shown in Figure 7.14, the magnetometer was often integrated into the direction-finding array of an AOA system. This avoided the difficult process of orienting the antenna array to magnetic North, significantly reducing the system deployment time.

Shipboard DF systems on large platforms could get their location and orientation references from the ships' navigation systems, which have been quite accurate for many years. The ship's inertial navigation system (INS) can be manually corrected by a highly trained navigator to provide long-term location and directional accuracy.

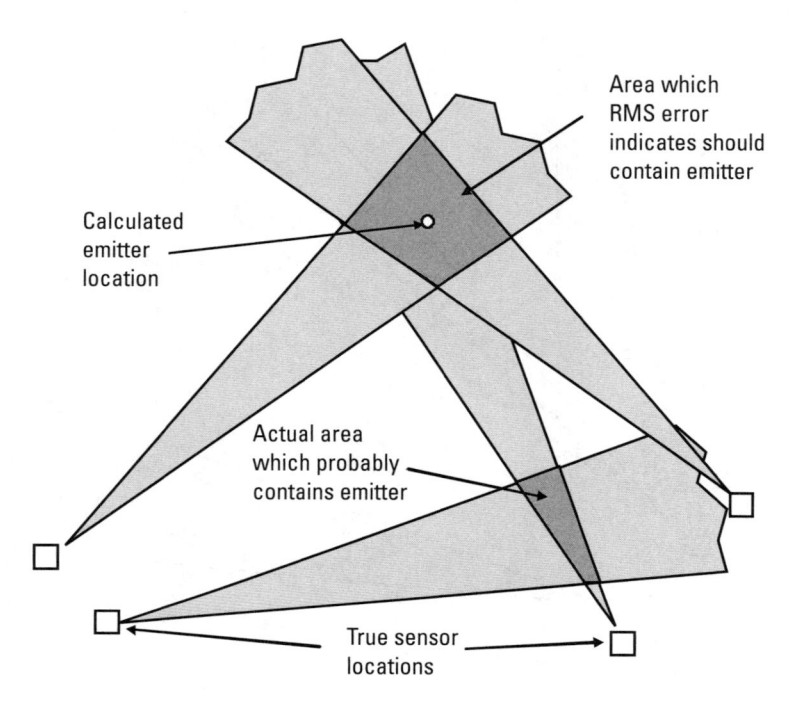

Figure 7.13 The impact of site location and directional reference accuracy on emitter-location accuracy.

Airborne DF systems, of course, also required that the location and orientation of each DF system be known and entered into the triangulation calculation. This was provided from the aircraft's INS, which required extensive initialization procedures before each aircraft mission. An INS derived its north reference from two mechanically spinning gyroscopes (oriented 90° apart) and its lateral location reference from three orthogonally oriented accelerometers, as shown in Figure 7.15. Each gyroscope can only measure angular motion perpendicular to its axis of rotation—hence the requirement for two gyroscopes to provide three-dimensional orientation. Each of the accelerometer outputs is integrated once to provide lateral velocity and a second time to provide location change (each in one dimension). The gyroscopes and accelerometers were mounted on a mechanically controlled platform within the INS which remained in a stable orientation as the aircraft maneuvered. After the aircraft left the compass rose on the airfield or the aircraft was launched from an aircraft carrier, the location and orientation accuracy decreased linearly with time because of the drift of gyroscopes

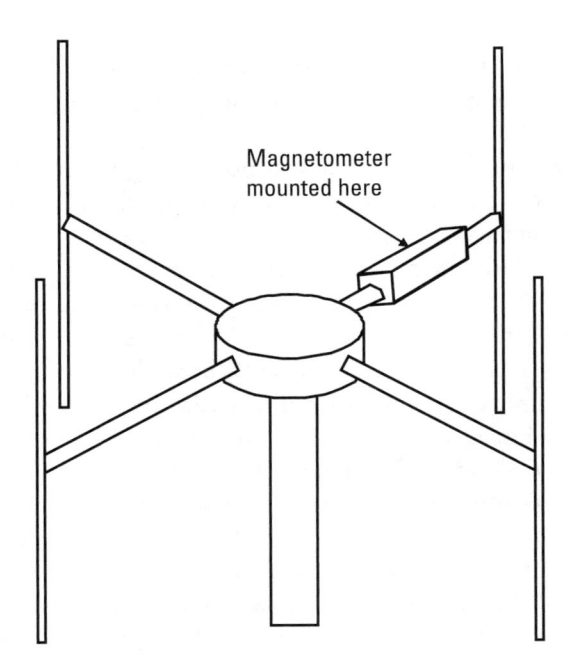

Figure 7.14 A magnetometer mounted in a direction-finding array.

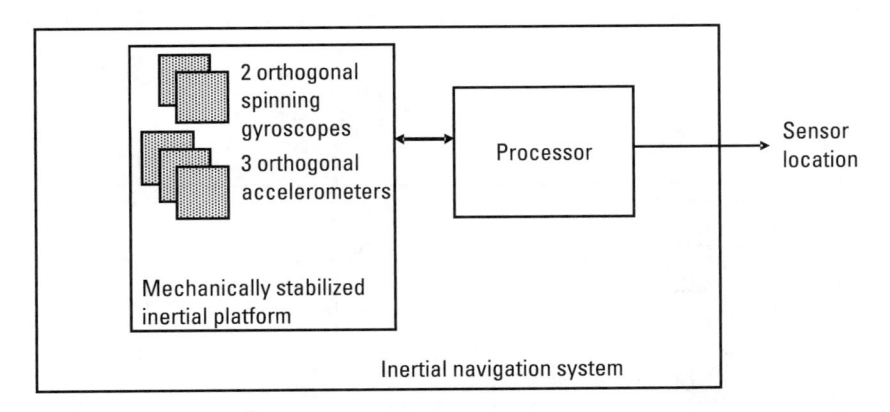

Figure 7.15 An older inertial navigation system required a mechanically stabilized inertial platform.

and the accumulated error of accelerometers. Hence, the accuracy of emitter location from airborne platforms was a function of the mission duration.

Also, effective airborne DF systems were constrained to deployment in large enough platforms to support INS installations (which were about two cubic feet in volume).

During the late 1980s, the global positioning system (GPS) satellites were placed in orbit and small, inexpensive, rugged GPS receivers became available. GPS has had a significant impact on the way we locate mobile assets. Now the location of small aircraft, ground vehicles, and even dismounted individuals can be automatically measured (electronically) with adequate accuracy to support emitter location. This allowed the many low-cost DF systems to provide significantly better location accuracy.

GPS has also had a significant impact on the way INS devices work. Because the absolute location can be directly measured at any time, INS location accuracy is no longer a function of mission duration. As shown in Figure 7.16, inputs from the inertial platform are updated with data from the GPS receiver. Location is measured directly by GPS, and angular updates can be derived from multiple location measurements.

Because of the development of new types of accelerometers and gyroscopes, and significant electronics miniaturization, the INS system can now be implemented in significantly less size and weight, and with no moving parts. Ring laser gyroscopes bounce a laser pulse around closed path (three precise mirrors). By measuring the time to get around the circular path, it determines angular velocity. The angular velocity is integrated to determine

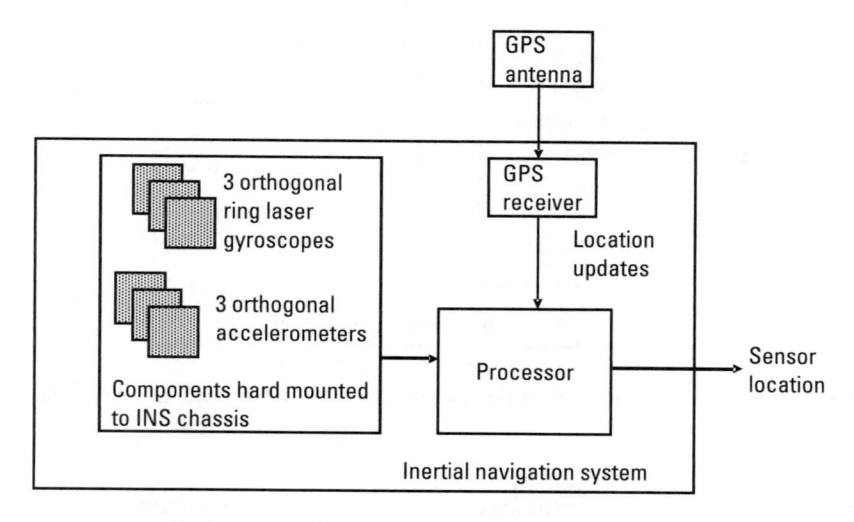

Figure 7.16 A GPS-enhanced inertial navigation system.

orientation. Three-ring laser gyroscopes are required to determine the three-axis orientation. Piezoelectric accelerometers have now replaced the old weight-on-a-spring type. There are also piezoelectric gyroscopes which measure angular velocity.

An additional value of GPS is to provide a very accurate clock at fixed or mobile emitter location sites. This clock function is required for the precision location techniques we will be discussing. The GPS receiver/processor synchronizes itself to atomic clocks in the GPS satellites. This has the effect of creating a virtual atomic clock in one printed circuit board plus an antenna. (Note that an actual atomic clock is "bigger than a bread box.") GPS has, therefore enabled the use of precision emitter location techniques in small platforms.

7.4 Moderate Accuracy Techniques

Since moderate accuracy systems are direction finders, their accuracy is most conveniently defined in terms of their RMS angular accuracy. A fairly good number for moderate accuracy is 2.5° RMS. This is the accuracy achievable in most DF approaches without calibration. We will be talking more about calibration later, but for now, "calibration" means systematically measuring and correcting errors in the measurement of angle of arrival of transmitted signals.

There are many moderate accuracy systems in use, and they are considered adequate for the development of electronic order of battle information. That is, they can locate enemy transmitters with enough precision to allow analysis of the types of military organizations present, their physical proximity, and their movements. This information is used by expert analysts to determine the enemy's order of battle and to predict the enemy's tactical intentions.

These systems are also relatively small, light, and inexpensive. In general, the higher the system accuracy, the more accurate site location and reference must be. This has been a significant problem in smaller scale (lower cost) systems. However, this has become much easier with the increasing availability of small, low-cost inertial measurement units (IMUs). Combined with GPS location reference, IMUs can provide adequate location and angle reference for moderate accuracy DF systems.

Two typical moderate accuracy techniques used for communications emitter location are Watson-Watt and Doppler.

7.4.1 Watson-Watt Direction Finding Technique

As shown in Figure 7.17, a Watson-Watt DF system has three receivers connected to a circularly disposed antenna array with an even number (four or greater) of antennas plus a reference antenna in the center of the array. The circular array has a diameter of about one-quarter wavelength.

Two of the outside antennas (opposite each other in the array) are switched to two of the receivers, and the center reference antenna is connected to the third receiver. In processing, the amplitude difference between the signals at the two outside antennas is referenced to (i.e., divided by) the amplitude of the signal at the center reference antenna. This combination of signals produces a cardioid gain pattern (gain versus direction of arrival) around the three antennas as shown in Figure 7.18. By switching another pair of opposite antennas into receivers 2 and 3, a second cardoid pattern is formed. At the moment of switching, we therefore have two points on the cardiod. After sequentially switching all of the opposite pairs a few times, the direction of arrival of the signal can be calculated.

The Watson-Watt technique works against all types of signal modulations and, without calibration, achieves about 2.5° RMS error.

7.4.2 Doppler Direction-Finding Technique

If one antenna is rotated around another antenna as shown in Figure 7.19, the moving antenna (A) will receive a transmitted signal at a different frequency from that received at the fixed antenna (B). As the moving

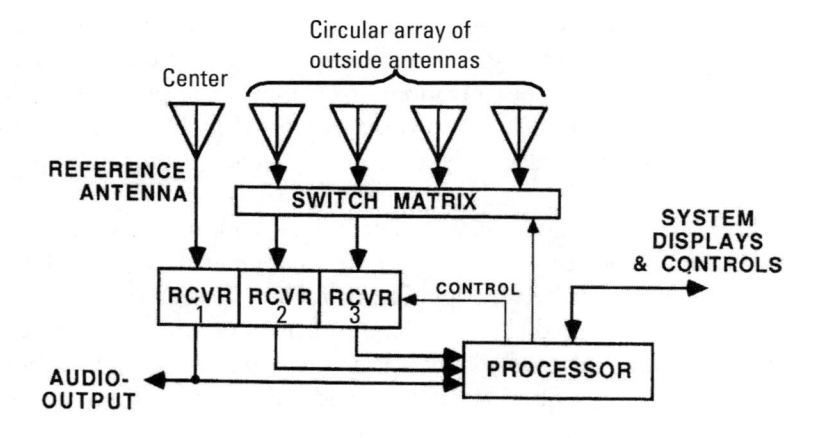

Figure 7.17 Watson-Watt DF system block diagram.

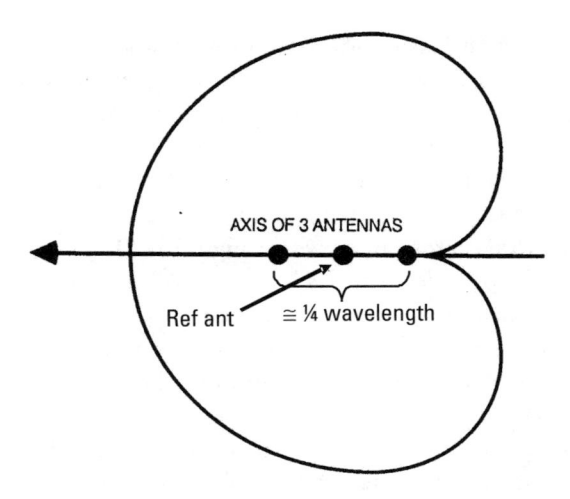

Figure 7.18 Difference of signals in two outside antennas normalized to reference antenna.

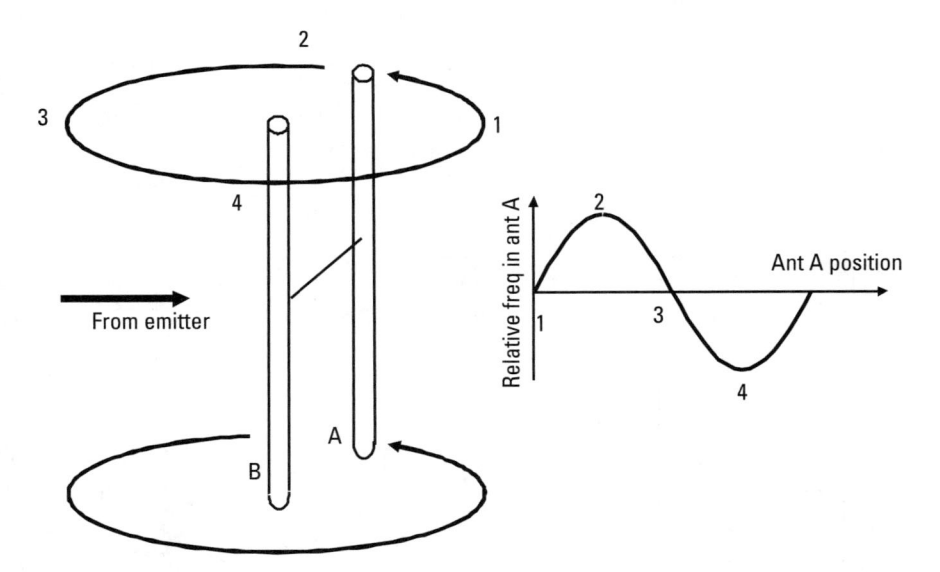

Figure 7.19 Doppler DF system concept.

antenna moves toward the transmitter, the receiving frequency will be increased by the Doppler shift. As it moves away, the frequency will be reduced. This frequency variation is sinusoidal, and can be used to determine the direction of arrival of the transmitted signal. Note that the emitter is in the direction at which the negative going zero crossing of the sine wave in this figure occurs.

In practice, multiple, circularly disposed antennas are sequentially switched into one receiver (A), while another receiver (B) is connected to a central antenna in the array as shown in Figure 7.20. Each time the system switches one of the outside antennas into receiver A, the phase change in the received signal is measured. After a few revolutions, the system can construct the sinusoidal variation of frequency (in antenna A versus antenna B) from the phase change data—and thus determine the angle of arrival of the transmitted signal.

The Doppler technique is widely used in commercial applications and can have as few as three outside antennas plus the central reference antenna. It typically achieves about 2.5° RMS accuracy. However, this technique has difficulty with frequency-modulated signals unless their modulation can be clearly separated from the apparent Doppler shifts of the sequentially switched outside antennas.

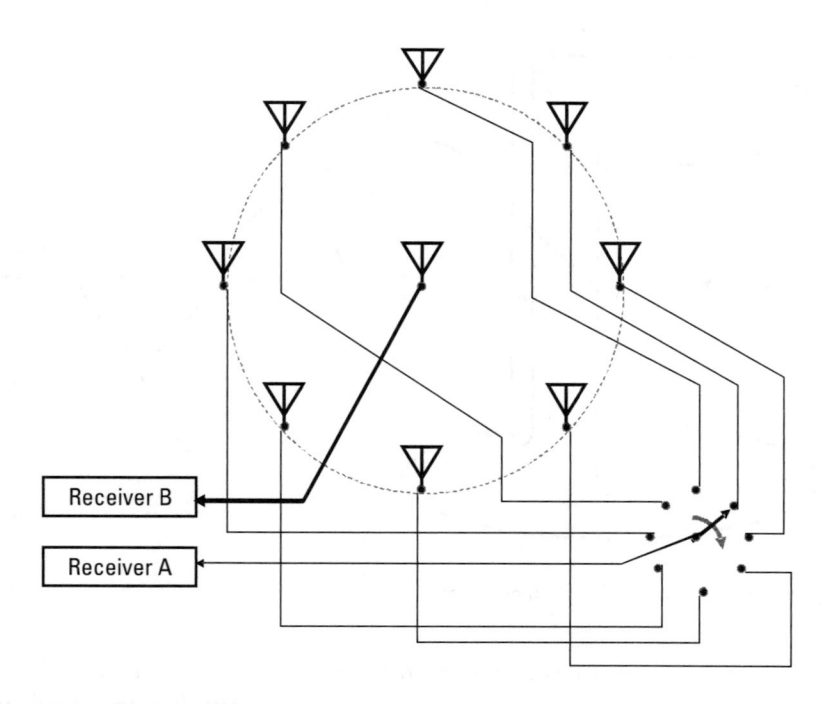

Figure 7.20 Doppler DF system.

7.5 High Accuracy Techniques

When we speak of high-accuracy emitter location techniques, we are generally talking about interferometer direction finding. Interferometers can generally be calibrated to provide on the order of 1-degreee RMS error. Some configurations provide better than that and some have less accuracy. The interferometer is a direction finder, determining only the angle of arrival of the signal. Emitter location is determined from one of the techniques (such as triangulation) discussed in Section 7.1.1.

We will begin by discussing single baseline interferometers and then will cover correlative and multiple baseline interferometers.

7.5.1 Single Baseline Interferometer

Although virtually all interferometer systems employ multiple baselines, the single baseline interferometer uses one baseline at a time. The presence of multiple baselines allows for the resolution of ambiguities. It also allows multiple, independent measurements to be averaged to reduce the impact of multipath and other equipment-based sources of error.

Figure 7.21 is a basic block diagram of an interferometric direction-finding system. Signals from two antennas are compared in phase, and the direction of arrival of the signal is determined from the measured phase difference. Remember that we characterize the transmitted signal as a sine wave traveling at the speed of light. One cycle (360-phase degrees) of the

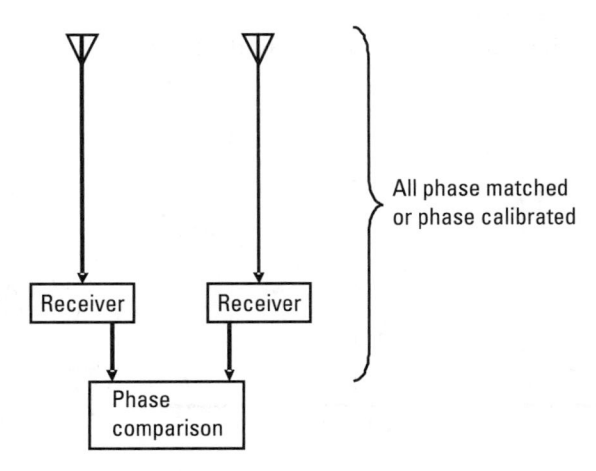

Figure 7.21 Interferometer functional diagram.

traveling sine wave is called the wavelength. The relation between the frequency of the transmitted signal and its wavelength is defined by the formula:

$$c = \lambda f$$

where
  c = the speed of light (3×10^8 m/s)
  λ = the wavelength (in meters)
  f = the frequency in cycles per second (units are 1/sec)

The interferometric principle is best explained by consideration of the interferometric triangle as shown in Figure 7.22. The two antennas from the figure form a baseline. It is assumed that the distance between the two antennas and their precise location are known. The "wave front" is a line perpendicular to the direction from which the signal is arriving at the direction-finding station. This is a line of constant phase for the arriving signal. The signal expands spherically from the transmitting antenna, so the wave front is actually a circular segment. However, since the baseline can be assumed to be much shorter than the distance from the transmitter, it is very reasonable to show the wave front as a straight line in this drawing. The precise location of the station is taken to be the center of the baseline. Since the signal has the same phase along the wave front, the phases at point A and point B are equal. Hence the phase difference between the signals at the two antennas (i.e., points A and C) are equal to the phase difference between the signal at points B and C.

The length of line BC is known from the formula:

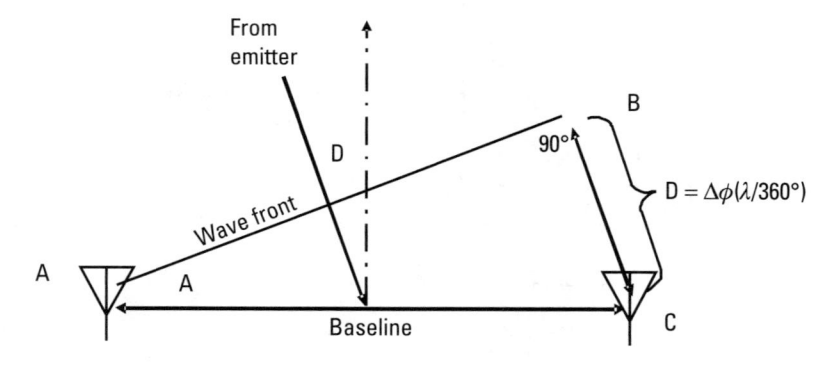

Figure 7.22 The interferometric triangle.

$$BC = \Delta\Phi(\lambda/360°)$$

where

$\Delta\Phi$ = the phase difference

λ = the signal wavelength

The angle at point B in the diagram is 90° by definition, so the angle at point A (call it angle A) is defined by:

$$A = \text{Arc} \sin(BC/AC)$$

where AC is the length of the baseline.

The angle of arrival of the signal is reported out relative to the perpendicular to the baseline at its center point, because the interferometer provides maximum accuracy at that angle. Note that the ratio of phase degrees to angular degrees is maximum here. By construction, you can see that angle D is equal to angle A.

Interferometers can use almost any type of antenna. Figure 7.23 shows a typical interferometer array which might be mounted on a metal surface—such as the skin of an aircraft or the hull of a ship. A horizontal array as shown would measure azimuth of arrival, while a vertical array would measure elevation angle of arrival. These antennas are cavity-backed spirals, which have a large front-to-back ratio, and thus provide only 180° of angular coverage. The spacing of the antennas in this array determine accuracy and ambiguity. The end antennas have a very large spacing, and thus provide excellent accuracy. However, their phase response is as shown in Figure 7.24. Note that the same phase difference (between the signals at the two antennas) can represent several different angles of arrival. This ambiguity is resolved by the two left-hand antennas, which are spaced not more than one-half wavelength apart, and thus have no ambiguity.

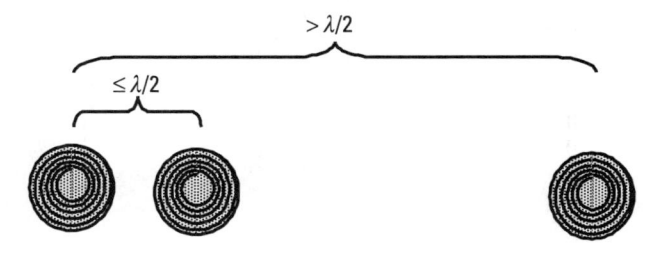

Figure 7.23 Interferometric array of antennas on a flat surface.

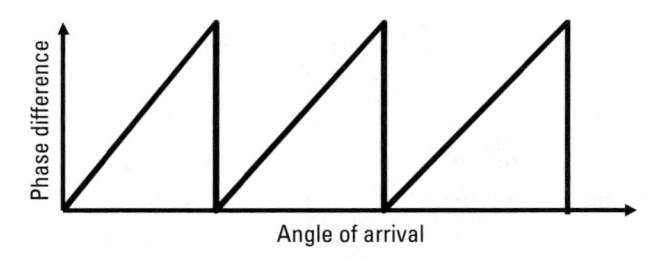

Figure 7.24 Phase versus angle-of-arrival for a long array.

Ground-based systems often use arrays of vertical dipoles as shown in Figure 7.25. To avoid the ambiguities shown in Figure 7.24, the antennas must be less than a half-wavelength apart. On the other hand, if the antennas are less than 1/10 wavelength apart, the interferometer is considered inadequately accurate. Thus a single array can provide direction finding only over a 5:1 frequency range. Some systems have multiple dipole arrays stacked vertically. Each array has different length dipoles with different spacing (smaller and closer dipoles used over higher frequency ranges). Note that the four antennas make six baselines as shown in Figure 7.26.

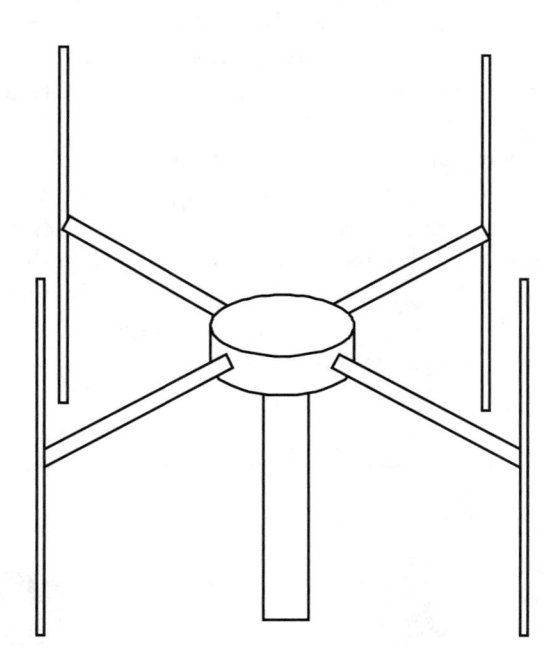

Figure 7.25 Typical ground-based interferometric antenna array.

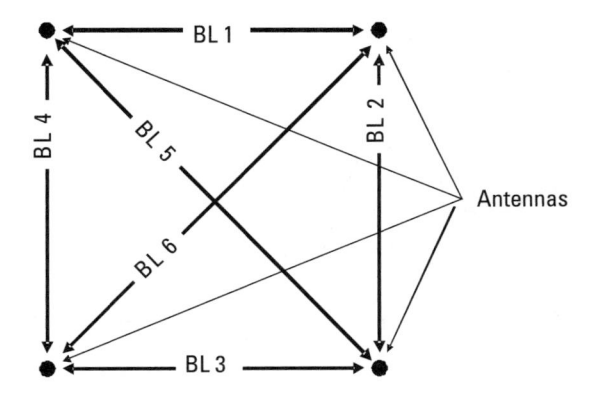

Figure 7.26 Six baselines in a four-antenna array.

Since these dipole arrays cover 360° of azimuth, the interferometer has a front-back ambiguity as shown in Figure 7.27, because signals arriving from either of the two angles shown would create the same phase difference. This problem is resolved as shown in Figure 7.28 by making a second measurement with a different pair of antennas. The correct angle of arrival is correlated in the two measurements, while the ambiguous angles of arrival do not correlate.

Figure 7.29 shows a typical interferometer DF system. The antennas are switched into the phase comparison two at a time, and the directions of arrival are measured. If there are four antennas, the six baselines are used sequentially. Often, each baseline is measured twice with the two antenna inputs switched to balance out any tiny differences in signal path length. The 12 angle-of-arrival results are then averaged and the direction-of-arrival reported out.

7.5.2 Multiple Baseline Precision Interferometer

Although it is typically applied only at microwave frequencies, the multiple baseline interferometer can be used in any frequency range as long as the length of the antenna array can be accommodated. As shown in Figure 7.30, there are multiple baselines, all greater than one-half wavelength. In the figure, the baselines are 5-, 14-, and 15-half wavelengths.

The phase measurements from all three baselines are used in a single calculation, using modulo arithmetic, to determine the angle of arrival and resolve all ambiguities. The advantage of this type of interferometer is that it can produce up to 10 times the accuracy of the single baseline interferometer.

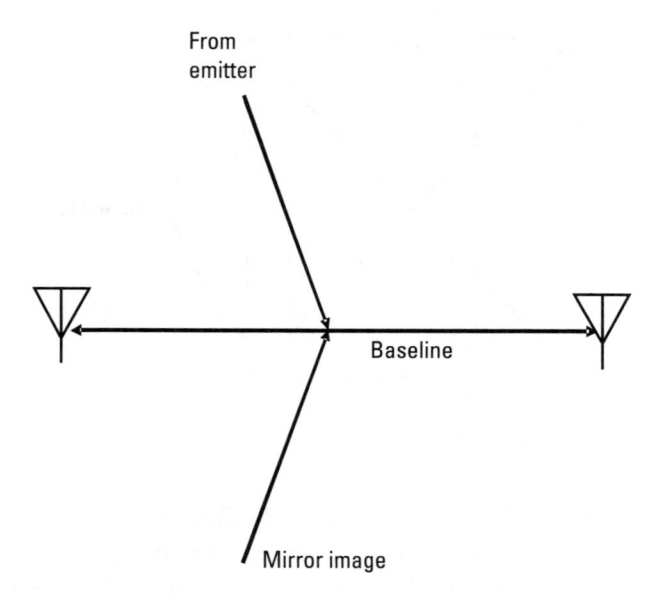

Figure 7.27 The front-back ambiguity.

The disadvantage at lower frequencies is that the arrays become extremely long.

7.5.3 Correlative Interferometer

The correlative interferometer system uses a large number of antennas, typically five to nine. Each pair of antennas creates a baseline, so there are many baselines. The antennas are spaced more than a half-wavelength apart—typically 1 to 2 wavelengths as shown in Figure 7.31. There are ambiguities in the calculations from all baselines. However, the large number of direction-of-arrival measurements allows a robust mathematical analysis of the correlation data. The correct angle of arrival will have a greater correlation value, and will be reported out.

7.6 Precision Emitter Location

Precision emitter location is generally thought of as providing adequate location accuracy for targeting. It is often described as allowing "first round fire for effect." This means that artillery can be fired at a location determined only by location of an important emitter. With the introduction of

Location of Communications Emitters 213

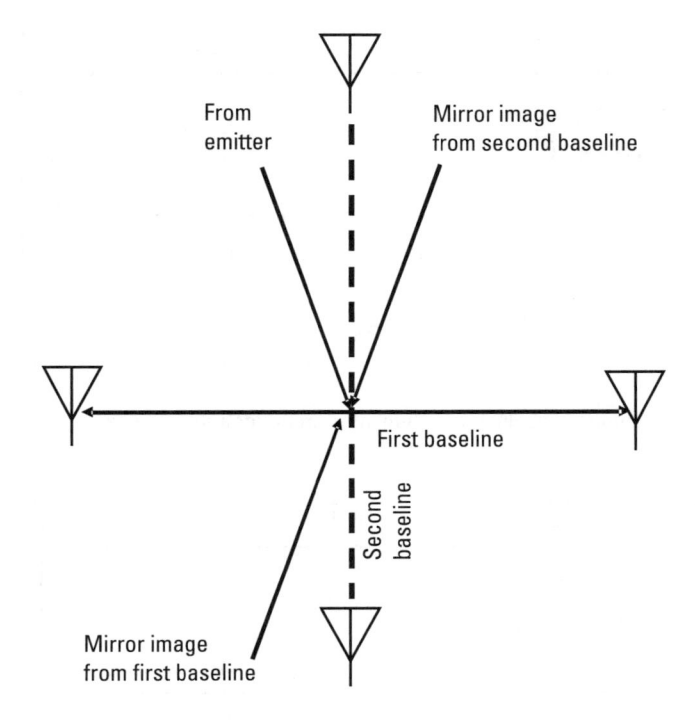

Figure 7.28 Ambiguity resolution with multiple baselines.

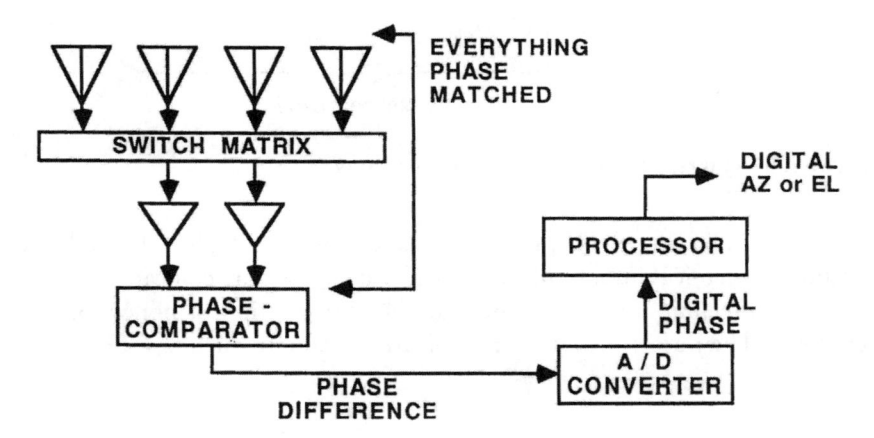

Figure 7.29 Interfometeric DF system block diagram.

GPS-guided missiles, it is possible to attack a distant target with great accuracy if the location is known with "1-burst radius" accuracy. There are also

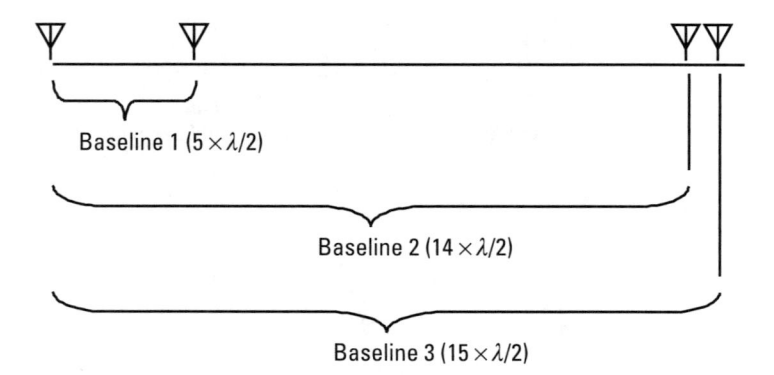

Figure 7.30 Multiple baseline precision interferometer antenna array.

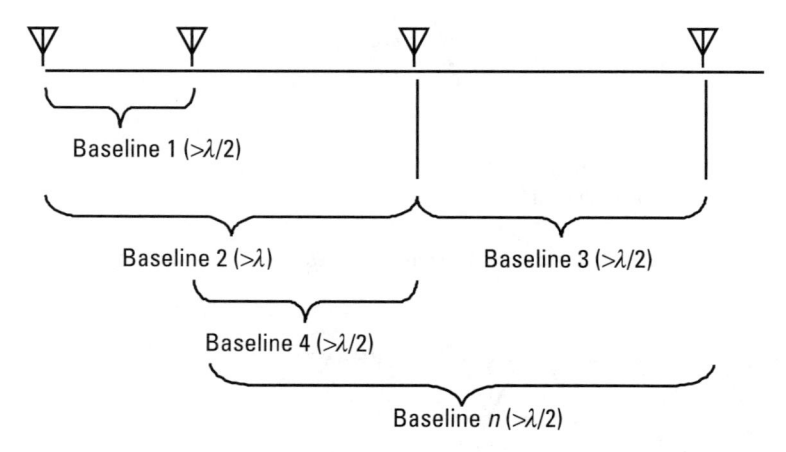

Figure 7.31 Correlative interferometer antenna array.

other situations in which precision location may be highly desirable, for example, when it is important to know if two emitters are colocated.

Two techniques provide precision emitter location of noncooperative emitters. They are time difference of arrival and frequency difference of arrival.

7.6.1 Time Difference of Arrival Method

A signal travels at the speed of light, so if we know when the signal left the transmitter and arrived at the receiver, we know the path length. When dealing with cooperative signals (such as GPS) or our own data link, coding on the signal can allow the determination of the time of departure. However, when dealing with hostile emitters, we have no way of knowing when the signal left the transmitter. The only information we can measure is when the signal arrives. However, by determining the difference between the times of arrival at two sites, we can know that the transmitting site is located along a hyperbolic curve. If the time difference of arrival is measured very accurately, the emitter location will be very close to the line, but since a hyperbola is an infinite curve, the location problem is not yet solved.

Figure 7.32 shows two sites receiving signals from a single transmitter. The two sites form a baseline. The area of uncertainty is the area which might contain the emitter of interest. Note that the difference between the two distances determine the time-of-arrival difference. Figure 7.33 shows a few of the infinite number of hyperbolas. Each curve represents a specific time-of-arrival difference, and is called an "isochrone."

TDOA Difference for Communication Signals

Pulsed signals have a very convenient time of arrival feature—the leading edge of the pulse—which can be recorded against an accurate local clock. Thus, the time difference of arrival at two locations is easily determined.

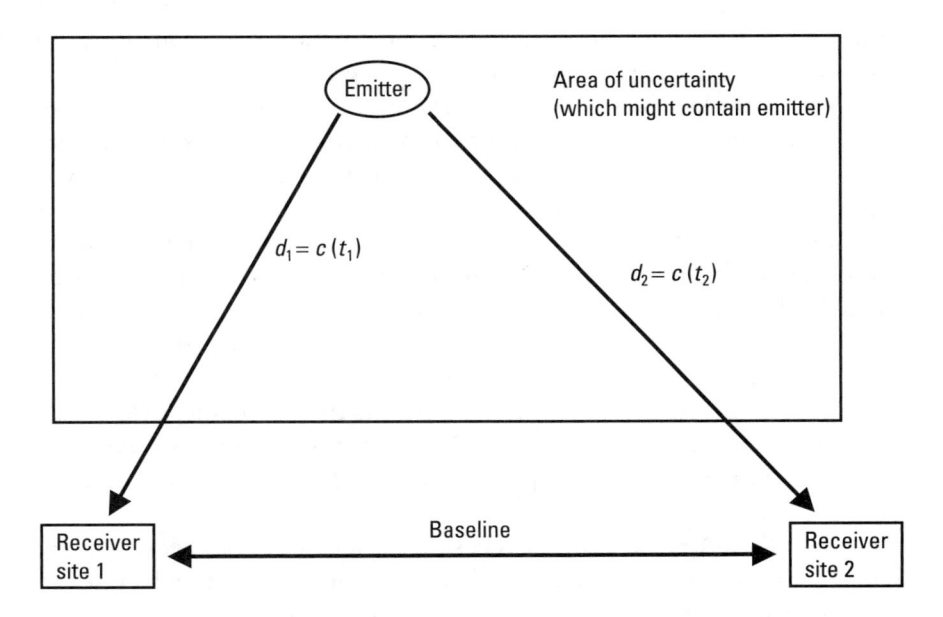

Figure 7.32 Emitter being received by two sites at which the differential time-of-arrival is determined.

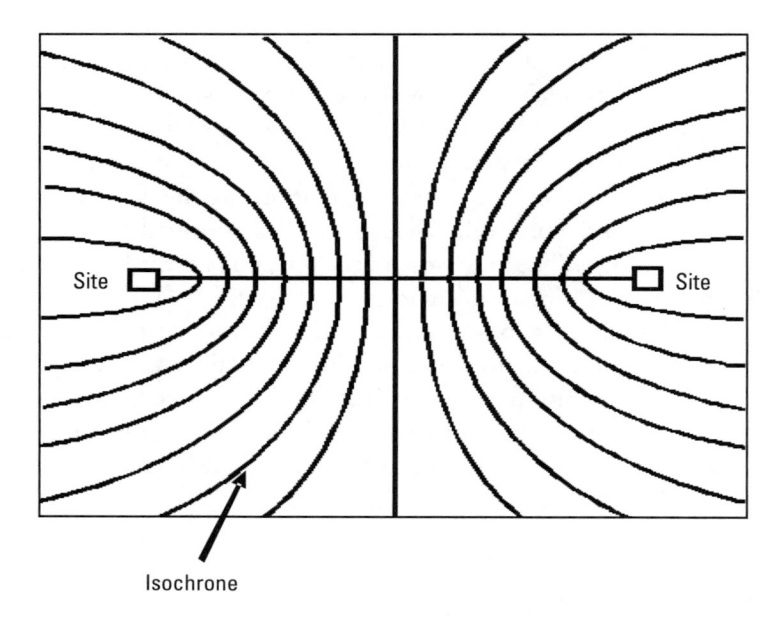

Figure 7.33 Hyperbolic contours of equal time difference of arrival.

However, most communication signals have continuous carriers (at the transmit frequency) and carry their information in the modulation of the frequency, amplitude, or phase of that carrier. The carrier repeats every wavelength (typically less than a meter), so the only attribute of the signal that we can correlate to determine the time of arrival at two receivers is the modulation. The time-of-arrival difference is determined by sampling the received signal many times with a varying time delay at one of the receivers. This time delay must be varied over sufficient range to cover the minimum to the maximum time difference possible over the area in which the emitter could be located. The samples are digitized, time coded, and sent to a common point at which the correlation between the two samples can be calculated.

The correlation changes as a function of the differential delay as shown in Figure 7.34. The correlation peak occurs at the differential delay value equal to the time difference of arrival. Note that this correlation curve has a fairly smooth top, but that the peak is typically determined to the order of 1/10 of the delay increments.

The TDOA process is relatively slow for analog signals because many samples must be taken, and requires significant data transmission bandwidth because many bits per sample are required for adequate location accuracy.

Location of Communications Emitters 217

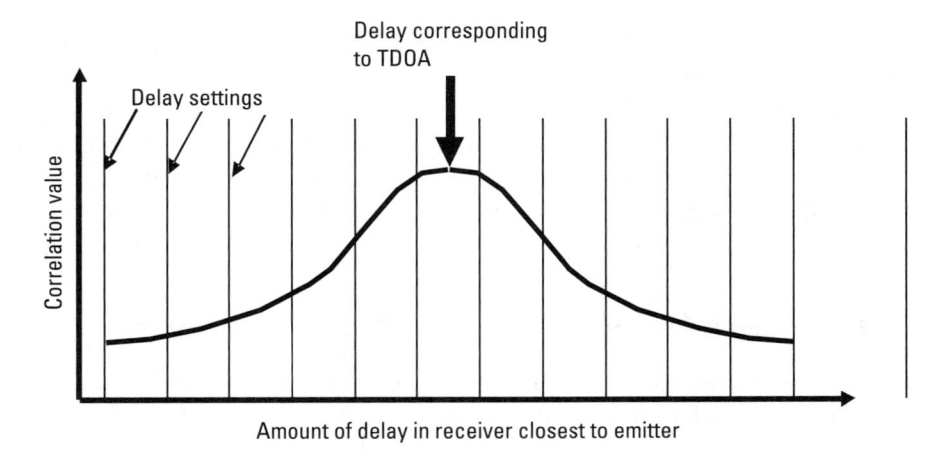

Figure 7.34 Correlation of analog signal received at two sites.

Location

Determining the actual location of the emitter requires a third receiver site so that there can be at least two baselines. As shown in Figure 7.35, each baseline forms a hyperbolic isochrone. These two hyperbolas cross at the emitter location. Since there are three sites, there is a third isochrone which would pass through the intersection of the other two.

In order to provide an accurate emitter location, it is necessary that the receiver site locations be accurately known. With the availability of GPS, accurate locations are available for small vehicles and even dismounted operators. If the receivers are moving, it is, of course, necessary to consider the

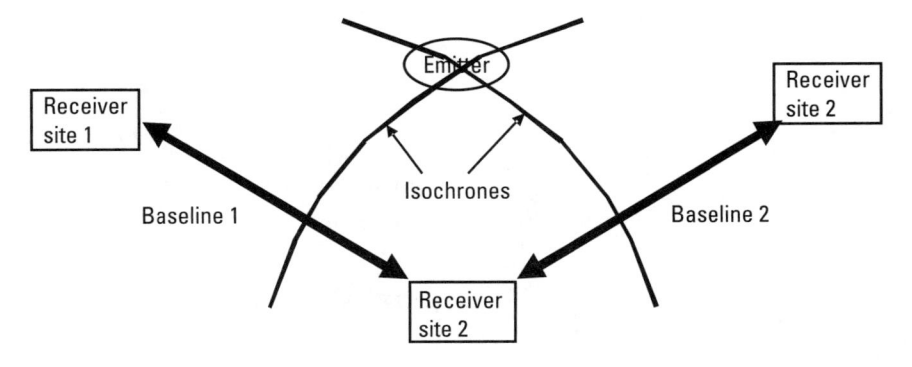

Figure 7.35 Location of emitter at intersection of isochrones from two.

instantaneous receiver locations when making the isochrone and emitter-location calculations.

The EEP and CEP can be calculated for TDOA systems from the "thickness" of the isochrone lines. That is, the uncertainty of the location of the isochrone caused by the various sources of error in the system. Because of the highly accurate time reference at each site, the location accuracy is typically specified as a few tens of meters.

7.6.2 Precision Emitter Location by Frequency Difference of Arrival

Frequency difference of arrival (FDOA) is one of the techniques for achieving precision emitter location. It involves the measurement of the difference between the received frequency at two moving receivers from a single transmitter (which is generally not moving). Because the difference in received frequency is caused by differences in Doppler shift, FDOA is also called differential Doppler (DD).

Frequency Difference of Arrival Method

First, consider the received frequency of a signal from a fixed transmitter if the receiver is moving. As shown in Figure 7.36, the received signal frequency depends on the transmitted frequency, the speed of the receiver, and the true spherical angle between the vector toward the transmitter and the velocity vector of the receiver. The received signal frequency is given by the formula

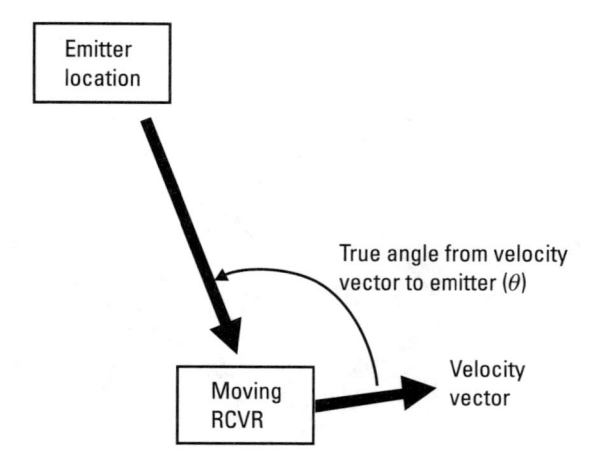

Figure 7.36 Received frequency in moving receiver.

$$F_R = F_T\left[1 + V_R \cos(\theta)/c\right]$$

where

F_R = the received frequency
F_T = the transmitted frequency
V_R = the speed of the receiver
θ = the angle from the receiver velocity vector to the transmitter
c = the speed of light

Now consider two moving receivers receiving the same signal from different locations as shown in Figure 7.37. The instantaneous locations of the two receivers form a baseline. The difference between the received frequencies at the two receivers is a function of the difference between θ_1 and θ_2 and the velocity vectors of the receivers. The difference between the two received frequencies is given by the formula:

$$\Delta F = F_T\left[V_2 \cos(\theta_2) - V_1 \cos(\theta_1)\right]/c$$

where

ΔF = the difference frequency
F_T = the transmitter frequency

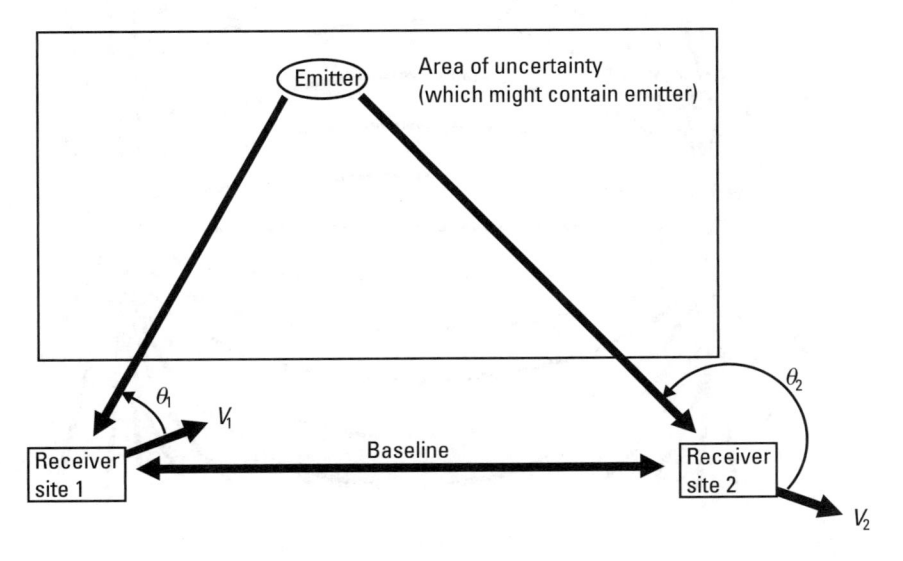

Figure 7.37 Received frequencies in two moving receivers.

V_1 = the speed of receiver 1

V_2 = the speed of receiver 2

θ_1 = the true spherical angle from the velocity vector of receiver 1 to the transmitter

θ_2 = the true spherical angle from the velocity vector of receiver 2 to the transmitter

c = the speed of light

There is a curving three-dimensional surface defining all of the possible transmitter locations which would produce the measured-frequency difference under the existing conditions. If we look at the intersection of this surface with a plane (e.g., the surface of the Earth) the resulting curve is often called an "isofreq." The two receivers can be moving in different directions at different speeds, and system computers can draw the correct isofreq curve for each velocity/geometry/frequency-difference condition. However, to simplify the visual presentation for us humans, Figure 7.38 shows a set of isofreq

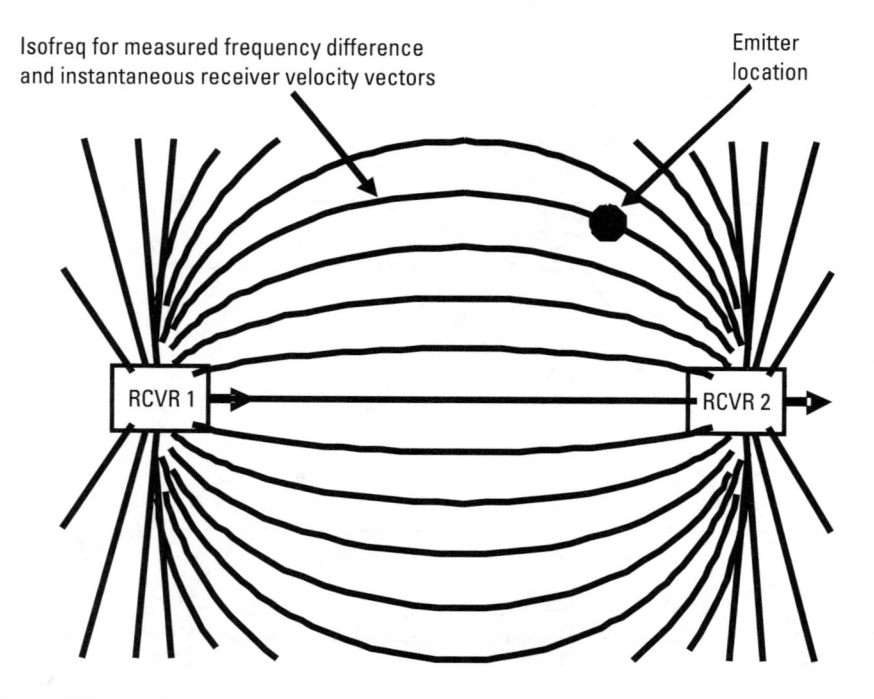

Figure 7.38 Isofreqs from baseline of two receivers moving in same direction at same speed.

curves for various frequency differences for two receivers going the same direction at the same speed (but not necessarily in a tail chase). Note that this set of curves fills all space. It looks like lines of magnetic flux (as though the two receivers were the ends of a bar magnet in a high school physics book).

Like TDOA, the frequency-difference measurement by two receivers does not define a location, it only defines a curve of possible positions (i.e., the isofreq). However, if the frequency difference is measured accurately, the transmitter location will be very close to the isofreq curve. With a third moving receiver, we have three measurement baselines, each of which can collect FDOA data and calculate isofreqs. The location of the transmitter can then be determined from the intersection of isofreqs from two baselines. There is, of course a third baseline which will create an isofreq which intersects the other two at the same location.

Like TDOA, the EEP and CEP of an FDOA system can be calculated from the thickness of the isofreq lines—that is the inaccuracy caused by the various sources of system error. Because the received frequencies at each site are measured against an extremely accurate clock, the location accuracy of FDOA systems are typically specified as a few tens of meters.

FDOA Against Moving Transmitters

There is a significant problem associated with the FDOA location of a moving transmitter (using moving receivers). The measured frequency differences are from the Doppler shift caused by the accurately known velocity vectors of the two receivers. If the transmitter is also moving, it causes Doppler shifts of the same order of magnitude as those from the moving receivers—but the transmitter velocity vector is not known. This brings another variable into the emitter-location calculation. While this should be mathematically resolvable, the required calculations (i.e., computer power and time required) are much more complex. Therefore, FDOA is generally accepted as being appropriate only for locating fixed or very slowly moving transmitters from moving airborne receivers.

7.6.3 Combined FDOA and TDOA

Since the measurement of frequency and time both require the presence of a highly accurate frequency reference, it is logical to perform both functions from the same two receivers. This is done in many precision location systems. Consider Figure 7.39, a set of isochrons (from TDOA) and a set of isofreqs (from FDOA), calculated for a single baseline of two receivers. You will note that the transmitter location is at the intersection of an isochron

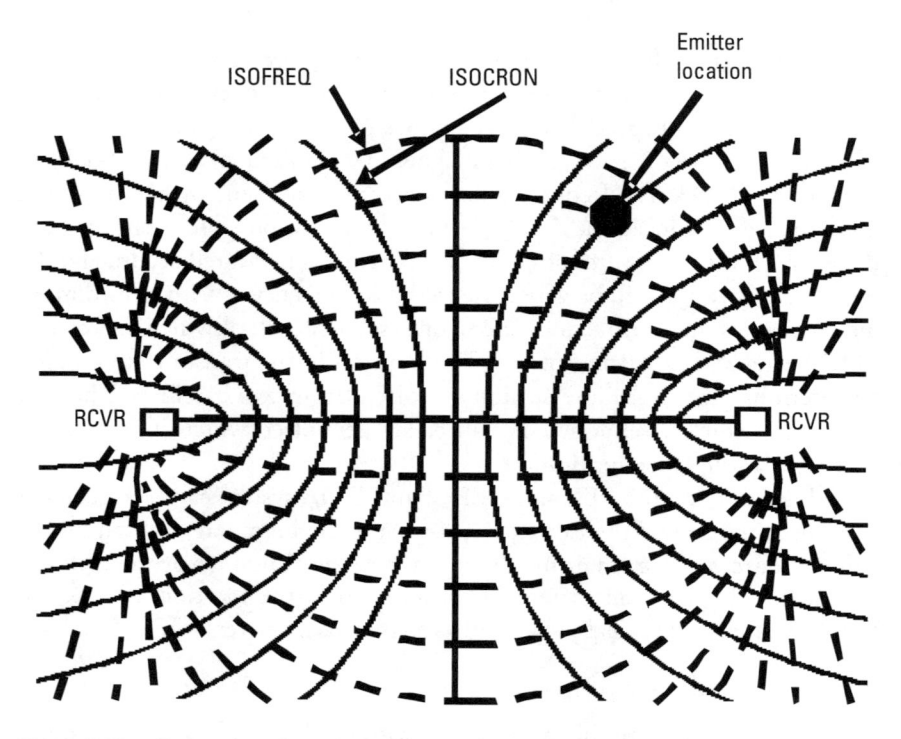

Figure 7.39 Emitter location using both TDOA and FDOA with two sites.

and an isofreq. Thus, a precision emitter location is determined from a single baseline of two receivers.

In practice, location systems typically use three or more platforms, so many solutions can be calculated—from TDOA or FDOA alone and by combined TDOA and FDOA. This multiplicity of solutions allows a more accurate ultimate result over a wide range of operational conditions.

7.7 Emitter Location—Error Budget

The most important measure of value of an emitter location system is location accuracy. In the specification of a system, it is necessary to account for all of the elements which contribute error. This is called the error budget. Some elements are common to more than one type of emitter-location approach, but many are associated with only one approach.

7.7.1 Combination of Error Elements

There are multiple sources of emitter-location error, some are random and some are fixed. In general, if sources of error are random and independent of each other, they are statistically combined. The total error is the square root of the sum of the squares of the components, as shown here:

$$\text{Total RMS error} = \text{SQRT}\left[\text{error}_1^2 + \text{error}_2^2 + \text{error}_3^2 + \text{error}_4^2 + \ldots \text{error}_n^2\right]$$

where there are n independent and random sources of error.

However, if the sources of error are not random, they must be summed directly.

Where a very accurate and complete measurement of system error can be made, for example on an instrumented range, it may be practical to offset all location measurements or direction-of-arrival measurements by the value of the statistical mean error. Then, the RMS error of the system will be equal to the standard deviation of the measured error data. Note that this assumes no significant site errors—but rather that the major sources of error are associated with the platforms. Direction-of-arrival systems on airborne platforms will usually have this characteristic, since reflections from the airframe can cause significant angle-of-arrival errors while multipath sources are far from the measurement system.

Impact of Reflections on AOA Error

Reflectors near the path from the target emitter to the AOA site cause errors by creating multipath. The AOA site measures the vector sum of the direct path component and all multipath components arriving at its antennas. As shown in Figure 7.40, reflectors near the target emitter cause multipath signals to arrive at relatively small offset angles which cause relatively small errors (typical in airborne systems). However, reflectors near the AOA site cause multipath signals which can arrive at relatively large angles. These reflections cause relatively large errors. Ground-based AOA systems are significantly impacted by close-in terrain. However all AOA systems have significant multipath errors from the vehicles (air or ground) on which they are mounted. Reflections from the sides (versus close or opposite to the signal AOA) cause the most significant errors. Thus it is desirable to locate emitter location systems in as clean an environment as possible.

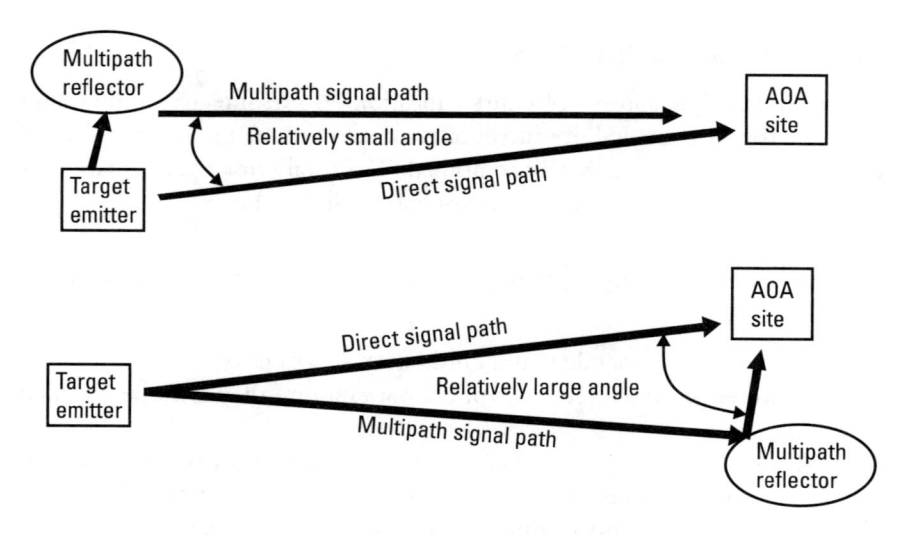

Figure 7.40 The impact of multipath reflectors on DOA measurement accuracy.

Error Related to Signal-to-Noise Ratio

It is typical to specify the location accuracy for a system receiving a strong signal; however, the system must typically handle much weaker signals. One way to specify the sensitivity of a direction-finding system is to take a series of measurements (usually 5 to 10) at increments of received signal strength. For strong signals, all AOA measurements will be very close (typically identical). Then as the signal strength decreases, the reduced signal-to-noise ratio will cause variations in the measured AOA. The system sensitivity is often stated as the received signal strength at which the standard deviation of these measurements equals 1°. It is also possible to calculate the RMS angular error component caused by any specified signal-to-noise ratio.

Calibration Errors

All AOA systems that achieve high accuracy are calibrated to remove the effects of fixed errors from antenna mounting geometry and from processing. The calibration involves measuring AOA on some sort of accurate range and the correction of measured data in later operations to remove the errors measured during calibration. The accuracy of the calibration data is an additional source of angular error.

7.8 Locating Spread Spectrum Emitters

As mentioned before, LPI communication techniques involve spreading the transmitted spectrum, so they are also called spread spectrum techniques. In general, any of the emitter location approaches described in this chapter can be used to locate spread spectrum transmitters. There are, however, some special considerations for each of the three spread spectrum techniques. Some aspects of the precision emitter-location techniques make their application to spread spectrum signals quite challenging.

In any case, the receiver or receivers used for emitter location must be able to detect and receive the target signals. Like the LPI search techniques described in Chapter 6, LPI emitter location is strongly impacted by intercept geometry. Clear line-of-sight and short-link distance will provide strong received signals with high signal-to-noise ratio. As you will see, some of the techniques suggested will only work with high-received signal-to-noise ratio.

This section describes techniques for the location of frequency hopping, chirp, and direct sequence spread spectrum emitters.

7.8.1 Locating Frequency Hoppers

As described in Section 2.4.2, a frequency hopping signal places all of its transmitted signal in a single information bandwidth, but pseudo-randomly hops to another transmit frequency every few milliseconds. (See Figure 7.41.) The hop period is the time the signal remains at one frequency.

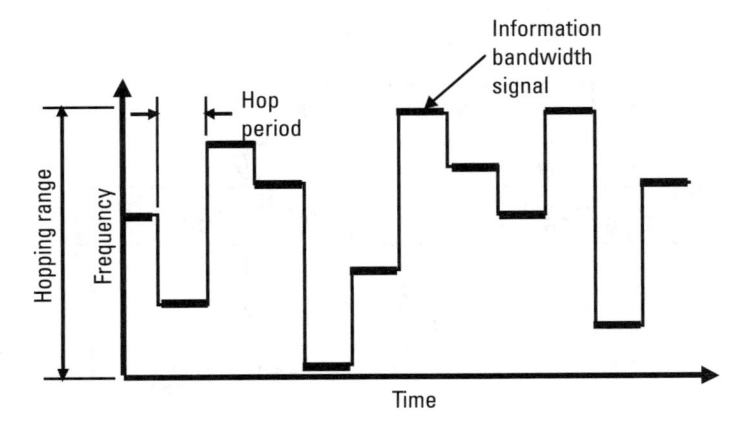

Figure 7.41 Frequency hopping signal.

A frequency hopper is the least challenging of the spread spectrum transmitters to locate, because it places all of its radiated power at one frequency for a period of milliseconds. The challenge is to determine that frequency before the transmitter hops to a different frequency.

There are two commonly used approaches to this problem. The approach used by many moderate cost systems is to use a simple sweeping receiver/direction finder to determine the direction of arrival of any encountered signal at a few of its hops during a transmission. The second is to use a fast digital receiver/direction finder to determine the direction of arrival during every hop.

At the end of this section, we will discuss some other approaches which might have promise in special situations.

Sweeping Receiver Approach

With this technique, each direction-finding station has a receiver with the general block diagram shown in Figure 7.42. The receiver usually sweeps at a high rate, pausing only long enough at each step to determine whether or not a signal is present. If there is signal power at a frequency, the receiver stops long enough to perform a DF on that signal. This type of DF system typically catches only a moderate percentage of the hops, but it keeps track of the direction of arrival each time it is able to make a measurement. After it gets some number of DF measurements at a single angle of arrival, it reports that there is a frequency hopping transmitter in that direction.

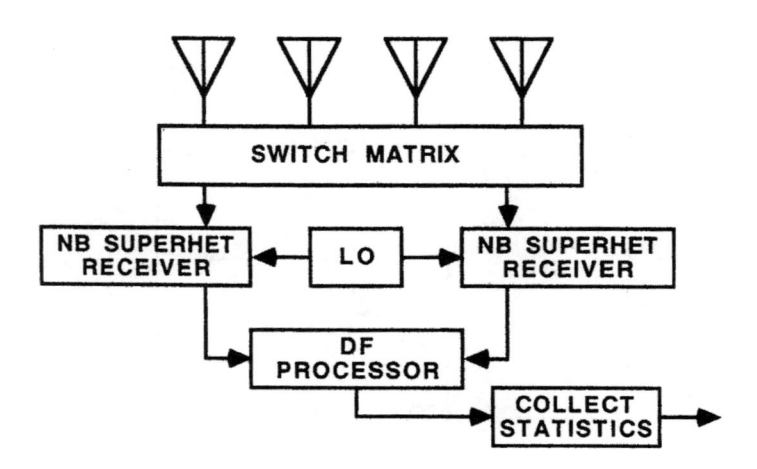

Figure 7.42 Sweeping frequency hopper DF system.

Data is collected in a computer file and is sometimes presented to an operator on a frequency versus angle-of-arrival display like that shown in Figure 7.43. Each dot on this display represents a received signal. Note that there are several intercepts at the same angle but at different frequencies. This is the characteristic of a frequency hopper. If some number of hits at the same angle of arrival are detected during a signal transmission (generally specified at a small number of seconds), a frequency hopper direction of arrival is reported. This same process is repeated at a second (and preferably a third) DF station so that triangulation can be performed to locate the transmitter. In multiple DF station systems, signal location versus frequency could be displayed to operators as shown in Figure 7.44.

As discussed in Chapter 2, the density of communication signals in a tactical military environment can be very high. This complicates the triangulation problem. Consider a very simple environment with two hopping emitters being measured by two direction-finding systems as shown in Figure 7.45. There are four possible emitter locations in this simple case, and in the real world it is a lot worse.

The solution most often applied is to cause the two systems to step together. Each frequency hopper can be at any frequency in its range at any given step, but if the two receivers are locked together, they will always look at the same frequency and will thus catch the same emitter on the same hop.

Fast Fourier Transform

The determination of the frequency of a hop with a digital receiver was discussed in Section 4.3.2, and the central importance of knowing the location

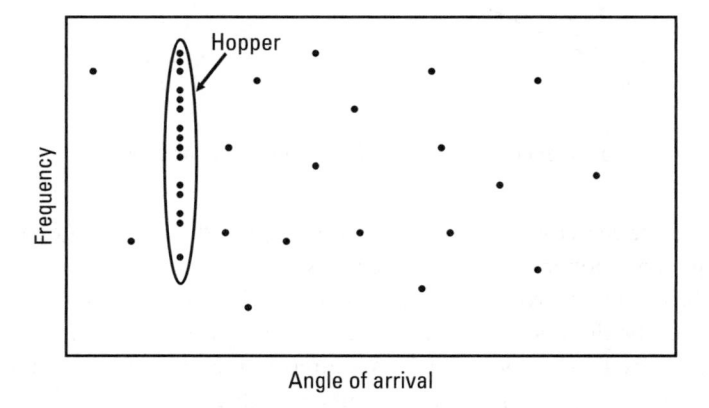

Figure 7.43 Display of emitter angle-of-arrival versus frequency.

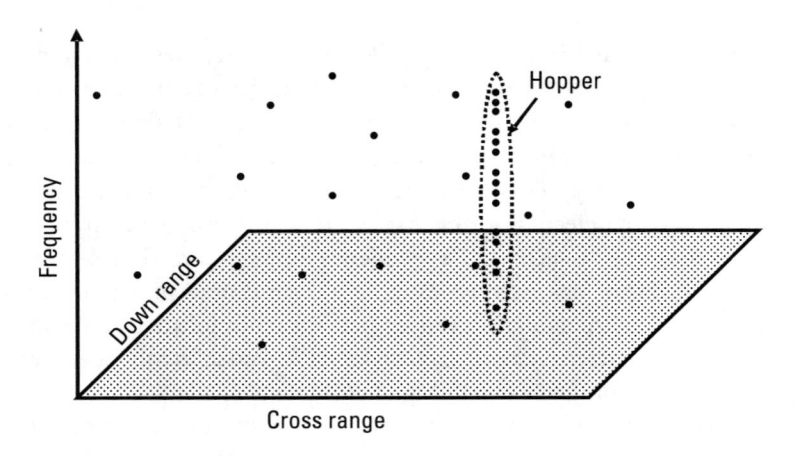

Figure 7.44 Display of emitter location versus frequency.

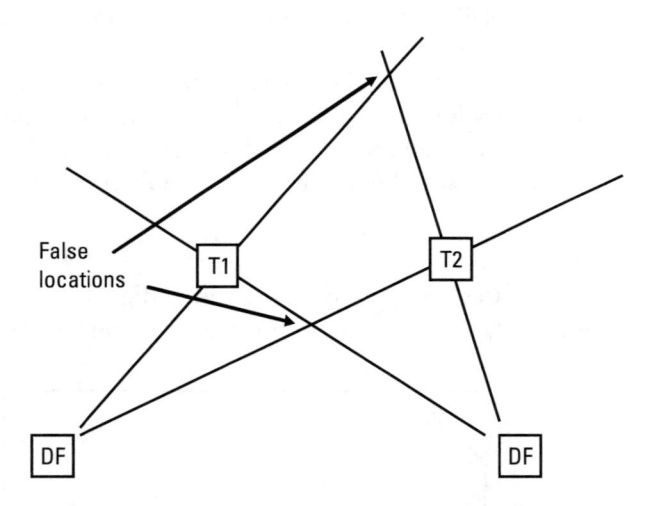

Figure 7.45 Location ambiguities with multiple frequency hoppers.

of the emitter was covered in Chapter 6. Here, we will discuss the location of the frequency hopper with digital receivers.

The digital receiver system discussed here is as shown in Figure 7.46. It determines the direction of arrival of signals using the interferometric principle as discussed in Section 7.5.1. An interferometer requires simultaneous measurement of signal phase at two antennas which form a baseline. A common practice for ground-based interferometers is to arrange four dipole

Location of Communications Emitters

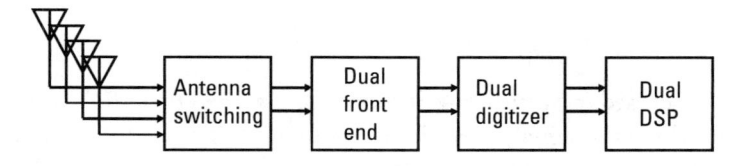

Figure 7.46 Direction finder with digital receivers.

antennas in a square array like that shown in Figure 7.25. As shown in Figure 7.26, these four antennas form six baselines. Each pair of antennas is sequentially switched into the two receiver channels, and is usually input a second time with the two antenna outputs reversed. Hence, 12-phase comparisons are made for each direction-of-arrival calculation.

If we make a 1,000-point FFT calculation for a 20-MHz IF bandwidth, we must collect 1,000 samples. To preserve the phase information, we must use I & Q digitization. As shown in Figure 7.47, I & Q digitization for this approach requires two parallel digitizers on each of two antennas which form the interferometric baseline. Since I & Q are one-fourth-wavelength apart, we have two samples per measurement (taken by parallel digitizers)—or 2,000 samples per IF bandwidth. A 2,000-point FFT provides a 1,000-channel analysis of the bandwidth. Thus, with 1,000 I & Q samples, we preserve the phase of any signal present in every 20-kHz frequency increment across the 20-MHz IF bandwidth. With three tuning steps, we cover the full 30- to 88-MHz hopping range of a VHF Jaguar V frequency hopping transmitter (with a little extra).

As described in Section 4.3.2, we will assume a sample rate of 40M samples/second (i.e., 25 ns per sample). Thus, we collect 1,000 I & Q samples in 25 μs. The three tuning steps require a total of 75 μs. This means that we have digitized the phase of any received signal across the whole hopping range, with a frequency resolution of 20 kHz in 75 μs. This is an acceptable channelization, since the Jaguar V has 25-kHz channel spacing.

We take 12 sets of samples (two per baseline), so the total phase data collection time is:

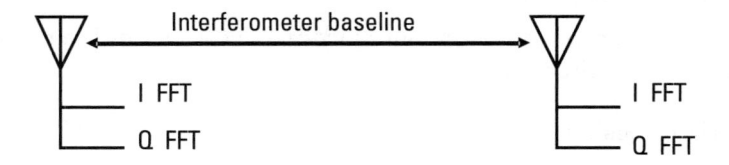

Figure 7.47 Digitization for collection of differential phase data.

$$12 \times 75 \,\mu s = 900 \,\mu s$$

As discussed in Section 6.4, it should be assumed that 10% of the channels over which the signal hops will be occupied—since we are sampling very quickly, There are 2,320 possible signal channels (58 MHz/25 kHz). If 10% are occupied, there will be 232 signals present.

Now, we have to perform direction-of-arrival calculations on the 232 signals assumed present in our data set. Processing a single direction-of-arrival measurement is assumed to take 100-DSP cycles. This is an experience-based estimate by the author. With 100-DSP cycles per DF calculation and 12 calculations per signal (two per baseline), there are 278,400-DSP cycles required. Using a 600 MFLOP/sec DSP, the whole set of direction-of-arrival measurements will require:

$$\text{DF Time} = 278{,}400 \text{ FLOPS}/600 \text{ MFLOPS/sec} = 464 \,\mu s$$

Adding this DF time to the 900 μs required to collect the phase data, the digital receiver can determine the direction of arrival of all signals present in the 58-MHz frequency range in 1.364 ms. A cooperating pair of these receiver systems could locate all signals. A hopper will be identified by its multiple frequencies at a single location. When the location of an enemy hopper transmitter is determined, a follower jammer could be assigned to the frequency associated with that location—thereby jamming every hop of the enemy signal. However, the enemy signal can only be identified from the transmitter location. Friendly frequency hoppers and many fixed-frequency transmitters operate in the same frequency range. Thus, we must determine the *frequency and direction of arrival* of each signal present during a small part of the hop. Better yet is to determine the location of each transmitter using two cooperative direction-of-arrival systems. Once the location of the enemy's frequency hopping transmitter is known, every hop of the frequency hopper can be jammed by tuning a jammer to the frequency of the signal from that location.

Special Receiver Approaches

These approaches are tempting, but practical considerations limit their usefulness to special circumstances.

Channelized Receiver

As described in Section 6.7, a channelized receiver with a channel at each hop frequency can detect the presence of a frequency hopping signal. By energy detection at the output of each channel, it can quickly determine the frequency of the signal in a small part of the hop dwell time. Then, a narrow bandwidth direction-finding system can be tuned to the hop frequency to determine the angle of arrival. This approach may be practical if there are not too many hop frequencies and the signal environment is not too dense.

Since many nonhopping signals can be expected to be present in a dense environment, this approach requires that the direction of arrival of every signal found in any channel be measured. Correlation with previous DOA measurements may be able to isolate the desired target signal location.

Compressive Receiver

A compressive receiver quickly determines the frequency of each signal present in its predetection bandwidth. The frequency is output as a digital signal which can be used to tune a narrowband DF system to the frequency of each new hop. Figure 7.48 is a generalized block diagram of such a system, which could use either a channelized or compressive receiver to measure frequency.

At lease one experimental system was developed to use a two channel compressive receiver as part of a DF system. The problem of sorting frequency hoppers from large numbers of nonhopping signals in a dense environment caused the experiment to be abandoned.

7.8.2 Chirp Emitters

If the chirped signal can be detected, most of the direction-finding techniques discussed earlier can be used to locate the transmitter. In general, the

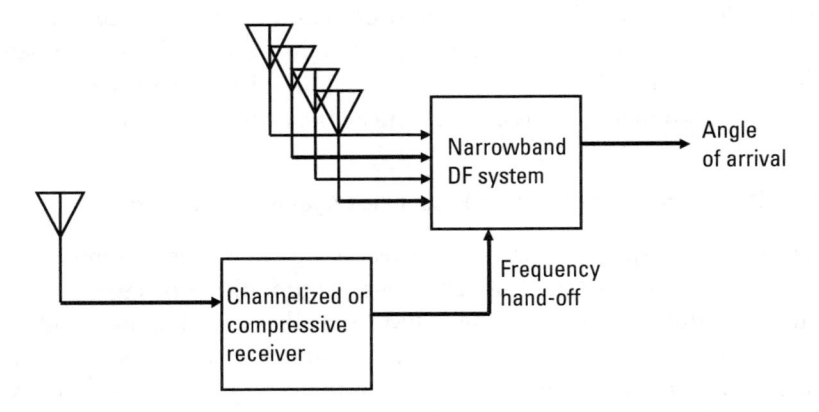

Figure 7.48 Narrowband DF system cued by wideband frequency measuring receiver.

implementation of the chosen technique must be such that intermittent reception of the carrier signal is sufficient to measure angle of arrival. Thus, techniques in which the signal is simultaneously received by two or more antennas seem most suitable. This has been done successfully with a Watson-Watt direction-finding system.

7.8.3 Direct Sequence Spread Spectrum Emitters

Location of a DSSS transmitter requires receivers which can detect the signal. Amplitude comparison approaches with multiple antennas seem the most appropriate. In general, the location of DS transmitters is fairly easy when strong signals are received, and very complicated for weak signals.

In Section 8.4.3, the nature of DSSS signals with short-spreading codes will be discussed. With such signals, it may be practical to receive one spectral line in a narrowband receiver. If so, any of the DF techniques discussed earlier may be practical.

When DSSS signals have longer spreading codes, the only practical detection approach may be energy detection. Two energy detectors with separate antennas as shown in Figure 7.49 could provide wide angle instantaneous DF. If a single energy detector operates on the output of a narrow beam directional antenna as in Figure 7.50, it can determine the angle of arrival of a "noise-like" signal such as DSSS. Two such systems can triangulate the emitter location.

Location with a Chip Detector

Chips (the digital bits used for spreading the signal) necessarily have highly predictable transition times which allow detection of the presence of the signal by implementation of a tapped delay line in software. Two chip detectors, each connected to an antenna may allow direction of arrival to be measured. Two such systems can triangulate the emitter location.

7.8.4 Precision Emitter Location Techniques Against LPI Emitters

The location of spread spectrum emitters using the precision emitter-location techniques is very challenging because communication signals use continuous modulations such as AM, FM, and phase modulation, and thus require significant correlation times that are much greater than hop times. The pseudo-random parameters of chirp and DSSS signals also make correlation very difficult to achieve. The significant problem is the subject of some current Ph.D. thesis efforts.

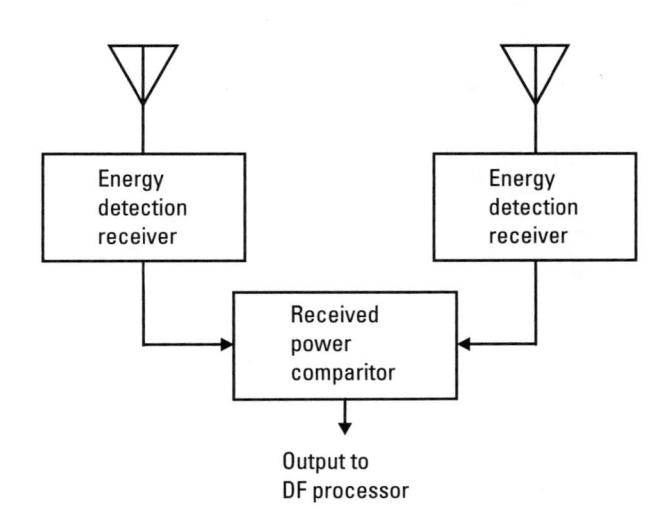

Figure 7.49 Energy detectors with multiple antennas.

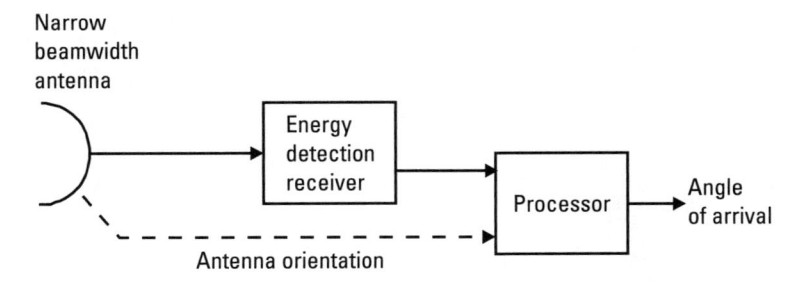

Figure 7.50 Energy detector with narrow beam antenna.

8

Intercept of Communications Signals

Intercept of a communications signal is generally considered to involve the demodulation of the signal by a hostile receiver to recover the information carried by that signal. This is also called communications intelligence (COMINT). However, communications ESM normally deals with signal externals such as frequency, emitter location, modulation, and others. This can be extremely useful for such electronic warfare activities as development of an electronic order of battle (EOB) from which the enemy's organization and perhaps intentions can be determined. Since many communications signals are encrypted, it may not be practical to recover the internals information—hence the value of the intercept may be limited to recovery and exploitation of signal externals.

This chapter deals primarily with the technical aspects of receiving and demodulating various types of hostile communications signals. In all cases, the propagation of the hostile signal from the transmitter to the intercept receiver follows one or more of the propagation modes discussed in Chapter 5. The intercept receiver sensitivity is determined by the techniques developed in Section 4.4. The techniques developed in this chapter will allow calculation of effective range for intercept as a function of the link parameters.

In Chapter 6, search for signals of interest was covered. If the search system has only one receiver, that receiver performs the intercept function after the signal of interest has been identified. In systems with search-and-monitor receivers, the intercept function as described in this chapter is performed by the assigned monitor receiver.

Also in this chapter, we will discuss the recovery of information carried by LPI communication signals.

8.1 The Intercept Link

Figure 8.1 shows a general intercept link. The received power in the intercept receiver is determined from the formula:

$$P_R = P_T + G_T - L + G_R$$

where

P_R = the signal strength into the intercept receiver from the intercept antenna (in dBm)

P_T = the transmitter output power (in dBm)

G_T = the gain of the transmitting antenna (in dB) in the direction of the intercept receiver

L = the propagation loss from the transmitter to the intercept receiver (in dB) using whatever propagation mode is appropriate

G_R = the gain of the intercept system receiving antenna (in dB) in the direction of the transmitter

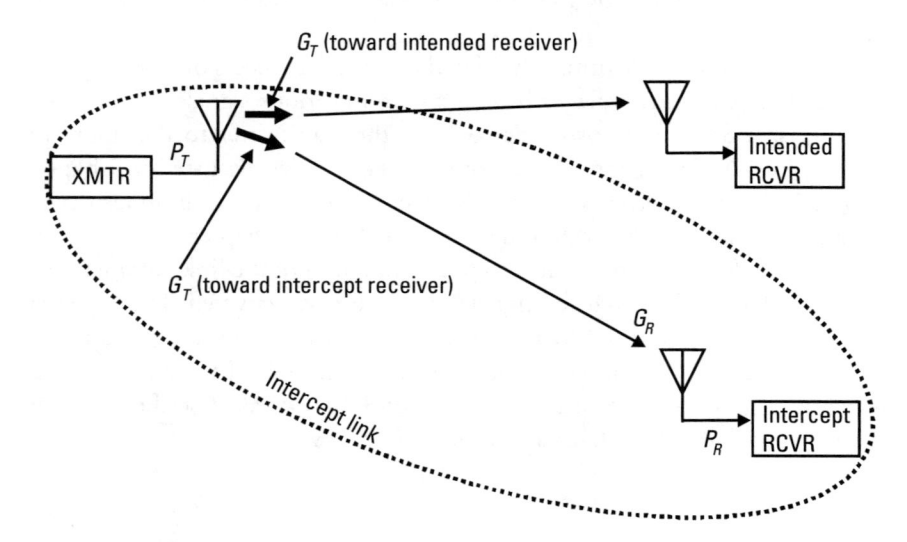

Figure 8.1 Intercept link.

Note that the transmitting antenna gain may not be the same in the direction of the intended receiver and the intercept receiver. If the received power is greater than the system sensitivity of the intercept receiver, intercept is possible. Ideally, the intercept receiver bandwidth should be matched to the modulation of the intercepted signal. This provides the best sensitivity and thus the maximum intercept range.

To determine the maximum range at which the intercept can be accomplished, set the received power to the sensitivity of the intercept receiver system. Then, solve for the range component of the propagation loss. This will be demonstrated in some of the following examples.

8.1.1 Intercept of a Directional Transmission

The situation shown in Figure 8.2 is the intercept of a data link by a hostile receiver. The transmitter has a directional antenna pointed toward the desired receiver, and the hostile receiver is not in the main lobe of the transmitting antenna pattern. The transmitter and receiver are both located on elevated terrain, so the receiving antenna is not illuminated by significant reflections from local terrain. This means that the propagation loss is determined from the line-of-sight model discussed in Section 5.4.

The received power in the intercept receiver is the transmitter power, increased by the transmit gain in the direction of the intercept receiver, reduced by the propagation loss and increased by the receiving antenna gain in the direction of the transmitter. Thus, the received power is calculated from the formula:

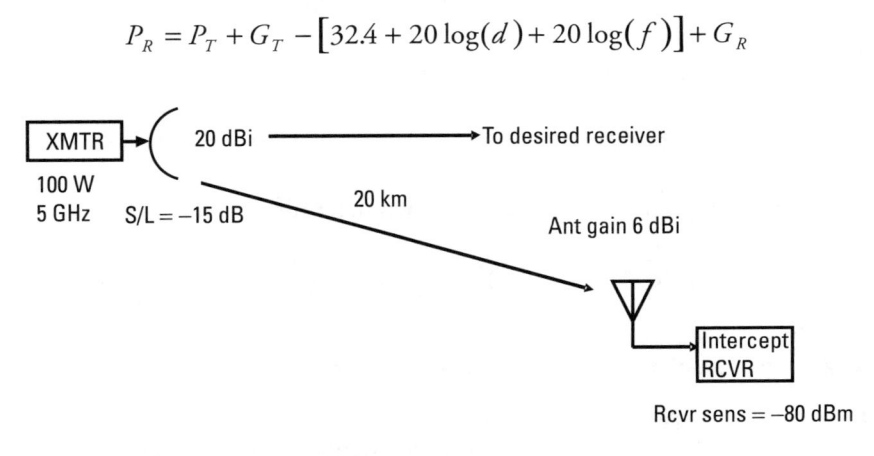

$$P_R = P_T + G_T - [32.4 + 20\log(d) + 20\log(f)] + G_R$$

Figure 8.2 Intercept of a directional transmission.

where

P_R = the received power

P_T = the transmitter power (in dBi)

G_T = transmit antenna gain (toward the receiver)

d = the link distance (in km)

f = the transmitted frequency (in MHz)

G_R = the receiving antenna gain (toward the transmitter) (in dBi)

The link transmitter outputs 100 watts (50 dBm) into its antenna at 5 GHz. The transmitting antenna has 20-dBi boresight gain and the receiver is located 20 km away in a −15-dB side lobe (i.e., 15-dB lower gain than the peak of the main beam). Thus, the transmit antenna gain for the intercept link is 5 dBi. The receiving antenna is oriented toward the transmitter and has 6-dBi gain. The received power in the intercepting receiver is calculated as:

$$P_R = +50 \text{ dBm} + 5 \text{ dBi} - [(32.4 + 26 + 74) \text{ dB}] + 6 \text{ dBi} = -71.4 \text{ dBm}$$

Since the receiver sensitivity is −80 dBm, the signal is successfully intercepted with 8.6 dB of margin. To determine the maximum range at which intercept could be accomplished, set the received power equal to the sensitivity and solve for the range component of the propagation loss from the following formula:

$$20 \log(d) = P_T + G_T - 32.4 - 20 \log(f) + G_R - S$$

where S = the receiver system sensitivity.

$$20 \log(d) = 50 + 5 - 32.4 - 74 + 6 - (-80) = 34.6$$

The maximum intercept range is then determined from:

$$d = \text{Antilog}\left[(20 \log(d))/20\right] = \text{Antilog}\left[1.73\right] = 53.7 \text{ km}$$

8.1.2 Intercept of a Nondirectional Transmission

In the intercept situation shown in Figure 8.3, the transmitter and receiver both are near the ground and have wide angular coverage antennas, therefore, they may be subject to either line-of-sight or two-ray propagation. The

Intercept of Communications Signals

1 watt ERP
whip antenna

Ant height 1.5m

XMTR
100 MHz

10 km

360° antenna
ant height 30m
gain = 2 dBi

Intercept RCVR

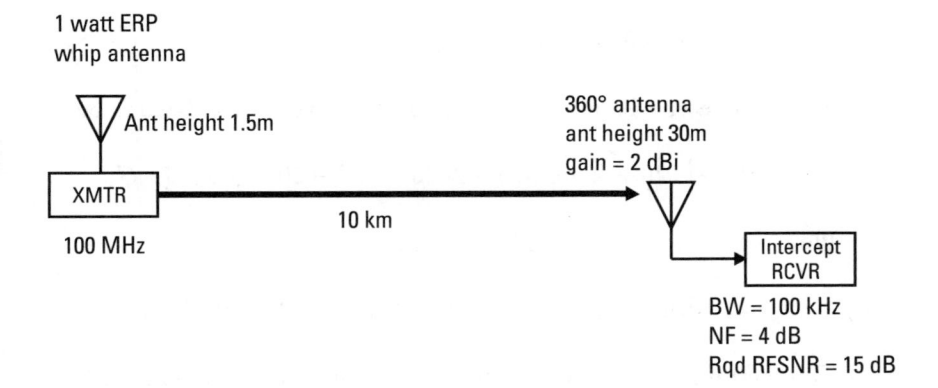

BW = 100 kHz
NF = 4 dB
Rqd RFSNR = 15 dB

Figure 8.3 Intercept of a nondirectional transmission.

proper propagation mode is determined by calculation of the Fresnel zone distance by the formula from Section 5.6:

$$FZ = (h_T \times h_R \times f)/24{,}000$$

where
  FZ = the Fresnel zone distance (in km)
  h_T = the transmit antenna height (in meters)
  h_R = the receiving antenna height (in meters)
  f = the transmitted frequency (in MHz)

If the transmitter to receiver path length is shorter than the Freznel zone distance, line-of-sight propagation applies. If the path is longer than the Fresnel zone distance, two-ray propagation applies.

The target emitter is a handheld push-to-talk system with a whip antenna, 1.5 meters from the ground. Note that the effective height of a whip antenna is at the bottom of the whip. The receiving antenna has 2-dBi gain. The power transmitted from the target emitter has effective radiated power of 1 watt (30 dBm) at 100 MHz. The Fresnel zone distance is:

$$(1.5 \times 30 \times 100)/24{,}000 = 188 \text{ meters}$$

The Fresnel zone distance is far less than the 10-km path distance, so two-ray propagation applies.

The propagation loss, from the formula in Section 5.5, is:

$$120 + 40\log(d) - 20\log(h_T) - 20\log(h_R)$$

Thus, the received power in the intercept receiver is calculated to be:

$$P_R = ERP - \left[120 + 40\log(d) - 20\log(h_T) - 20\log(h_R)\right] + G_R$$

Plugging in the values from Figure 8.3,

$$P_R = 30\,\text{dBm} - [120 + 40 - 3.5 - 29.5] + 2\,\text{dB} = -95\,\text{dBm}$$

To determine whether or not the signal is successfully intercepted, we must calculate the sensitivity of the receiver using the technique presented in Section 4.4:

$$Sens = kTB + NF + Rqd\ RFSNR$$

where

Sens = the receiver sensitivity in dBm

NF = the receiver noise figure in dB

Rqd RFSNR = the required predetection signal-to-noise ratio in dB

Remember that the sensitivity is the minimum signal strength that a receiver can receive and still do its job.

$$kTB = -114\,\text{dBm} + 10\log(\text{bandwidth}/1\,\text{MHz}) = -124\,\text{dBm}$$

The receiver system noise figure is given as 4 dB and the required RFSNR is given as 15 dB, so:

$$Sens = -124 + 4 + 15 = -105\,\text{dBm}$$

Since the signal is received at a level 10 dB above the receiver system's sensitivity level, the intercept receiver has achieved a 10-dB performance margin.

The maximum intercept range is determined from the formula:

$$40\log(d) = ERP - 120 + 20\log(h_T) + 20\log(h_R) + G_R - S$$

Plugging in the values from Figure 8.3 and the sensitivity above,

$$40 \log(d) = 30 - 120 + 3.5 + 29.5 + 2 - (-105) = 50$$

The maximum intercept range is then determined from:

$$d = \text{Antilog}\left[(40 \log(d))/40\right] = \text{Antilog}\left[1.25\right] = 17.8 \text{ km}$$

8.1.3 Airborne Intercept System

In Figure 8.4, the intercept system is located in a helicopter which is 50 km from the enemy transmitter at an altitude above local terrain of 1,000 meters. The target emitter is a handheld 400-MHz transmitter transmitting 1-watt ERP at 400 MHz. The bottom of its whip antenna is 1.5 meters above the local terrain.

First we need to calculate the Fresnel zone distance for the intercept link using the formula given above:

$$FZ = (h_T \times h_R \times f)/24{,}000 = (1.5 \times 1{,}000 \times 400)/24{,}000 = 25 \text{ km}$$

Since the transmission path is longer than the Fresnel zone distance, two-ray propagation occurs, so:

$$P_R = ERP - \left[120 + 40 \log(d) - 20 \log(h_T) - 20 \log(h_R)\right] + G_R$$

The received intercept signal strength is:

$$P_R = 30 \text{ dBm} - \left[120 + 68 - 3.5 - 60\right] \text{ dB} + 2 \text{ dBi} = -92.5 \text{ dBm}$$

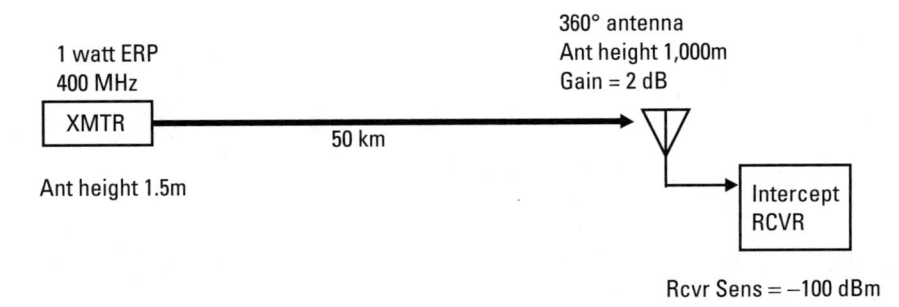

Figure 8.4 Intercept of ground transmission by an airborne receiver.

Since the sensitivity is –100 dBm, the intercept is successful with 7.5 dB of margin.

The maximum intercept range is determined from:

$$40 \log(d) = ERP - 120 + 20 \log(h_T) + 20 \log(h_R) + G_R - S$$

Plugging in the values from Figure 8.4 and the sensitivity above,

$$40 \log(d) = 30 - 120 + 3.5 + 60 + 2 - (-100) = 75.5$$

The maximum intercept range is then determined from:

$$d = \text{Antilog}\left[(40 \log(d))/40\right] = \text{Antilog}[1.89] = 77.2 \text{ km}$$

8.1.4 Nonline-of-Sight Intercept

Figure 8.5 shows intercept of tactical communication emitter across a ridge line 11 km from the emitter. In this problem, the direct line distance from the transmitter to the intercept receiver is 31 km, the transmit antenna height is 1.5m, and the intercept antenna height is 30m. The transmit signal has 1-watt ERP at 150 MHz and the receiving antenna has 12-dBi gain (G_R).

As discussed in Section 5.7, the link loss is the line-of-sight loss (ignoring the terrain interference), plus a knife-edge diffraction (KED) loss factor. If the ridge rises 210m above the local ground (assuming flat earth) it will be 200 meters above the line-of-sight between the two antennas.

The line-of-sight loss, using the formula from Section 5.4, is:

$$32.4 + 20 \log D + 20 \log f$$

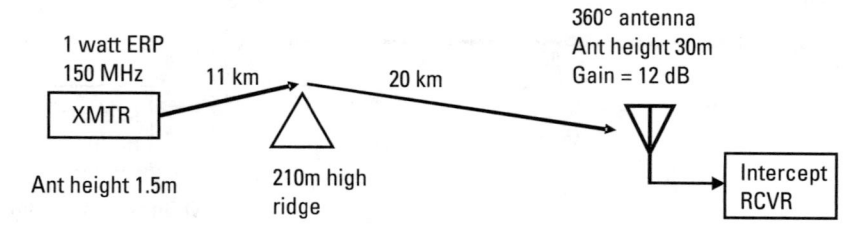

Figure 8.5 A nonline-of-sight ground intercept.

Note that we use a capital D here for full link distance to avoid confusion with the lowercase d used in the KED loss determination.

LOS loss = 32.4 + 20 log(31) + 20 log(150) = 32.4 + 29.8 + 43.5 = 105.7 dB

To determine the KED loss (as explained in Section 5.7), we first calculate d from the formula:

$$d = \left[\sqrt{(2)}/(1+(d_1/d_2))\right]d_1$$

where
  d is the distance term entered into the KED loss nomograph
  d_1 is the transmitter-to-ridge distance
  d_2 is the ridge-to-receiver distance

For this problem, $d = [\sqrt{(2)}/1.55]\,11 = 10$ km, but remember we could also just set $d = d_1$ for a slightly less accurate KED determination.

Figure 8.6 is the nomograph from Section 5.7 used to calculate knife-edge diffraction (KED) loss. It shows that the values from this problem ($d=10$ km, $H=200$m, $f=150$ MHz) will cause 20-dB KED loss. Thus, the total link loss is:

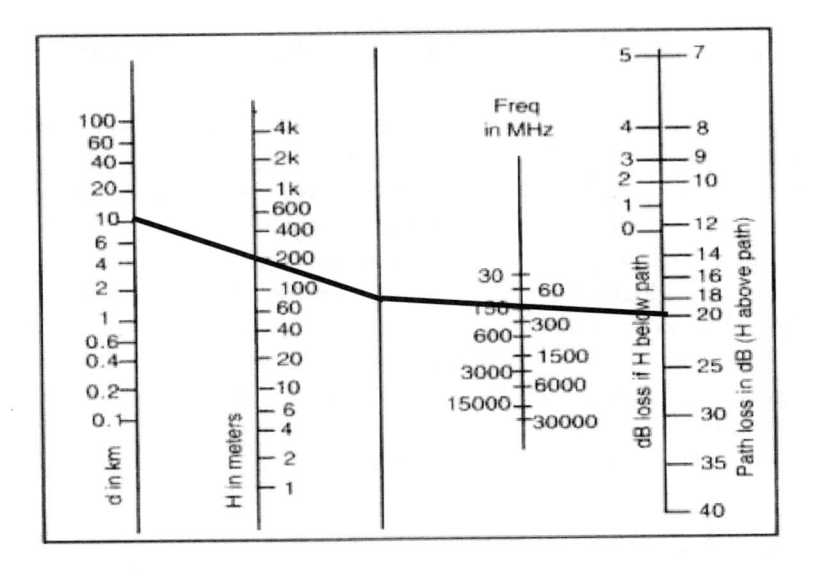

Figure 8.6 KED nomograph for nonline-of-sight intercept.

$$\text{LOS loss} + \text{KED Loss} = 105.7 \text{ dB} + 20 \text{ dB} = 125.7 \text{ dB}$$

The received power in the intercept receiver is then:

$$P_R = ERP - \text{Loss} + G_R = 30 \text{ dBm} - 125.7 \text{ dB} + 20 \text{ dB} = -83.7 \text{ dBm}$$

The maximum intercept range can be determined by setting the received signal power equal to the intercept receiver sensitivity. Remove the KED loss (20 dB) from the total acceptable loss and solve for the range at which the LOS loss is the calculated value. However, the ridgeline complicates the process. Moving the intercept receiver or the target transmitter farther from the ridge (to the maximum range) will change the KED geometry and thus the KED loss. The new KED loss value will change the acceptable maximum line-of-sight loss. Thus the maximum range will need to be recalculated. This will again change the KED geometry. If this process is repeated for a few cycles, which is fairly easy to do with a computer, the maximum range can be calculated to acceptable accuracy. Of course, you must remain aware of the KED restriction described in Section 5.7 that the receiver be at least as far from the ridgeline as the transmitter—which may be the factor determining the maximum intercept range.

8.2 Intercept of Weak Signal in Strong Signal Environment

In most situations, a communication intercept system must be able to intercept a weak signal in the presence of strong signals. Thus, the system dynamic range is of primary importance. Since automatic gain control (AGC) optimizes the receiver operation to the strongest signal, it is almost a universal rule that an intercept system cannot have AGC, even though it may be required to receive signals over a range of 100 dB or more. Ideally, the intercept receiver instantaneous dynamic range is wide enough to handle all signals of interest. If that is impractical, it may be appropriate to have a switched gain control (or input attenuation) to allow extended dynamic ranged (noninstantaneous).

Consider the following intercept system:

- Its sensitivity is −110 dBm.
- Its instantaneous dynamic range is 80 dB.

This means that the system can receive and demodulate a signal at −110 dBm in the presence of an undesired signal at −31 dBm.

8.3 Intercept of LPI Signals

The operation of LPI signals is discussed in detail in Chapter 2. In this section, we will build on that information to discuss the recovery of modulation from such signals.

8.3.1 Intercept of Frequency Hoppers

Because the frequency hopping signal changes to a new, pseudo-randomly selected frequency every few milliseconds, the only practical way to recover its information is to tune to the new frequency as soon as possible. If the hopping code is available, for example in a commercial system with few and rather short hopping patterns, this will be straight forward. Just try the possible patterns until the signal is present. However, in a secure military system, the code will be protected and will not repeat for an extended period. Thus, a digital receiver with fast Fourier transform analysis will be required to identify the new frequency in a small part of the hop period. Further, in order to follow a single signal, the emitter location must be determined for each hop. The intercept receiver can then be tuned to the frequency of the signal at the known target transmitter location.

In Chapter 6, the process of finding the frequency with an FFT processor was discussed. Then, in Chapter 7, the location of the emitter was discussed. Both of these processes take time. During this delay time, the information in the threat signal goes on. If a system like that shown in Figure 8.7 is practical, the intercept channel is delayed by enough to allow the

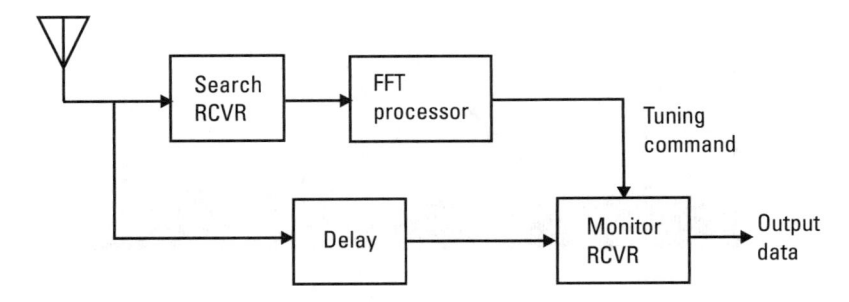

Figure 8.7 Hopper intercept system with delayed channel.

intercept receiver to receive all of the target signal. Note that digital RF memories can provide several milliseconds of delay.

The frequency range over which a frequency hopper operates can be adjusted to avoid spectral areas of high interference. The active channels in a hostile hopper can be determined by spectral analysis as shown in Figure 8.8.

8.3.2 Intercept of Chirped Signals

To intercept a chirped signal (i.e., recover the information it carries), it is necessary to generate a more or less continuous output of the signal's modulation. The obvious way to do this is to provide a sweeping receiver which has a tuning slope the same as that of the chirped transmitter, and to somehow synchronize the receiver sweep to the signal sweep.

If the tuning slope pattern can be calculated from a series of carrier frequency intercepts, generation of the correct receiver-tuning curve is straight forward. If the pseudo-random sweep synchronization scheme can be solved, the sweep-to-sweep timing can be predicted. Another approach is to digitize the full chirp range and do curve fitting in software to recover the modulation with some process latency.

Either way, recovering the modulation of a chirp signal with pseudo-random selection of slope or sweep synchronization is technically challenging. If a delayed channel as shown in Figure 8.9 is practical, analysis of the output of the search receiver can be performed to determine the pseudo-random sweep synchronization delays and to generate a tuning signal (for the monitor receiver) that is synchronized to the intercepted signal. If the pseudo-random sweep synchronization code length is short, this process

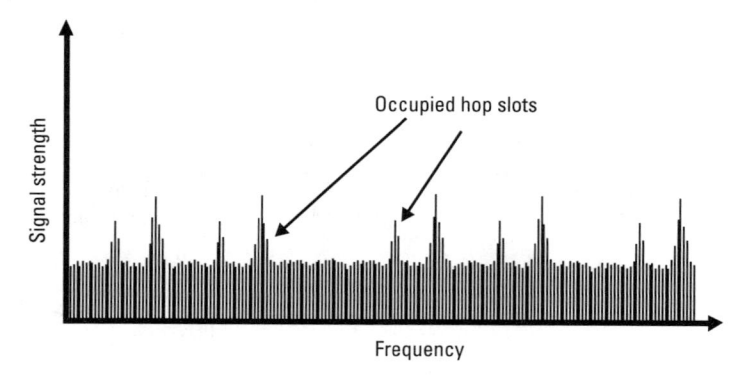

Figure 8.8   Spectrum of frequency hopping signal.

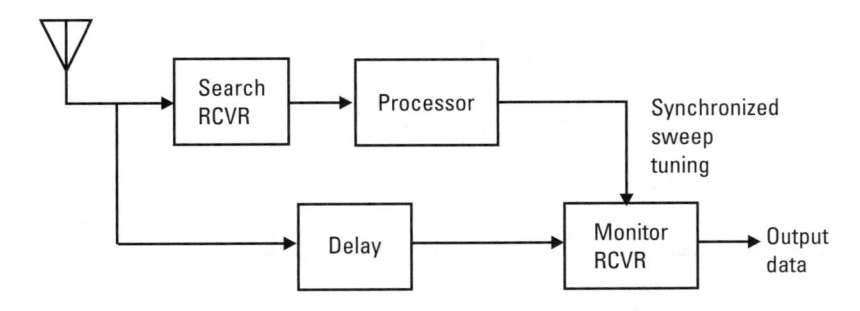

Figure 8.9 Chirp intercept system with delayed channel.

is fairly straightforward. For longer codes, massive computer capability would be required.

8.3.3 Intercept of Direct Sequence Spread Spectrum Signals

Like all spread spectrum signals, DS signals are hard to intercept (i.e., to recover the transmitted information). If the spreading code is known, it can be applied to despread the signal. If there are several publicly known codes, they can be tried sequentially until the signal is present in its nonspread form.

Secure, Long Spreading Codes

For DSSS signals with secure, long spreading codes, intercept is quite challenging, but there are approaches that may be practical if the intercept receiver is close enough to the transmitter to provide adequate received signal strength.

An interesting characteristic of the direct sequence spread spectrum signal is that it has the nonspread modulation on each of the spectral lines of the spread signal. As shown in Figure 8.10, the spectrum of the spread signal has a null at the chip rate (converted to Hz) from the carrier frequency. There are $2n$ distinct lines in the main lobe of the spectrum, where n is the number of bits in the modulating code. For example, if the chip rate is 1 Mbps and the code length is 31, there will be 62 spectral lines separated by 32.258 kHz. Figure 8.11 shows the modulation on the individual lines. If the code is short enough that the spectral lines are far enough apart to keep the modulations from interfering with each other, the unspread signal can be recovered by tuning a receiver to one line. This, of course, requires that the receiver have enough sensitivity to receive one line. Remember that the signal power is divided among the spectral lines by the spreading factor.

f_0 = Carrier frequency
f_c = Chip rate

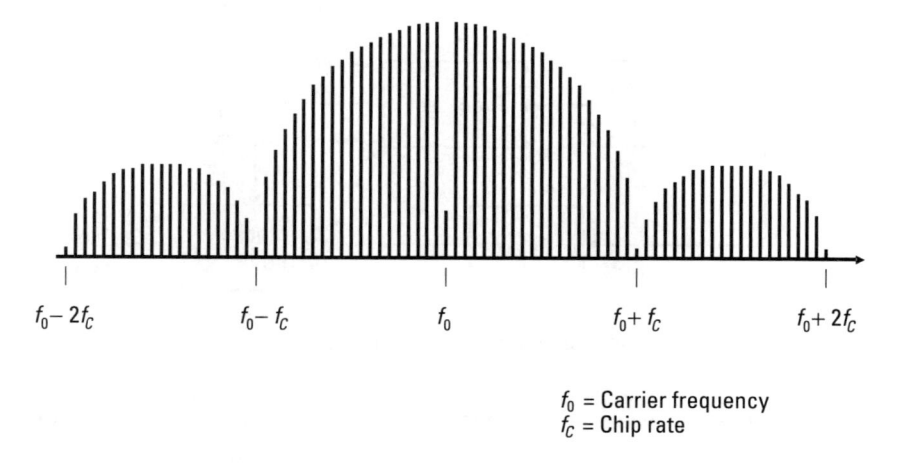

Figure 8.10 DSSS spectrum showing spectral lines.

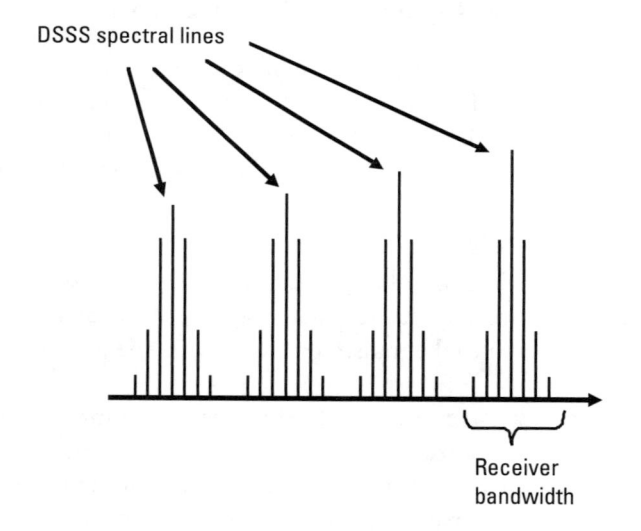

Figure 8.11 DSSS spectral lines with modulation.

Another approach is to determine the spreading code by analysis and design a despreader. The chip rate is determined from the spread signal spectrum as shown in Figure 8.12. The RF modulation can be determined by viewing the waveform on an oscilloscope. Figure 8.13 is an example of the appearance of a digital RF modulation on an oscilloscope. This will typically be at a lower IF frequency. The modulation shown in the figure is binary

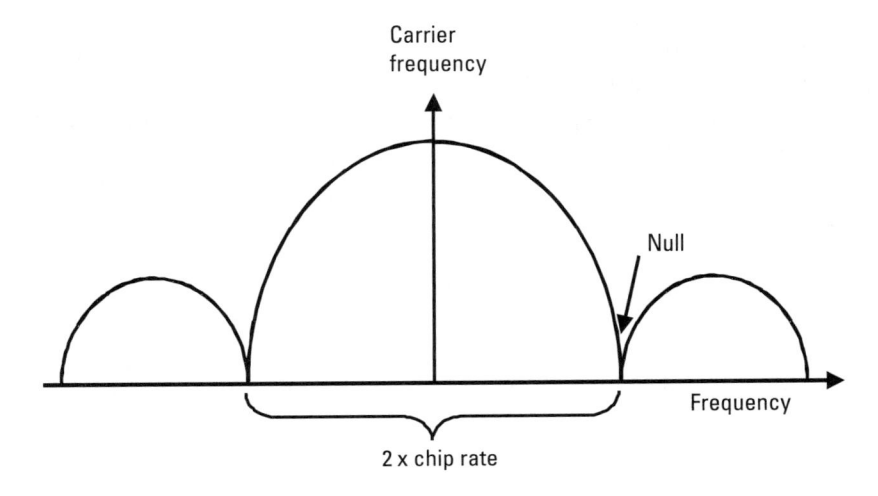

Figure 8.12 DSSS spectrum.

Figure 8.13 Oscilloscope display of BPSK modulation.

phase shift keying (BPSK). The oscilloscope is synchronized to the signal with a positive going synch.

A series of bits is captured in a computer register and analyzed to determine the code length and sequence. The code will be a maximal length code as discussed in Section 2.4.1. Once the code is determined, a shift register with feedback loops (discussed in Section 2.4.2) can be designed to reproduce the code. Then the signal can be despread.

This approach is fairly straight forward if the spreading code is short. For longer codes, it would require massive computer capability.

Intercept of Cell Phone Transmissions

Since the TDMA and CDMA schemes used by cell phones use publicly known structures, it is presumed that they will be available to the intercepting party. They are, of course, most easily implemented by acquiring the commercial equipment used by cell phone companies. To intercept a cell phone transmission, it is necessary to tune a receiver to the proper channel frequency and apply the appropriate timing slot or spreading code to recover the information in the targeted transmission.

The legal and/or political considerations associated with this practice are beyond the scope of this book.

9

Communications Jamming

The purpose of communication is to move information from one location to another. All of the following types of transmitted signals are considered communication:

- Voice communication;
- Computer-to-computer communication;
- Command links;
- Data links;
- Weapon-firing links;
- Cell phones.

The purpose of communication jamming is to prevent the transfer of information. Communication jamming requirements depend on the signal modulation, the geometry of the link, and the transmitted signal power.

Figure 9.1 shows the communication jamming geometry. Whereas a typical radar has both the transmitter and the associated receiver at the same location, a communication link—because its job is to take information from one location to another—always has its receiver in a different location from that of the transmitter.

Note that you can only jam the receiver. Of course, communication is often done using transceivers (each including both transmitter and receiver), but only the receiver at location B in the figure is jammed. If transceivers are

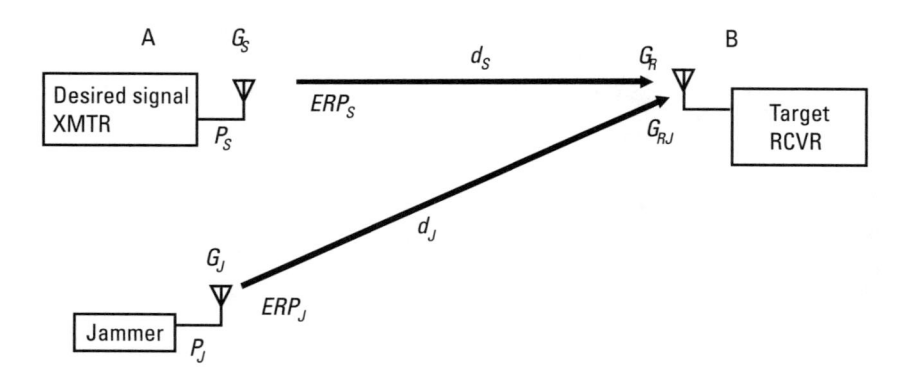

Figure 9.1 Communication jamming geometry.

in use and you want to jam the link in the other direction, the jamming power must reach location A.

There are some important communications cases in which transceivers are not used—for example in UAV links as shown in Figure 9.2. This figure shows the data link (or "downlink") being jammed. Again, you jam the receiver.

9.1 Jammer-to-Signal Ratio

The real test of jammer effectiveness is the thoroughness with which information flow is stopped. This latter factor is commonly tested by having a trained communicator read a text while a trained listener copies down what he or she heard. As the jamming gets worse, the percentage of correct words decreases. One problem with this test is that the human brain allows an operator to pull information from low-quality signals by pattern recognition. We know our language and therefore can complete words or sentences from past

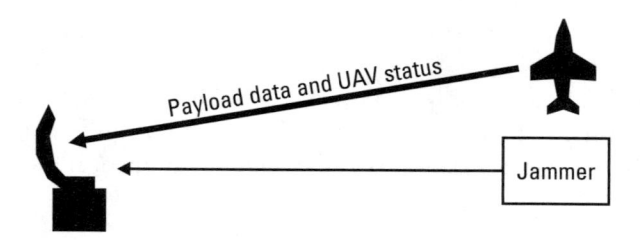

Figure 9.2 UAV link jamming geometry.

experience. One way to overcome this problem is to change key words in the text each time a message is sent—and count only those words for accuracy. Another way is to use random lists of words.

The mechanism by which a communication jammer interferes with communication is injecting an undesired signal into the target receiver along with any desired signals that are being received. The undesired signal must be strong enough that the receiver cannot recover the required information from desired signals. The ratio of the jamming signal (in the receiver) to the desired signal (in the receiver) is called the jamming-to-signal ratio (J/S). This ratio is commonly stated in dB. The required J/S for effective jamming depends on the transmitted modulation, but the calculation of J/S for any modulation can be done using the following approach.

The jammer-to-receiver link and the desired transmitter-to-receiver link in Figure 9.1 can employ any of the propagation models explained in Chapter 5. They do not have to have the same propagation model. For this reason, the jamming-to-signal ratio formulas in this section include general terms for loss.

The formula for communication J/S is:

$$J/S = ERP_J - ERP_S - L_J + L_S + G_{RJ} - G_R$$

where

J/S = the ratio of the jammer power to the desired signal power at the input to the receiver being jammed (in dB)

ERP_J = The effective radiated power of the jammer (in dBm)

ERP_S = The effective radiated power of the desired signal transmitter (in dBm)

L_J = The propagation loss from the jammer to the receiver (in dBi)

L_S = The propagation loss from the desired signal transmitter to the receiver (in dB)

G_{RJ} = The receiving antenna gain in the direction of the jammer (in dBi)

G_R = The receiving antenna gain in the direction of the desired signal transmitter (in dBi)

In many cases the receiving antenna of the target receiver has 360° azimuth coverage. Examples of such antennas are whips and monopole antennas on aircraft. When the receiver has a 360° antenna, the communications J/S equation simplifies to the following:

$$J/S = ERP_J - ERP_S - L_J + L_S$$

Since the receiving antenna has the same gain toward the jammer and the desired signal transmitter, the two antenna gain terms cancel out.

The effective radiated power terms are the sum of the appropriate transmitter output power (in dBm) and the transmit antenna gain in the direction of the target receiver (in dBi).

The appropriate propagation loss terms (for jammer and desired signal links) depend on the frequency and propagation geometry. Table 9.1 shows the propagation model to use. Note that the selection of propagation loss model for the jamming and desired signal links are completely independent.

Each of these propagation models is explained in Chapter 5, and calculation techniques are presented for each.

9.1.1 Other Losses

Although the spreading loss is the dominant factor, and J/S equations are usually expressed this way, the jamming and desired signal propagation paths also have atmospheric losses and can be subject to nonline-of-sight or rain losses. If there are large differences between the two distances or line-of-sight conditions, these calculations should be made, and the J/S ratio adjusted accordingly. Note that atmospheric and rain losses are covered in Section 5.8.

9.1.2 Stand-In Jamming

Stand-in jamming is the placement of a jammer close to the target receiver as shown in Figure 9.3. The effect is to reduce the jammer-to-receiver distance, increasing the J/S by the square (or fourth power) of the reduced distance.

Table 9.1
Appropriate Propagation Model

Signal and Environment	Propagation Model	
High frequency and/or link path distant from the ground and/or narrow antennas.	Line-of-sight	
Signal near the ground or water and frequency below microwave.	Link distance less than Fresnel zone	Line-of-sight
	Link distance greater than Fresnel zone	Two-ray
Signal path passes near ridgeline or hill.	Line-of-sight plus knife-edge diffraction loss.	

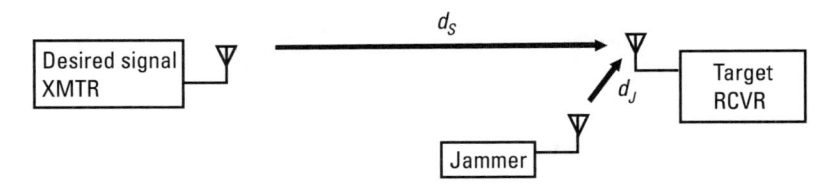

Figure 9.3 Stand-in jamming.

This technique has the advantage of allowing decreased jamming power to achieve the same J/S. A derivative advantage is that friendly receivers—which are presumably much farther from the jammer than the target enemy receiver—will not be jammed. This prevents fratricide, or the unintentional jamming of friendly communications.

Stand-in jamming techniques include emplaced jammers, jamming payloads on UAVs, and artillery-delivered jammers.

Stand-in jamming can be particularly advantageous against spread spectrum communication where jamming from long distances is difficult because of the requirement to overcome processing gain in the receiver.

9.2 Digital Versus Analog Signals

When jamming analog modulated communication signals, it is normally necessary to achieve a high J/S ratio (10 dB is generally considered adequate). It is also generally necessary to jam with 100% duty cycle. This is necessary because a receiver operator has a significant ability to listen "adaptively." In all analog communication, we can "fill in the blanks" in a low-quality transmission from the context. This is particularly true in tactical military communication where important information is sent in fairly rigid formats. Examples are the standard five-paragraph operations order and the phonetic alphabet.

When jamming digitally modulated communications signals, we attack the signal by trying to make it unreadable to the digital demodulator. We can either interfere with the synchronization or produce bit errors. Since the synchronization tends to be quite robust, the basic approach is to create bit errors.

As shown in Figure 9.4, there is a nonlinear relationship between the signal-to-noise ratio into a digital receiver and the percentage of bit errors present in the digital output it produces. Communication theory textbooks contain families of these curves—one for each type of modulation technique

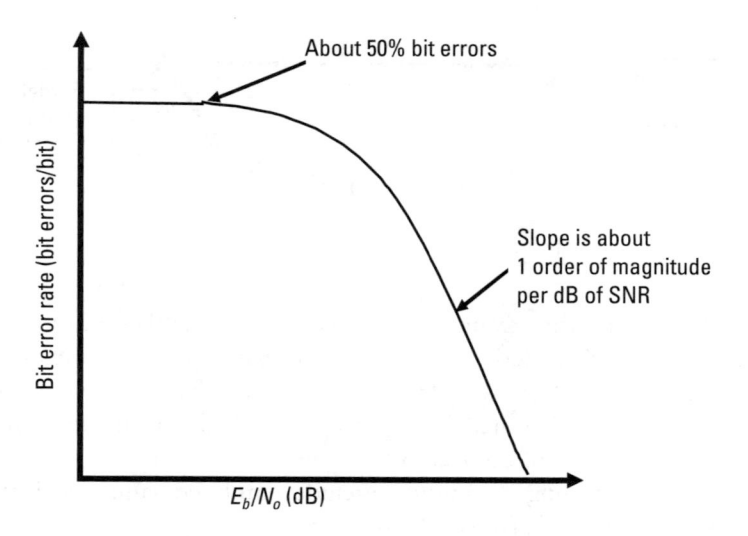

Figure 9.4 Bit error rate versus E_b/N_o.

that is used to carry the digital data. The signal quality in this diagram is in terms of bit error rate versus E_b/N_o. This value is related to the predetection signal-to-noise ratio by the formula:

$$E_b/N_o = RFSNR(\text{bandwidth}/\text{bit rate})$$

where RFSNR is the predetection signal-to-noise ratio.

All of the curves for various modulations have the basic shape shown in this typical example. The top of the curve flattens out at 50% bit errors. This is logical when you think about it—50% errors is as bad as it gets in a digital signal. If the bit error rate goes above 50%, the output becomes more coherent with the transmitted message. All of the curves reach this 50% point at about 0-dB signal-to-noise ratio (i.e., signal = noise). This means that regardless of the type of modulation used, if the noise level (or the jamming level) is equal to the received signal level, the percentage of bit errors does not increase when the jamming level is increased. Since increased jamming-to-signal ratio reduces the predetection signal to "noise" ratio, the impact of jamming on the bit error rate can be shown against a typical digital modulation as in Figure 9.5. This figure shows bit error rate as a function of J/S ratio (which increases to the left). Note that when J/S reaches 0 dB, the bit error rate is very close to 50%, and increased J/S does not significantly increase the bit error rate. Thus, in general, it is assumed that the received

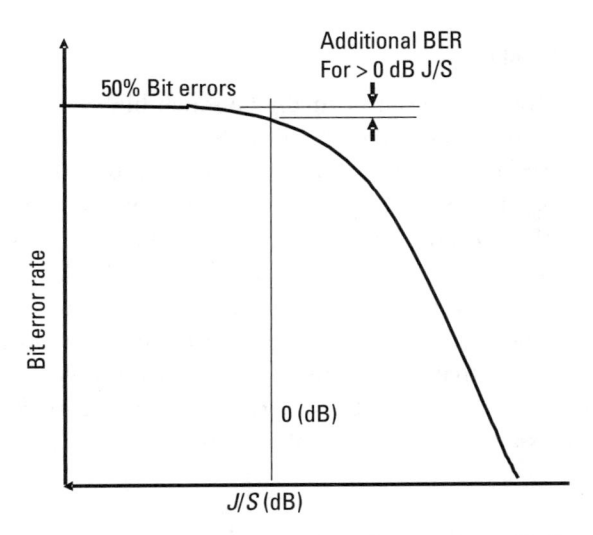

Figure 9.5 Bit error rate versus jamming-to-signal ratio.

signal quality of a digital signal is not further reduced by J/S values greater than unity (i.e., 0 dB). It is important to realize that this 0 dB applies at the point in the circuitry where the digital waveform is recovered from the RF modulation. In spread spectrum signals which allow processing gains in the receiver, the actual J/S is reduced by the processing gain.

Some types of information (for example remote control commands) require extremely low bit error rates, while voice communication is much more tolerant of errors.

Further, if the signal is (in the short term) unreadable a third of the time, it is considered useless. This means that a digital signal need only be jammed to a J/S of 0-dB one-third of the time, while an analog signal requires positive J/S 100% of the time. In some literature, 20% is stated as the required jamming duty cycle for complete jamming of digital signals, but the 33% value is more commonly accepted. This 33% must be applied over time periods less than the syllabic rate in speech or its equivalent in digital data.

Note that the presence of error-correction codes can increase the required jamming duty cycle against digital signals, because they correct some of the jamming-caused bit errors.

For reasons that are discussed in Section 2.4 dealing with specific types of LPI signals, all LPI signals can be expected to carry their information in digital form under most circumstances.

9.2.1 Pulse Jamming

The peak power of a pulse can normally be much higher than the constant power from a continuous signal transmitter. Since we only need to jam a digital signal one-third of the time, a 33% duty factor pulse will be adequate for effective jamming. The enhanced peak power improves the J/S. As shown in Figure 9.6, equal total transmitter energy creates three times as much effective radiated power if the duty cycle is reduced to one-third.

Error-correction codes allow the receiver to correct some percentage of bit errors, and some codes can be designed to replace whole blocks of erroneous data. This means that pulse jamming, which causes blocks of data with many bit errors may not be effective. In general, if error-correction codes are present, it is necessary to jam digital signals with a large jamming duty cycle.

9.3 Jamming Spread Spectrum Signals

Low probability of intercept (LPI) communication signals spread their energy (pseudo-randomly) across a wider frequency range than required to just carry their information from the transmitter to the receiver. Thus, they are also called spread spectrum signals. The minimum bandwidth required for a communication transmission is the "information bandwidth." The "transmission bandwidth" is the frequency over which the signal is spread or rapidly tuned.

The desired receiver for a spread spectrum signal has despreading capability that is synchronized to the spreading circuitry in the transmitter—allowing the receiver to process the signal in its original, unspread form.

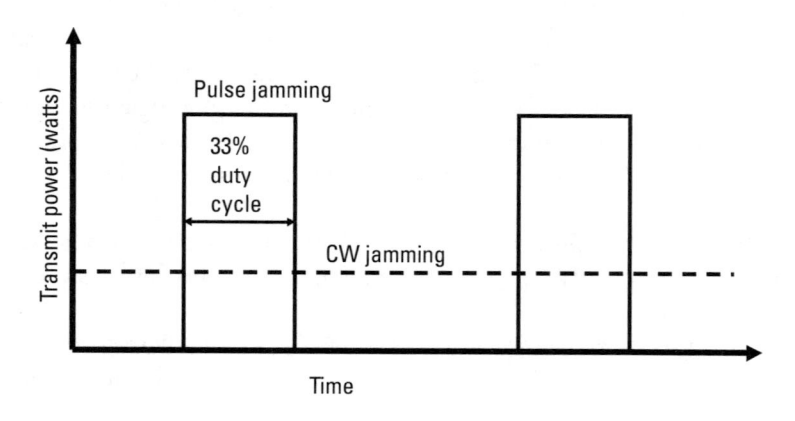

Figure 9.6 Pulse jamming.

A hostile receiver does not have the synchronized despreading capability. Thus, signal intercept jamming and emitter location are greatly complicated. You will recall from Chapter 4 that the noise power in a receiver is proportional to its effective bandwidth. Thus, the noise power in a hostile receiver with enough bandwidth to receive the spread spectrum signal will be high enough to hide the signal.

As discussed in Chapter 2, there are three basic types of spread spectrum signals: frequency hopping, chirp, and direct sequence. Each type spreads the signal. However, the nature of the power versus frequency versus time distribution for each type of modulation gives it different vulnerability to jamming.

Spread spectrum signals are subject to the same jamming equations as any other signals, but the ability of the cooperative receiver to collapse the spectrum gives it a "processing gain" that reduces the effectiveness of the jamming. In general, the processing gain advantage is the same as the spreading ratio (i.e., the transmission bandwidth/the information bandwidth). It is also defined (in DSSS signals) as the code rate (used for spreading) divided by the data rate. Another applicable term is "jamming margin," which is defined by the following formula.

$$M_J = G_P - L_{SYS} - SNR_{OUT}$$

where

M_J = the jamming margin (in dB)

G_P = the processing gain (in dB)

L_{SYS} = the system losses (in dB)

SNR_{OUT} = the required output signal to noise ratio

9.3.1 Partial Band Jamming

This is a jamming technique that optimizes the jammer performance against some kinds of spread spectrum signals. As the name implies, partial band jamming covers only part of the full frequency range of the spread spectrum signal as shown in Figure 9.7. As stated above, spread spectrum signals can be expected to be digital. Thus, optimum J/S is about 0 dB and the jamming duty cycle can be relatively small. Since frequency hopping and chirp signals selectively occupy portions of their frequency spectrum, partial band jamming is a way to create low duty cycle jamming at the optimum J/S.

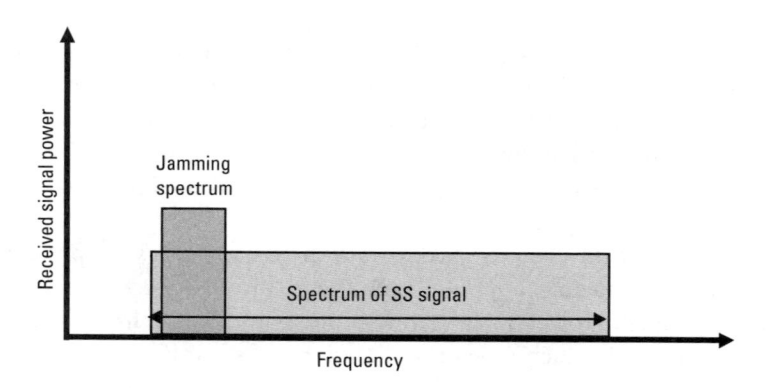

Figure 9.7 Partial band jamming.

Partial Band Jamming Calculation

With this technique, it is necessary to determine the level of the desired signal in the receiver. Then, the jammer power is spread over the maximum frequency range that will allow jammer power in the receiver to equal the desired signal power at each hop frequency as shown in Figure 9.8. The jamming power is generally spread over enough of the hopping channels so that the J/S in each jammed channel is 0 dB. If there is a secondary spreading technique, the 0 dB applies after the processing gain has been applied.

Figure 9.9 shows a jamming geometry. In this example, it is assumed that the transmitter and target receiver have whip antennas or some other antenna type with 360° azimuth coverage. The problem is more complex when the jammed link uses directional antennas, but can still be solved. This discussion assumes that the locations of the desired signal transmitter and target receiver are known because the jammer is supported by an emitter location system. Typically communication nets use transceivers, so every station can be located when it transmits. A measurement of desired signal strength at the jammer receiver will allow calculation of the effective radiated

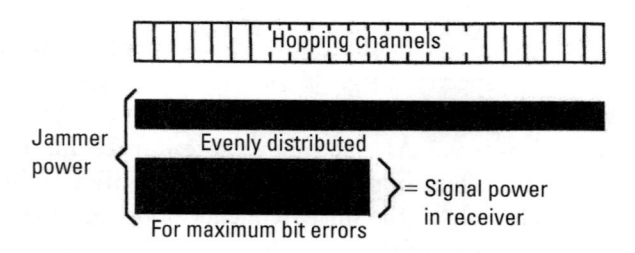

Figure 9.8 Maximum bit errors per channel.

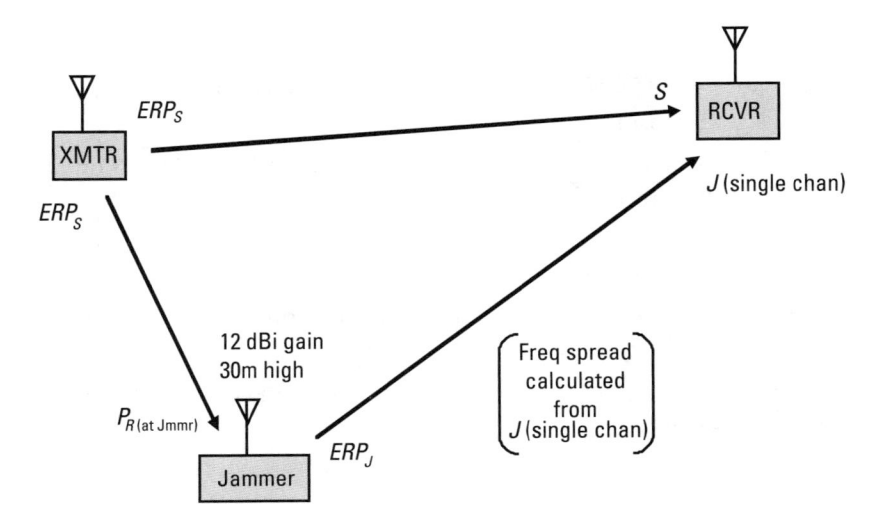

Figure 9.9 Partial band jamming geometry.

power. The ERP of the desired signal transmitter is the received signal strength (adjusted for the jammer receiving antenna gain) increased by the spreading loss by the formula:

$$6ERPs = P_R + L_{SJ} - G_{RJ}$$

where
ERP_s = the effective radiated power of the desired signal transmitter (in dBm)

P_R = received power at jammer receiver from desired signal transmitter (in dBm)

L_{SJ} = the propagation loss from the desired signal transmitter to the jammer receiver (in dB)

G_{RJ} = the jammer's receiving antenna gain in the direction of the desired signal transmitter (in dBi)

In this and subsequent steps, the loss is calculated using the appropriate formulas from Chapter 5.

Next, calculate the received power from the desired signal transmitter at the target receiver by the formula:

$$S = ERPs - L_S + G_R$$

where

> S = the received signal strength at the target receiver from the desired signal transmitter
>
> ERP_S = The effective radiated power of the desired signal transmitter (in dBm)
>
> L_S = The propagation loss from the desired signal transmitter to the receiver (in dB)
>
> G_R = The receiving antenna gain in the direction of the desired signal transmitter (in dB)

Now, calculate the received power at the target receiver from the jammer by the equation:

$$J = ERP_J - L_J + G_{RJ}$$

where

> S = the received signal strength at the target receiver from the desired signal transmitter (in dBm)
>
> ERP_J = the effective radiated power of the jammer (in dBm)
>
> L_J = the propagation loss from the jammer transmitter to the receiver (in dB)
>
> G_{RJ} = the receiving antenna gain in the direction of the jammer (in dBi)

Now, subtract the strength of the received desired signal (S) from the strength of the jamming signal (J) to determine the total J/S in the target receiver.

You have now calculated the single channel J/S, which is the J/S which would be achieved if the signal were not spectrum spread. However, since the frequency hopper or chirp signal moves its bandwidth across a wide frequency range, the jamming signal must be spread to provide optimum J/S (i.e., 0 dB) over as wide a frequency range as possible.

This process starts by converting the single channel J/S of hopping signals or the information bandwidth J/S of chirped signals back to non-dB form by the technique in Section 1.4. Then multiply that non-dB ratio by the information bandwidth of the target signal. The resulting bandwidth is the bandwidth over which 0-dB J/S can be supported. Numerical examples will be given later for specific hopping and chirp signal parameters.

9.3.2 Jamming Frequency Hop Signals

If a narrow bandwidth jamming signal is applied to a frequency hopping receiver, it will be received only when the receiver happens to hop to that frequency. This will cause the jamming effectiveness to be significantly reduced. For example, if a CW jamming signal is applied to a Jaguar V receiver (which hops randomly over a maximum of 2,320 25-kHz wide channels between 30 and 88 MHz) the receiver will only see the jamming signal 0.043% of the time. Alternately, if the jamming signal is spread over the whole 2,320 channel frequencies, the J/S per channel will be reduced by 33.65 dB. Thus, more sophisticated jammers are required for frequency hopping signals. Either partial band or follower jamming can be used.

Partial Band Jamming

For effective partial band jamming of a frequency hopper, the jammer should provide 0-dB J/S in at least 33% of the hopping channels. To illustrate this way to jam, the technique described in Section 9.3.1 can be applied to the Jaguar V frequency hopping signal. Figure 9.10 shows the jamming geometry, which assumes that we know the locations of the desired signal transmitter and the target receiver.

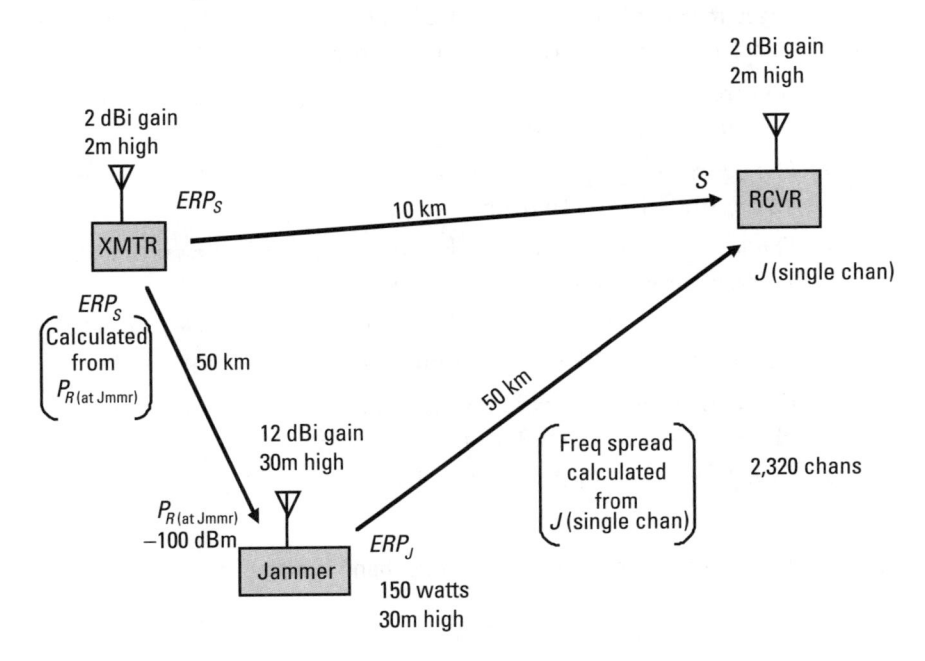

Figure 9.10 Partial band jamming of Jaguar V.

We see from the diagram that the distance from the jammer to the desired signal transmitter and the target receiver are both 50 km, while the distance between the desired signal transmitter and the target receiver is 10 km. The jammer has a 12-dB antenna on a 30-m mast, and the both the desired signal transmitter and the target receiver have whip antennas 2m from the ground. The average transmitted frequency is 59 MHz. The received signal in the jammer's receiver is −100 dBm and the jammer transmitter outputs 150 watts. $G_{RJ} = G_R$ since the receiving antenna is a whip. All of the loss calculations use the techniques described in Chapter 5.

Calculate the Single Channel J/S

- The Fresnel zone from the transmitter toward the jammer in km is (product of antenna heights in m × frequency in MHz)/24,000 = (2 × 30 × 59)/24,000 = 148 meters. Since this is less than 50 km, we use the two-ray propagation loss model.
- The link loss is 120 + 40 log(50) − 20 log(2) − 20 log(30) = 152.5 dB.
- The ERP_S is P_R + Loss − G_J = −110 + 152.5 − 12 = 30 dBm = 1 watt.
- The Fresnel zone distance from the desired signal transmitter to the target receiver is (2 × 2 × 59)/24,000 = 10 meters, much shorter than 10 km, so two-ray propagation loss applies.
- Loss from desired signal transmitter to target receiver is 120 + 40 log(10) − 20 log(2) − 20 log(2) = 148 dB.
- The desired signal power in the target receiver (S) is $ERP_S - L_S + G_R$ = 30 − 148 + 2 = −116 dBm.
- The loss from the jammer to the target receiver is the same as from the desired signal transmitter to the jammer = 152.5 dB.
- The jammer power is 150 watts which is 51.8 dB, with the 12-dB antenna gain, the ERP of the jammer is 63.8 dBm.
- The jammer signal power in the target receiver (J) is $ERP_J - L_J + G_{RJ}$ 63.8 − 152.5 + 2 = −86.7 dBm.
- $J/S = J - S$ = −86.7 dBm −(−115 dBm) = 29.3 dB.

Calculate the ideal frequency spread for partial band jamming:

- 29.3 dB is a non-dB ratio of 851.
- The partial band jamming bandwidth is 851 × 25 kHz = 21.3 MHz.

Evaluate the jamming effectiveness:

- 851 of 2,320 channels are jammed at 0 dB J/S.
- This is 36.7% which is more than 33%, so the jamming will be effective.

Partial and jamming does not require great jammer sophistication, but it does require enough jammer power to jam over some number of channels. It is a technique that is available to a great number of deployed jammers. One disadvantage is the danger of fratricide against friendly frequency hopping communications. If the jammed portion of the enemy spectrum is constant, the enemy will avoid the jammed frequencies, so the partial jamming band must be randomized.

Follower Jamming

Follower jamming jams only at the frequency used for each hop of an enemy transmitter. Thus, it has minimal interference with friendly communications.

The frequency hopper selects each hop frequency by a pseudo-random process, so there is no way to predict the next frequency. However, if we can measure the frequency during a small portion of the hop, we can set a jammer to that frequency for the balance of the hop. Then, we only have to provide enough jamming power to apply 0-dB J/S to the signal. To achieve 33% jamming duty factor, a rather sophisticated receiver must determine the frequency within 67% of the hop time after the hopping synthesizer has settled as shown in Figure 9.11. Since the settling time is expected to be about 15% of the hop period, that means that the frequency measurement and jammer set on must be accomplished in 57% of the hop period. For example, if the hopping rate is 100 hops/sec, 570 milliseconds is available for search and set-on.

Follower jamming requires a digital receiver with a fast Fourier transform (FFT) processor. Such a receiver was described in Section 4.3.2. The process was further defined in Section 7.8.1 when the location of frequency hoppers with the digital receiver was described. To understand the full implications of follower jamming, it is necessary to discuss the operational environment. There will be many signals present, most likely including both enemy and friendly frequency hoppers. The only way to identify a specific hopper that is to be jammed is by its location. Thus, two digital receiver emitter locators must work together to triangulate each signal in the

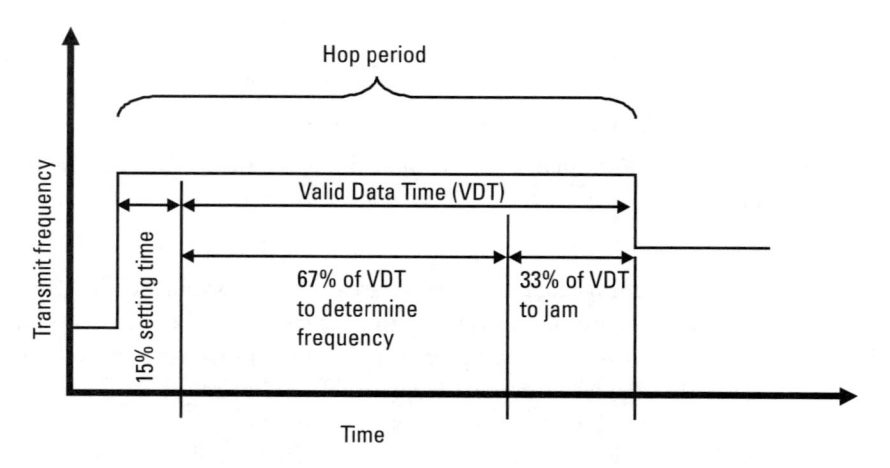

Figure 9.11 Follower jammer timing.

environment. There will then be a computer file with the frequency and location of each signal in the environment. The follower jammer is set to the frequency of the signal at the location of the targeted emitter. Figure 9.12 is a three-dimensional representation of the frequency environment space (geographical location and frequency). A large number of signals are typically present—randomly spread through the signal space. The figure includes representation of two frequency hopping signals. Note that each has a large number of frequencies coming from a single location.

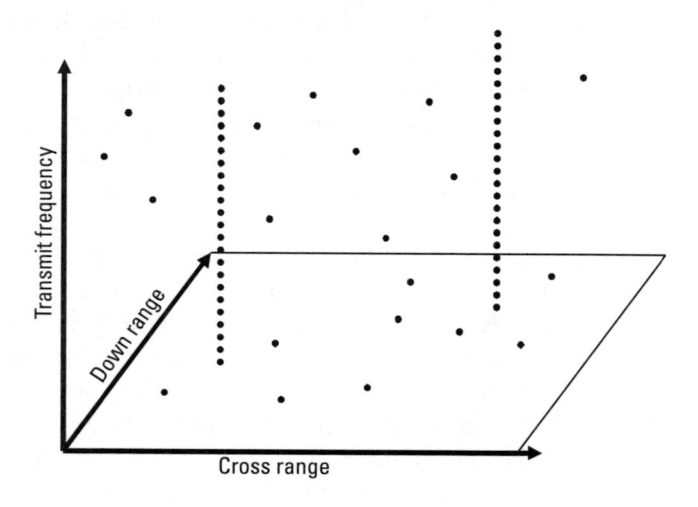

Figure 9.12 Signal environment with frequency hoppers.

Fratricide is avoided when follower jamming is applied because there is only a very small likelihood (0.043%) that a friendly emitter will be hopping at the same frequency as the targeted emitter.

9.3.3 Jamming Chirp Signals

In Chapter 2, the frequency versus time curve for chirped signals is described. The sweeping rate is so high that the signal cannot be seen in a receiver with the information bandwidth of the signal. (This requires a dwell time in-band of 1/bandwidth.) The desired receiver sweeps in synchronization with the transmitter, and can therefore use bandwidth near the information bandwidth. The start time of each sweep can be random and/or the slope of the sweep can be nonlinear to make it difficult for a hostile receiver to synchronize with the transmitter.

The frequency range of the chirp signal can be determined by a spectrum analyzer, but in order to perform "follower" jamming, the exact slope and start time of the sweep would need to be known. Since the sweep may have a nonlinear slope, there is an additional complication. However, if the chirp transmitter uses a predictable sweep synchronization, the sweep slope can be determined by a digital receiver with FFT and a jammer can be synchronized to the sweep to create a follower jammer.

Again, with predictable sweep synchronization, a digital RF memory (DRFM) could measure the first sweep and generate subsequent sweeps to match the chirped signal.

Partial band jamming, as described in Section 9.3.1 can be applied to the chirped signal. Since the chirped signal must be digital to avoid drop-outs during the sweep fly-back and pseudo-random delays in the sweep start time, 0-dB J/S and 33% jamming duty cycle are expected to be adequate.

If the chirp excursion is 5 MHz and the information bandwidth is 25 kHz, jamming bandwidth of 1.65 MHz would provide 33% jamming duty factor. To achieve 0-dB J/S in every information bandwidth across this jamming range requires an additional 18.2 dB of jammer power as compared to a single channel matched-tuned jammer.

9.3.4 Jamming Direct Sequence Spread Spectrum Signals

The direct sequence spread spectrum (DSSS) signal varies from the frequency hopping and chirp signals in that it continuously occupies a spread frequency spectrum as described in Section 2.4.4. It is generated by applying a secondary digital modulation with a very high bit rate. Since the bandwidth of the signal is proportional to the bit rate, the signal energy is spread over the much wider bandwidth. This, in turn, reduces the signal strength versus frequency as shown in Figure 9.13. In the figure, the lower (spread) signal has only five times the bit rate of the normal digital signal (called the "chip rate.") A normal DSSS signal would have a much higher bit rate ratio, typically 100 to more than 1,000, so the spreading and resulting reduction of signal density (versus frequency) would be much greater.

Like the other spread spectrum signals, the DSSS signal carries its information in a digital modulation. Thus, it can be jammed with a duty factor of 33% and a J/S of 0 dB after despereading. However, the DSSS receiver provides a significant processing gain by removing the pseudo-random spreading signal. If the spreading factor is 1,000, the despreading circuitry will reduce the level of any signal which does not have the correct pseudo-random code (at exactly the correct phase) by 30 dB.

If a CW signal is input to the DSSS receiver, it will be reduced by the spreading rate (converted to dB) as shown in Figure 9.14. Note that the jammer signal is reduced relative to the desired signal because the jammer does not have the benefit of the processing gain. Thus, the actual J/S (or effective J/S) is reduced by the processing gain (which is the spreading factor

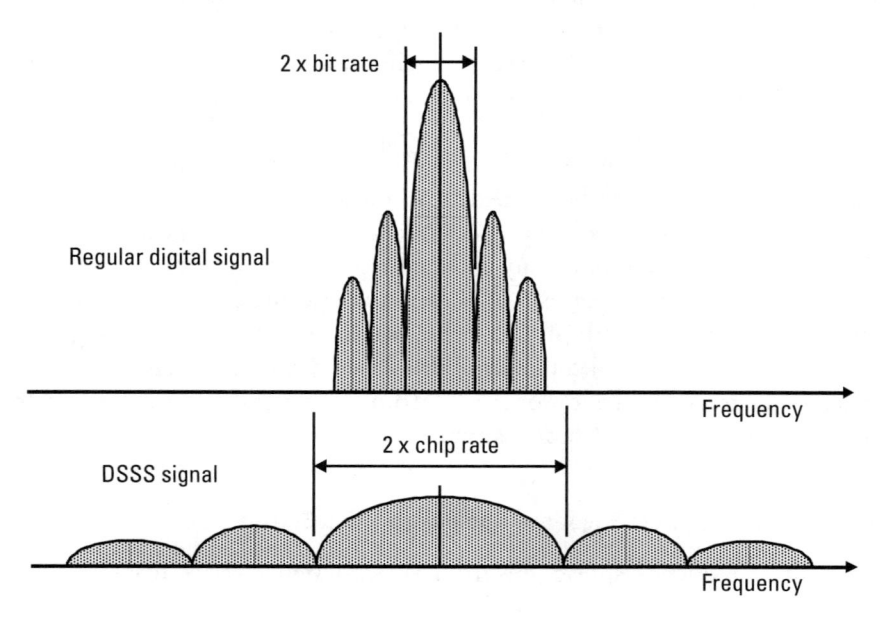

Figure 9.13 Frequency spectrum of normal and DSSS signals.

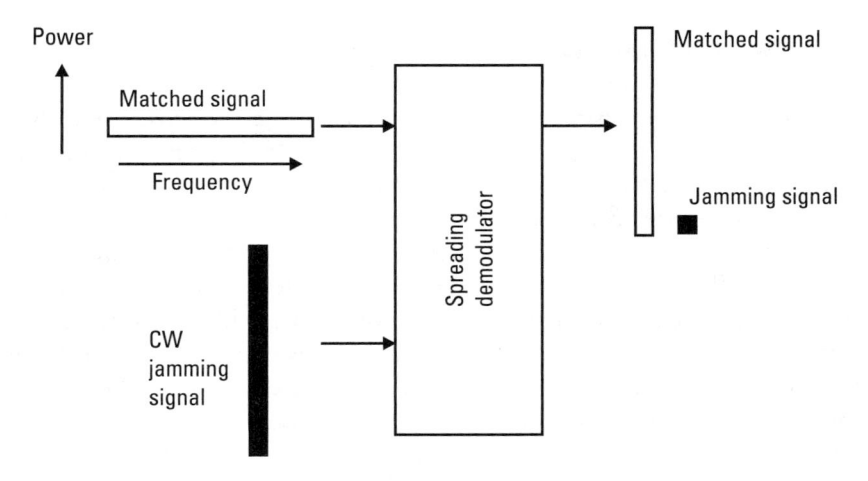

Figure 9.14 Narrowband jamming through despreader.

converted to dB). However, a CW oscillator is a relatively simple device, so it may be practical to provide a CW jammer with enough extra power to overcome the despreading processing gain and still provide 0 dB J/S.

Since the jamming needs to be present only 33% of the time, a pulse jammer can typically provide a significant increase in available power. Also, it may be practical to employ a stand-in jammer which, from a very short range, would provide very large J/S since the jamming signal attenuation is reduced by the square or fourth power of the decreasing range, depending on the propagation mode. The simplicity of a CW or pulse jammer makes it an excellent candidate for a remotely emplaced and operated stand-in jammer.

9.3.5 Jamming of Combined Mode Spread Spectrum Signals

Combined mode spread spectrum signals significantly increase the challenge to jammers. A very common multiple technique approach is DSSS with frequency hopping. A pulse CW jammer at the operating frequency will effectively jam the signal. But, the CW signal needs to be at the correct frequency. Thus, an approach to quickly measuring the operating frequency needs to be used; then the pulse jammer needs to be set to the correct frequency. This may require the use of a multiple channel energy detector as described in Chapter 6, since a normal digital receiver may not have enough signal to determine the frequency with an FFT processor.

Another approach, if there are only a few hop frequencies, would be to set a pulse jammer to each of the hop frequencies.

9.4 Impact of Error-Correction Coding on Jamming

Since we jam digital signals to create bit errors, error-detection coding directly attacks the effects of the jamming. Thus, additional jamming power may be required to overcome the error correction function.

The more error correction bits that are added to a signal, the greater the increase in J/S required. An excellent example of the mechanism is the widely used 31/15 Reed Solomon block code. It can correct up to 8-bad bytes of each 31 bytes sent. It is typically employed along with interleaving that assures that not more than 8-sequential bits are likely to be jammed. If the protected signal is sent as a frequency hopper, there would be 31-bytes per hop and the bytes would be spread over at least four hops, so not more than 8 adjacent bytes would be carried by a single hop. When partial band jamming is employed, the error correction code will be able to correct many of the bit errors caused by the jamming. This can have the net effect of allowing acceptable output (post correction) bit error rates in the presence of 11-dB J/S. Thus, when forward error correction is present, the jamming duty cycle must be increased.

9.4.1 Jamming Cell Phones

Cell phone systems and how they operate were discussed in Section 2.4.6. Since the entire cell phone network is controlled from the mobile system control (MSC) it is clear that with adequate access to the computers in this system some sophisticated approaches could be used to stop communication by specific users or locations. This, of course, depends on political considerations—which are beyond the scope of this book. Thus, we will focus only on the jamming of radio links to and from the cell phones.

If the cell system is analog, each conversation is on a single frequency channel, so if that channel is known, normal communication jamming techniques can be used against either the uplink or the downlink.

If the cell system is digital, it will use either time division multiple access (TDMA) or code division multiple access (CDMA) to allow multiple conversations on each RF channel. This means that the jammer can jam the whole RF channel using normal communication jamming techniques. However, if specific conversations are to be jammed, it will be necessary to use modulation equipment like that used by the cell phone system to jam in the appropriate time slot or with the appropriate code.

One of the big issues in jamming cell phones is interference with nonmilitary cell phone traffic. In troubled parts of the world where asymmetrical warfare is common, wired telephone service may be so difficult to establish that most of the population depends on cell phones. Thus, jamming whole bands or even whole RF channels may cause significant political problems (which are beyond the scope of this book). The following examples consider only the (much easier) technical part of the problem.

9.4.2 Jam the Uplink

Consider an example of uplink jamming. We will jam a single RF channel of a GSM system as shown in Figure 9.15. The operating frequency is 1,800 MHz. The cell tower is 50m high. The cell phone has 1 watt (30 dBm) effective radiated power (ERP), is 1m high and is 1 km from the cell tower. The jammer has 100 watts (50 dBm) ERP, its antenna is 3m high, and it is 2 km from the cell tower. Since the cell tower is the receiver, the jamming link is from jammer to tower, and the desired signal link is from the cell phone to the tower.

First, you need to determine the propagation modes that apply to the desired signal and jamming links. Using your slide rule, you can quickly determine that the Fresnel zone distances are:

- Cell phone to tower: 3.8 km;
- Jammer to tower: 11.3 km.

This means that line-of-sight propagation applies to both links (since the link distances are less than the Fresnel zone distances). The two link losses can be quickly calculated with your slide rule as:

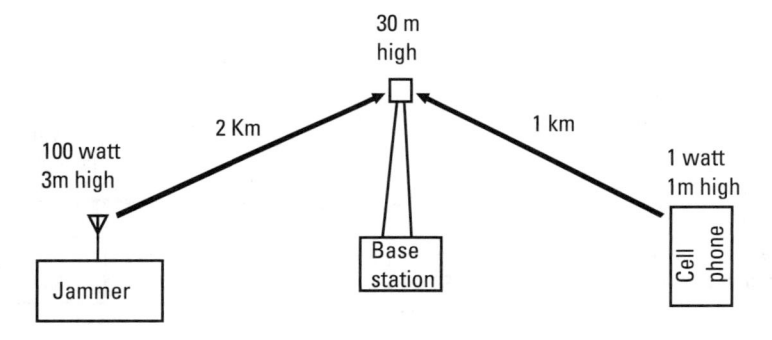

Figure 9.15 Jamming cell phone uplink.

- Cell phone to tower: 97.6 dB;
- Jammer to tower: 103.8 dB.

From Section 9.1:

$$J/S = ERP_J - ERP_S - L_J + L_S$$

So J/S is calculated as:

$$J/S = 50 - 30 - 103.8 + 97.6 = 13.5 \text{ dB}$$

Since the GSM cell phone system is digital, only 0 dB is required, so there is a comfortable margin in the jamming effectiveness.

9.4.3 Jam the Downlink

Now, consider downlink jamming as shown in Figure 9.16. In this case, the desired signal link is from the tower to the cell phone and the jamming link is from the jammer to the cell phone. The desired signal ERP is now the 50 watts (47 dBm). Again, the first step is to calculate the Fresnel zone distances to determine the propagation modes. Using your slide rule, you can quickly determine that the Fresnel zone distances are:

- Tower to cell phone: 3.8 km;
- Jammer to cell phone: 220 meters.

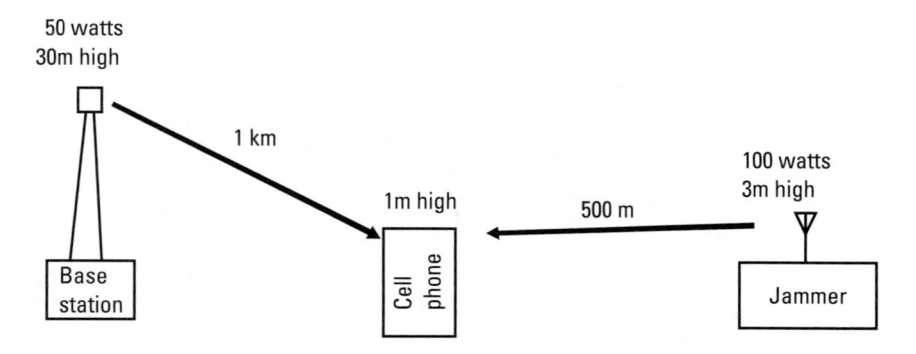

Figure 9.16 Jamming cell phone downlink.

This means that the desired signal uses line-of-sight propagation while the jamming link uses two-ray propagation. The desired signal link (tower to cell phone) has the same loss as in the uplink jamming case above (97.6 dB). However, the jamming link not only has two-ray propagation, but is shorter than 1 km, so it is off the scale for the slide rule. You can calculate the jamming link loss from the formula:

$$\text{Loss} = 120 + 40\log(0.5) - 20\log(1) - 20\log(2)$$
$$= 120 - 12 - 0 - 6 = 108 \text{ dB}$$

The jamming-to-signal ratio is calculated from:

$$J/S = ERP_J - ERP_S - L_J + L_S$$

So J/S is calculated as:

$$J/S = 50 - 47 - 108 + 97.6 = -7.4 \text{ dB}$$

Unfortunately, this is not adequate J/S to prevent cell phone communication. If 7.4-dB more jamming power is applied (i.e., 550 watts ERP), the required 0 dB J/S would be achieved.

Appendix A
Problems with Solutions

This appendix is a series of problems covering all of the subject matter in this book involving calculations. The problems cover:

- The trade-offs in antenna parameters;
- Receiver system sensitivity;
- Receiver system dynamic range;
- One-way link equation (line-of-sight);
- One-way link equation (two-ray);
- Fresnel zone;
- Knife-edge diffraction;
- Effective range;
- Circular error probable from RMS error;
- Circular error probable from elliptical error probable;
- Communication jamming;
- Partial band jamming of frequency hopping signals;
- Jamming of direct sequence spread spectrum signals;
- Cell phone jamming.

Problem 1: Antenna gain and beamwidth

From Section 3.7

This problem involves determining the gain and beamwidth of a symmetrical parabolic dish antenna as a function of its diameter, its efficiency, and the operating frequency.

The antenna is 3-ft in diameter, has 30% efficiency, and is operating at 6 GHz.

What is its boresight gain and 3- and 10-dB beamwidths?

Solution

Set the antenna scales on side 1 of the slide rule with 6 GHz aligned with 3 feet as shown at point A in Figure A.1. Read the gain at 30% efficiency at point B on the same figure (~30 dB).

Read the 3-dB beamwidth (4°) and 10-dB beamwidth (7.3°) at points C and D, respectively, on the same figure.

Note that the gain (in dB) can also be determined from the formula in Section 3.7.

$$\text{Gain (at 55\% efficiency)} = -42.2 + 20 \log(\text{diameter in m})$$
$$+ 20 \log(\text{frequency in MHz})$$
$$= -42.2 + 20 \log(0.91) + 20 \log(6{,}000)$$
$$= -42.2 - 0.8 + 75.6 = 32.6 \text{ dB}$$

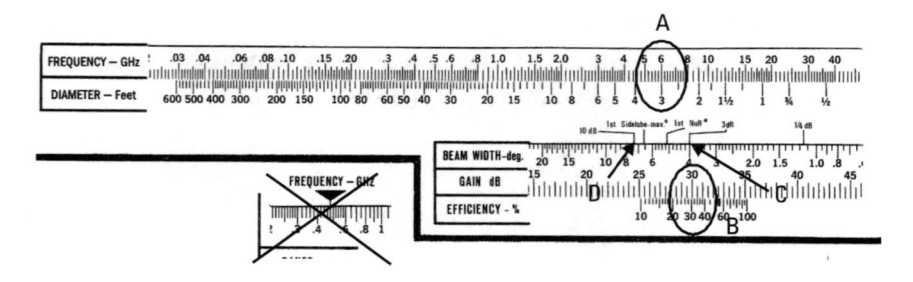

Figure A.1 Slide rule setting for Problem 1.

From Table 3.4, a 35% efficient antenna has 2-dB less gain: 32.6 − 2 = 30.6 dB

Problem 2: Gain of nonsymmetrical antenna

From Section 3.7
Find the gain of a nonsymmetrical parabolic dish antenna from its dimensions and efficiency.

Solution

As shown in Figure A.2, the vertical beam width is 40°, the horizontal beam width is 3°, and the efficiency is 50%.
The gain (not in dB) is 29,000/(vert BW × Hor BW) = 29,000/(40 × 3) = 241.7 if the efficiency were 55%.
From Table 3.4, the gain is reduced by 0.4 dB for the reduced efficiency, so the gain is 23.4 dB.

Problem 3: Antenna gain and beamwidth versus frequency

From Section 3.7
Use slide rule to determine the gain and 3-dB beamwidth at 2-, 6- and 18-GHz for an antenna with 30-dB gain at 8 GHz and 40% efficiency.

Solution

The gain and 3-dB beamwidth values are as shown in Table A.1.

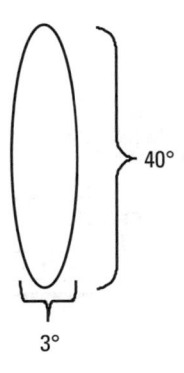

Figure A.2 Antenna beamwidths for Problem 2.

Table A.1
Gain and Beamwidth Versus Frequency

Frequency	Boresight Gain	3-dB Beamwidth
2 GHz	18 dB	18.2°
8 GHz	30 dB	4.6°
18 GHz	37.2 dB	2°

First, set the slide rule so that 30 dB is next to 40% efficiency as shown at point A on Figure A.3. Then note that the antenna diameter at 8 GHz is a tiny bit less than 2 ft as shown at point B on the same figure.

Now move the slide so that 2 GHz is at the same antenna diameter (about 2 ft) as shown at point A on Figure A.4. Then read the gain at 40% efficiency at point B (18 dB) and the 3-dB beamwidth at point C (18.2°).

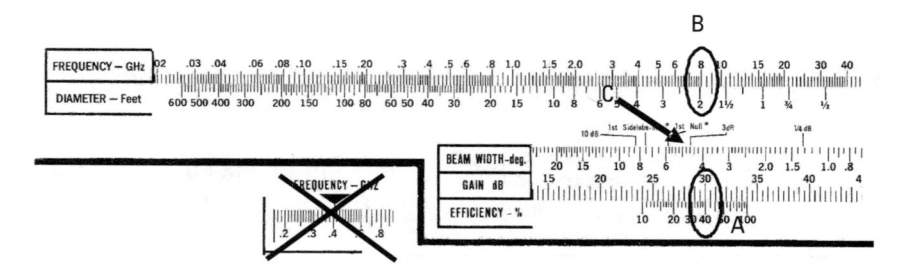

Figure A.3 First slide rule setting for Problem 3.

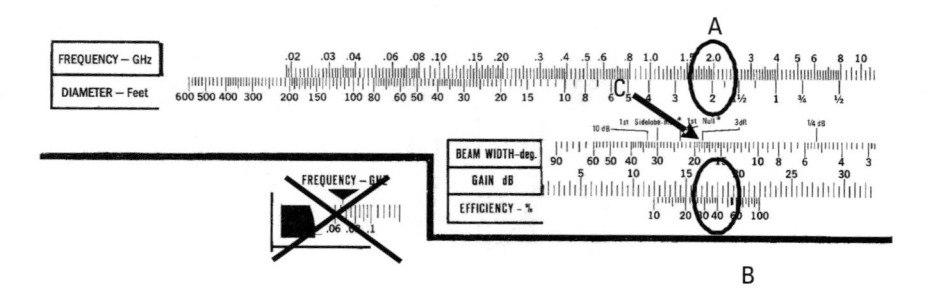

Figure A.4 Second slide rule setting for Problem 3.

Now move the slide so that 18 GHz is at 2-ft antenna diameter as shown at point A on Figure A.5. Then read the gain at 40% efficiency at point B (37.2 dB) and the 3-dB beam width at point C (2°).

Problem 4: Receiver system sensitivity

From Section 4.4

A receiver system has an effective bandwidth of 100 kHz, and requires 15-dB predetection signal-to-noise ratio (RFSNR). It has the block diagram shown in Figure A.6. What is the system sensitivity?

Solution

First, we must determine the system noise figure. Note that the loss before the preamplifier is 1 dB. The preamplifier gain is 20 dB and its noise figure is 3 dB. The loss following the preamplifier is 10 dB and the receiver noise figure is 11 dB.

From Figure A.7, the degradation factor is 2 dB. This makes the system noise figure:

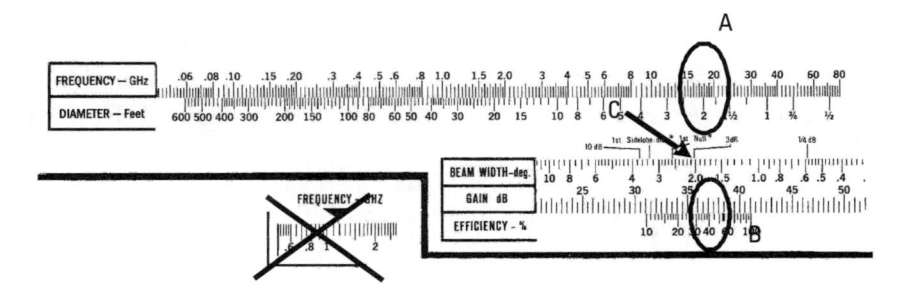

Figure A.5 Third slide rule setting for Problem 3.

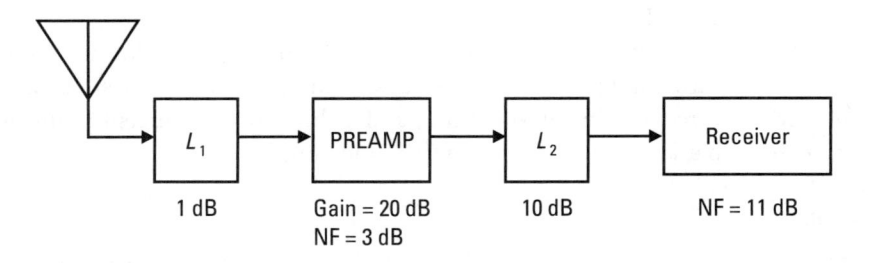

Figure A.6 Receiver system block diagram for Problem 4.

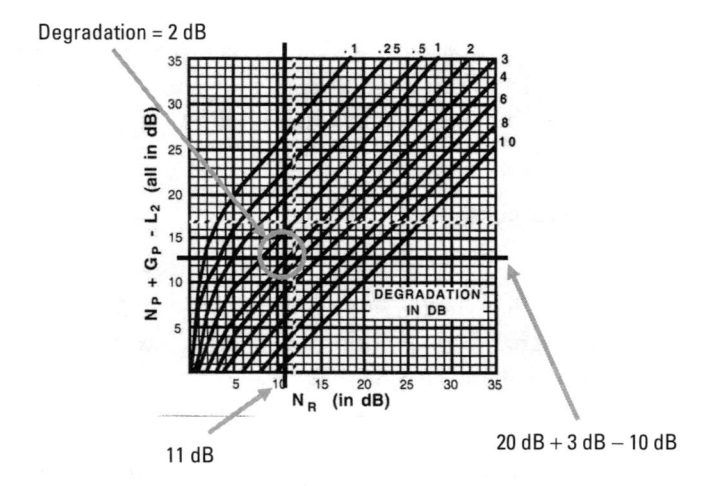

Figure A.7 Noise figure degradation chart for Problem 4.

The system noise figure is $L_1 + N_P +$ degradation factor $= 1$ dB $+ 3$ dB $+ 2$ dB $= 6$ dB

$$kTB \text{ is } -114 + 10\log(BW/1\,\text{MHz}) = -114 + 10\log(100\,\text{kHz}/1\,\text{MHz})$$
$$= -114 - 10 = -124\,\text{dBm}$$

The receiver system sensitivity is kTB + Noise Figure + Required RFSNR

$$= -124\,\text{dBm} + 6\,\text{dB} + 15\,\text{dB} = -103\,\text{dBm}$$

Problem 5: Analog receiver dynamic range

From Section 4.5.2

When a receiver has a sensitivity level of -90 dBm (at the output of the preamplifier after adjustment for all upstream gains and losses), a second-order intercept point of $+60$ dBm, and a third-order intercept point of $+30$ dBm, what is the receiver system dynamic range?

Solution

Figure A.8 shows the intercept chart for the preamplifier. Note that the second-order line crosses the fundamental line at $+60$ dBm and the

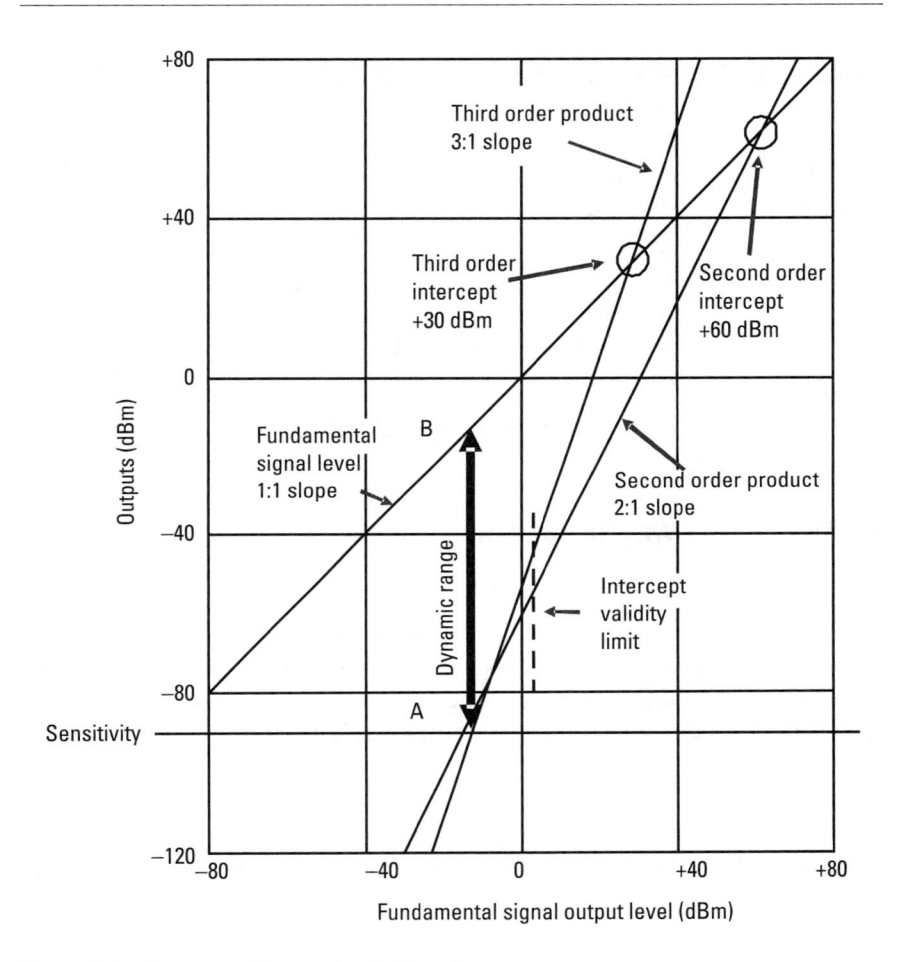

Figure A.8 Intercept diagram for Problem 5.

third-order line crosses at +30 dBm. The sensitivity line is drawn at −90 dBm. This means that the system sensitivity (at the output of the antenna) + the gain of the preamplifier − any losses between the antenna and the preamplifier sum to −90 dBm.

In this problem, we assume that the second-order spurs are eliminated by filtering, so the system dynamic range is limited by third-order spurs.

Draw a vertical line from point A where the third-order line crosses the sensitivity line—up to the fundamental line at point B. The length of this line is 75 dB (from −15 dBm to −90 dBm) so the receiver system dynamic range is 75 dB.

Problem 6: Digital receiver dynamic range

From Section 4.5.3

What is the dynamic range of a digital receiver which has 12-bit digitizing?

Solution

$$\text{Dynamic range} = 20 \log(2^n) \text{ where } n = 12$$

$$\text{Dynamic range} = 20 \log(4{,}095) = 72.2 \text{ dB}$$

Problem 7: One-way link equation

From Section 5.2

What is the received power in the link shown in Figure A.9 where:

The transmitter power = 1 watt, the transmitting antenna gain = 2 dBi, the receiving antenna gain (toward the transmitter) = 2 dB, and the link loss (by whatever propagation mode is appropriate) is 100 dB?

Solution

$$P_R = P_T + G_T - \text{Loss} + G_R = 30 \text{ dBm} + 2 \text{ dB} - 100 \text{ dB} + 2 \text{ dB} = -66 \text{ dBm}$$

Problem 8: Line-of-sight propagation loss

From Section 5.4

What is the propagation loss between two isotropic antennas (far from the ground) if the frequency is 500 MHz and the range is 75 km?

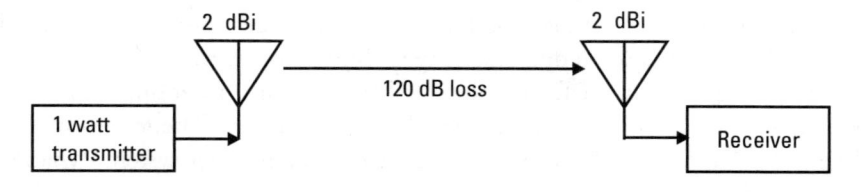

Figure A.9   One-way link for Problem 7.

Solution

$$\text{Loss (LOS)} = 32.4 + 20 \log(\text{distance in km})$$
$$+ 20 \log(\text{frequency in MHz})$$
$$= 32.4 + 20 \log(75) + 20 \log(500)$$
$$= 32.4 + 37.5 + 54.0 = 123.9 \text{ dB}$$

You can also determine this loss from the nomograph in Figure A.10. Draw a line from point A (500 MHz) to point B (75 km).

Note that the line crosses the center index at point C (a little less than 124 dB).

You can also determine this loss from side 1 of the slide rule as shown in Figure A.11.

Move the slide until 0.5 GHz is centered in the window at point A.

Read the propagation loss (a little less than 124 dB) at 75 km (point B).

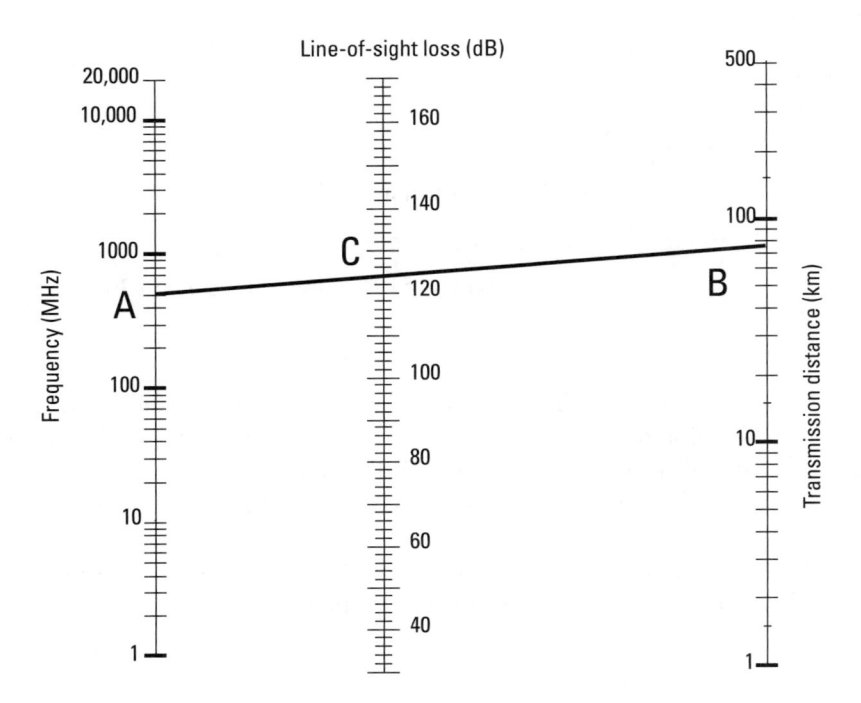

Figure A.10 Line-of-sight loss nomograph for Problem 8.

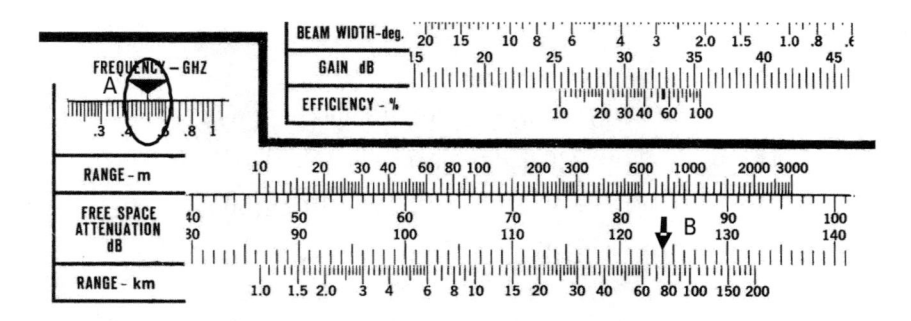

Figure A.11 Slide rule setup for Problem 8.

Problem 9: Two-ray propagation loss

From Section 5.5

What is the propagation loss between a transmit antenna 2m from the ground and a receiving antenna 20m from the ground, separated by 15 km. In this problem, it is given that the two-ray propagation model is appropriate.

Solution

$$\text{Loss (two-ray)} = 120 + 40 \log(\text{distance in km})$$
$$- 20 \log(\text{transmit ant height in m})$$
$$- 20 \log(\text{receive ant height in m})$$
$$= 120 + 40 \log(15) - 20 \log(2) - 20 \log(20)$$
$$= 120 + 47 - 6 - 26 = 135 \, \text{dB}$$

You can also determine the loss from the nomograph in Figure A.12.

Draw a line from 2 (point A) to 20 (point B). then draw a line from point C through 15 km (point D) to 135 dB (point E).

You can also determine the loss from side 2 of the slide rule as shown in Figure A.13.

Move the slide so that 15 km is next to 2 meters as shown at point A. Read the attenuation (135 dB) at 20m (point B).

Appendix A

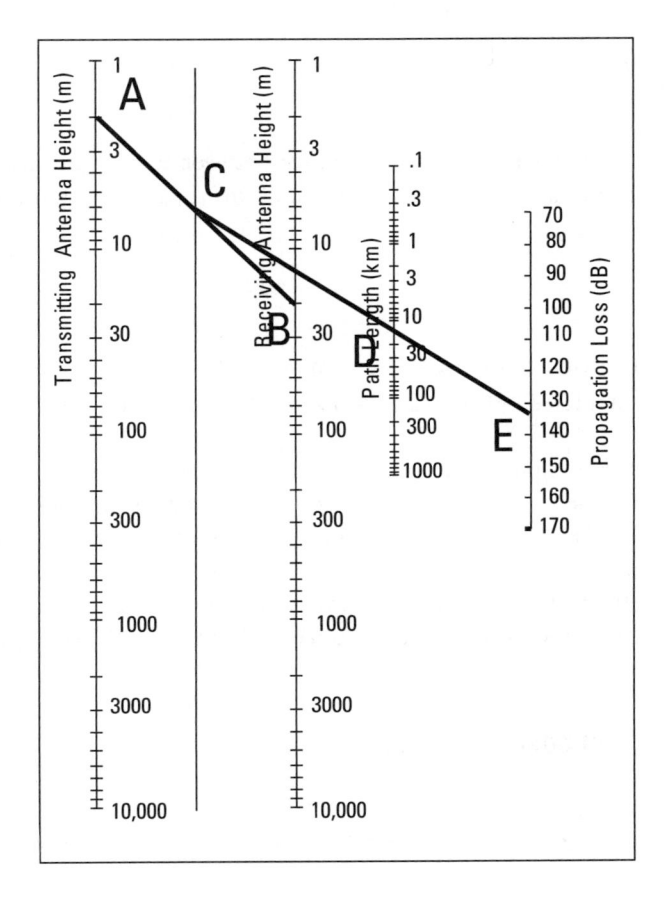

Figure A.12 Nomograph for Problem 9.

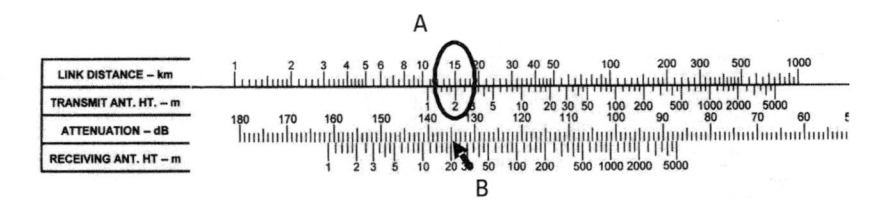

Figure A.13 Slide rule setup for Problem 9.

Problem 10: Fresnel zone

From Section 5.6

Determine the Fresnel zone distance between a transmit antenna 2m above the ground and a receiving antenna 200m from the ground if the frequency is 400 MHz.

Solution

Fresnel zone distance in km = (h_T in meters × h_R in meters × frequency in MHz)/24,000 = (2 × 200 × 400)/24,000 = 6.7 km

This means that if the link is shorter than 6.7 km, line-of-sight loss applies, and if the link is longer than 6.7 km, two-ray loss applies.

You can also determine the Fresnel zone from the slide rule as shown in Figure A.14.

Move the slide so that 2m is next to 200m at point A.

Then, read the Fresnel zone distance (6.7 km) at 400 MHz (point B).

Problem 11: Atmospheric loss

From Section 5.8

A 15-GHz link at sea level is 50 km long. What is its atmospheric loss?

Solution

On Figure A.15, draw a vertical line from 15 GHz (point A) up to the atmospheric loss line at point B, then draw a horizontal line left to the ordinate at 0.04 dB per km (point C).

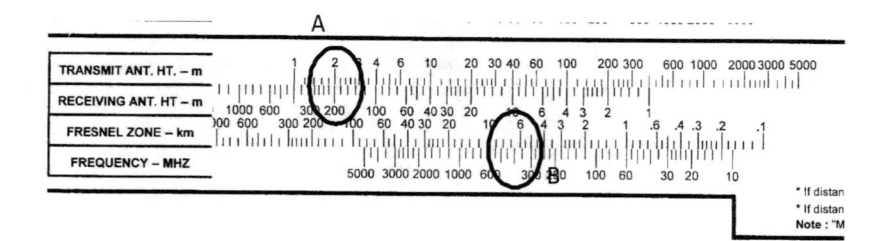

Figure A.14   Slide rule setup for Problem 10.

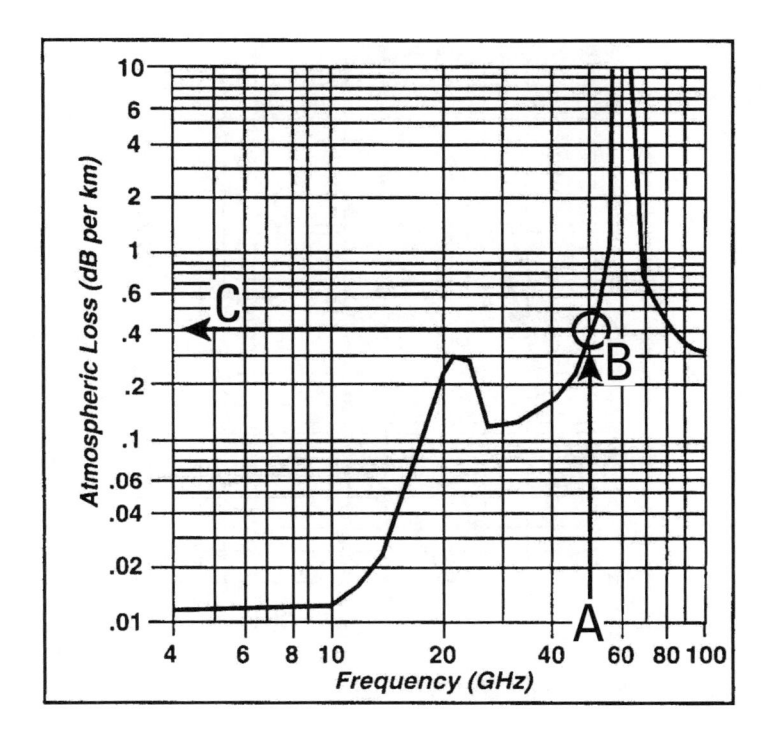

Figure A.15 Atmospheric loss for Problem 11.

The atmospheric loss is 0.04 dB/km × 50 km = 2 dB.

Problem 12: Rain loss

From Section 5.8
 A 50-km link at 20 GHz goes 50 km through moderate rain. What is its rain loss?

Solution

From Table 5.3 you can determine that moderate rain (4 mm/hr) causes loss as shown by line C in Figure A.16. Draw a vertical line from 20 GHz (point X) up to curve C (point Y), then left to 0.3 dB/km (point Z).

 The rain loss is 0.3 dB/km × 50 km = 15 dB.

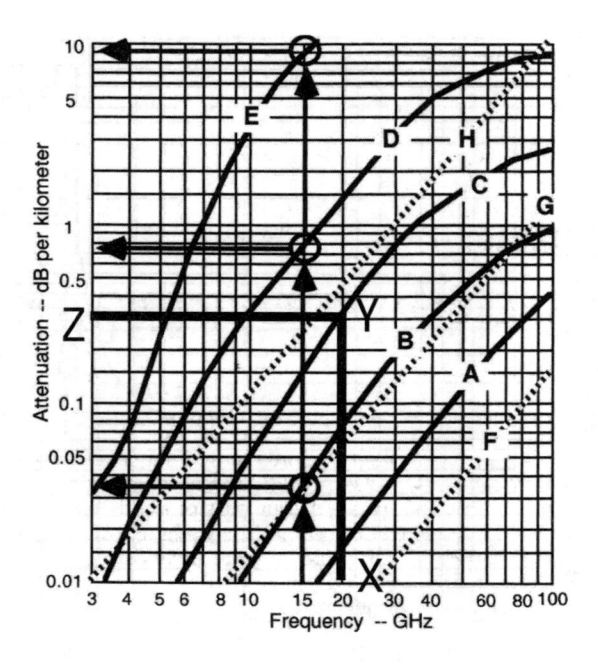

Figure A.16 Rain loss for Problem 12.

Problem 13: Satellite link loss

From Section 5.10

As shown in Figure A.17, a ground-based receiver with a 15-dB antenna pointed toward a low-orbit satellite is to intercept signals from the satellite at 5 GHz. The satellite is 4,000 km from the ground receiver at an elevation angle of 5° above the local horizon. The expected 0° isotherm height is 4 km, and the intercept needs to be successful when there is heavy rain present.

What sensitivity does the ground receiver need to have to receive a satellite signal that has 100 watts ERP?

Solution

The sensitivity needs to be equal to the received signal strength which is:

$$P_R = \text{ERP} - \text{LOS loss} - \text{atmospheric loss} - \text{rain loss} + \text{receiving antenna gain}$$

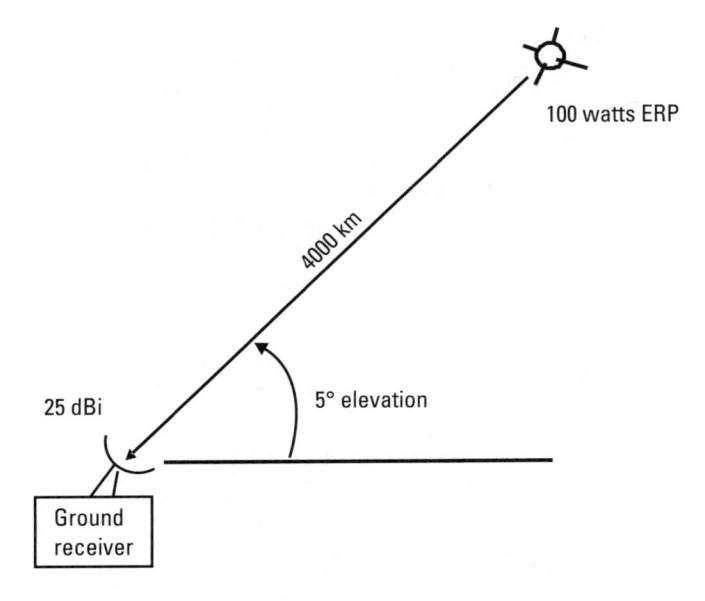

Figure A.17 Satellite link for Problem 13.

Atmospheric loss = 0.2 dB from the chart in Figure A.18 [draw vertical line from 5 GHz (point A) to 5° loss line (point B) and left to 0.2 dB (point C)].

Rain loss geometry is determined from Figure A.19. The path from the ground station to the 0° isotherm is 4 km / sin 5° = 4 km / 0.087 = 45.9 km.

The rain loss per km is 0.045 dB/km from Figure A.20 [draw vertical line from 5 GHz (point X) up to curve D (heavy rain) at point Y, then left to 0.045 dB/km at point Z].

The rain loss is 0.045 dB/km × 45.9 km = 2 dB.
The receiving antenna gain = 15 dBi.
The received signal strength—hence the required sensitivity is:
ERP − LOS loss − atmospheric loss − rain loss + receiving antenna gain
= 50 dBm − 178.4 dB − 0.2 dB − 2 dB + 15 dBi = −115.4 dBm.

Problem 14: Effective range (LOS link)

From Chapter 5

At what range can a receiver with −80-dBm sensitivity and a 3-dBi antenna receive a 1-watt ERP signal at 400 MHz? The antennas are far

290 EW 103

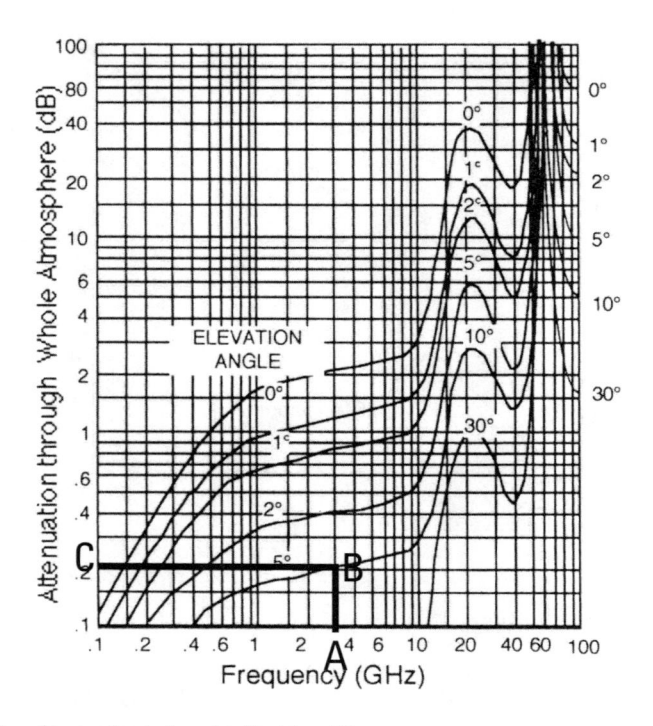

Figure A.18 Atmospheric loss for Problem 13.

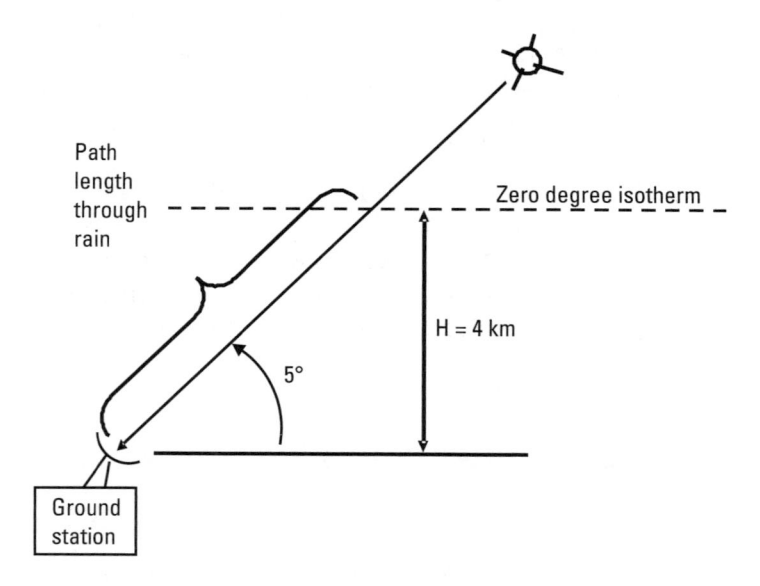

Figure A.19 Rain loss geometry for Problem 13.

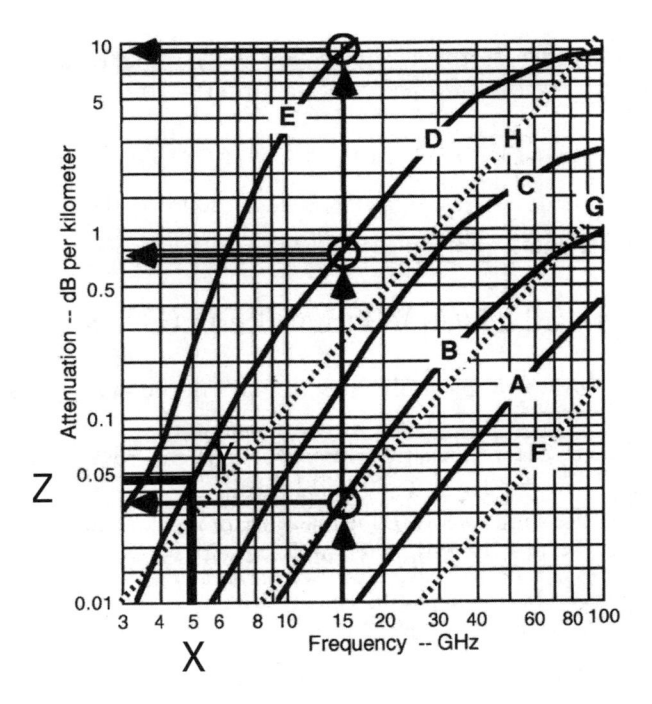

Figure A.20 Rain loss for Problem 13.

enough from the ground that line-of-sight propagation can be assumed. See Figure A.21.

Solution

From Section 5.4:

$$P_R = ERP - 32.4 - 20\log(d) - 20\log(F) + G_R$$

Figure A.21 Link diagram for Problem 14.

Remember that d is the distance in km and F is the frequency in MHz. Set P_R equal to the sensitivity and solve for the 20 log(d) term.

$$20 \log(d) = ERP - 32.4 - 20 \log(F) + G_R - Sens$$
$$= +30 - 32.4 - 20 \log(400) + 3 - (-80)$$
$$= 30 - 32.4 - 52 + 3 + 80 = 28.6$$

$$d = \text{Antilog}\left[20 \log(d)/20\right] = \text{Antilog}\left[28.6/20\right] = 26.9 \text{ km}$$

You can also work this problem with side 1 of the slide rule as shown in Figure A.22.

First isolate the propagation loss, which is the ERP + the receiving antenna gain – $Sens$ = 30 + 3 –(–80) = 113 dB.

Then set the slide rule to 0.4 GHz at point A. Find the range at which the attenuation is 113 dB (point B) and you will see that the range is 26.9 km.

Problem 15: Effective range (two-ray link)

From Chapter 5

At what range can a receiver with –100-dBm sensitivity and a 3-dBi antenna that is 5m above the ground receive a 400 MHz, 10-watt ERP signal with an antenna height of 2m? See Figure A.23.

Solution

First, you should check the Fresnel zone to determine the appropriate propagation mode.

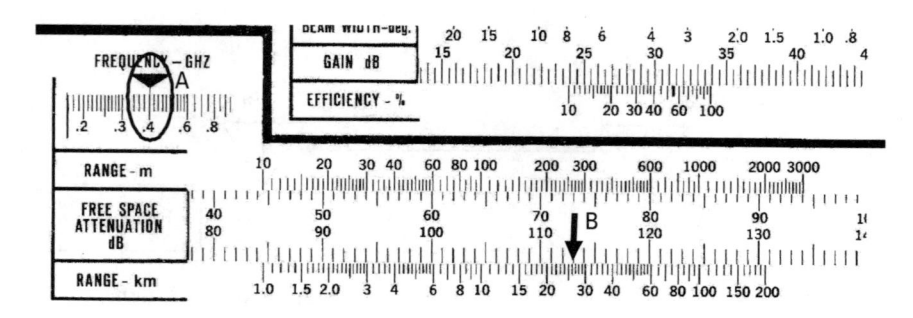

Figure A.22 Slide rule setup for Problem 14.

Appendix A

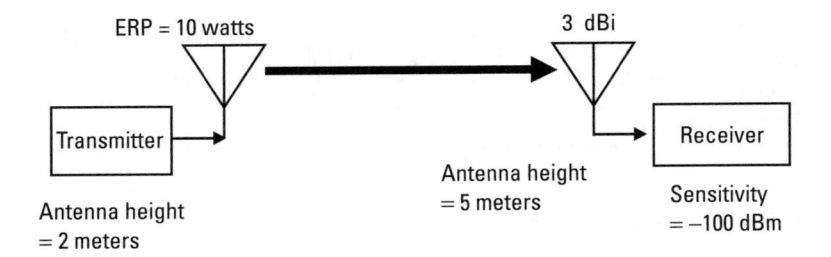

```
ERP = 10 watts              3 dBi
       ▽                     ▽
       |                     |
   ┌───┴────┐            ┌───┴────┐
   │Transmitter│─────────→│ Receiver │
   └────────┘            └────────┘
                Antenna height
  Antenna height   = 5 meters      Sensitivity
   = 2 meters                      = –100 dBm
```

Figure A.23 Link diagram for Problem 15.

$FZ = (2 \times 5 \times 400)/24{,}000 = 167$ m, so the propagation mode is almost certainly two-ray. If the effective range were to fall below 167 m, you would need to redo the problem.

From Section 5.5, the received power is:

$$P_R = ERP - 120 - 40\log(d) + 20\log(h_T) + 20\log(h_R) + G_R$$

Remember that the distance is in km and the heights of the two antennas are in meters.

Set the received power equal to the sensitivity and solve for the $40\log(d)$ term

$$\begin{aligned}
40\log(d) &= ERP - 120 + 20\log(h_T) \\
&\quad + 20\log(h_R) + G_R - Sens \\
&= 40 - 120 + 20\log(2) + 20\log(5) + 3 \\
&= 40 - 120 + 6 + 14 + 3 + 100 = 43
\end{aligned}$$

$$d = \text{Antilog}\,[40\log(d)/40] = \text{Antilog}\,[43/40] = 11.9 \text{ km}$$

You can also work this problem on side 2 of the slide rule, as shown in Figure A.24.

First, determine the propagation loss, which = $ERP - Sens +$ receiving antenna gain

$$= 40 + 100 + 3 = 143 \text{ dB}$$

First, set 143-dB attenuation at the receiving antenna height (5 m) at point A.

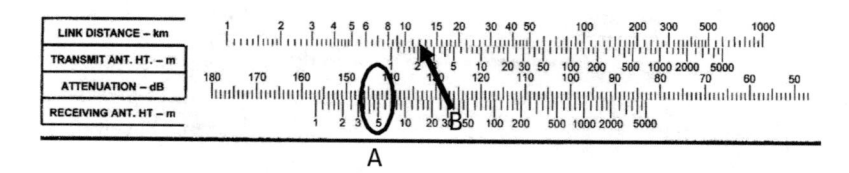

```
| LINK DISTANCE – km      |
| TRANSMIT ANT. HT. – m   |
| ATTENUATION – dB        |
| RECEIVING ANT. HT – m   |
```

                                    A

Figure A.24 Slide rule setup for Problem 15.

Then read the range at the transmitting antenna height (point B) You will see that the range is 11.9 km.

Problem 16: Probability of intercept in narrow-band search

From Chapter 6

A receiver has 200-kHz bandwidth and is swept to find a 25-kHz-wide signal between 30 and 88 MHz. The receiver can determine signal energy in its bandwidth within 100 microseconds, and can analyze any signal found within 1 millisecond, including its modulation parameters and the actual signal frequency. What is the probability that it will find a particular signal of interest (identifiable from its measured parameters) within one-half second if 5% of the channels are occupied? Use 50% bandwidth overlap when sweeping for signals.

Solution

First, how many signals are present? 58 MHz is covered (88 – 30) with 25-kHz signal channels. That makes 2,320 possible signal locations. 5% of 2,320 = 116 signals present.

How long will the system spend analyzing signals found? 116 × 1 ms = 116 ms.

How long will the receiver spend searching for signals? 58 MHz contains 290 200-kHz bandwidths. With 50% overlap, it takes 580 steps to cover the frequency range. The receiver must pause 100 microseconds on each step to determine if any signals are present on that tuning step: 580 steps × 100 microseconds/step = 58 ms. So, the receiver must spend 58 ms sweeping.

Thus, it takes 116 + 58 = 174 ms to search the whole range and analyze all signals found. Thus, the desired signal will be found in 174 ms if it is present. The probability of intercept is 100%.

Another way to look at the probability of intercept is to compare the total frequency range the receiver could cover to the frequency range of the signal of interest.

With 5% occupancy, there is an average of 2 signals per MHz. (58 MHz/25 kHz) × 0.05 = 2.

It takes 10 tuning steps to cover a MHz with 50% overlap. (1 MHz/200 kHz) × 2 = 10.

Each tuning step requires 100 μs. This means that it takes 1 ms to sweep 1 MHz (0.1 ms/step × 10 steps/MHz).

Each signal found requires 1 ms of analysis time, so analyzing the two signals found requires 2 ms per MHz.

Thus it takes 3 ms to cover 1 MHz. In one-half second, the system could search and analyze all signals present in 166.7 MHz (500 ms/3 ms). The probability of intercept (POI) is the total frequency range covered divided by the possible frequency range of the signal = the probability of intercept = 287%. However, we have chosen a case in which the POI is greater than 100%, so the answer is 100%.

NOTE ON PROBLEMS 17 THROUGH 23:

In these solutions, formulas are used to calculate Fresnel zones and propagation losses. You could just as well use the slide rule or Chapter 5 propagation nomographs to calculate these values. On the slide rule, you will often find that the Fresnel zone distance is off the scale to the right, which means that the Fresnel zone distance is less than 100m—this means that if the link is longer than 100m, two-ray propagation applies.

Problem 17: Communication J/S

From Section 9.1

Consider the jamming geometry shown in Figure A.25.

You are jamming an enemy push to talk net 50 km from your jammer with 5 km between stations. The target transceivers use whip antennas. The figure shows a typical target link. You, of course, jam the receiver (i.e., the transceiver which is receiving at the moment).

Solution

The transmitter to receiver FZ is (1.5 × 1.5 × 400)/24,000 = 37.5m, so propagation is two-ray.

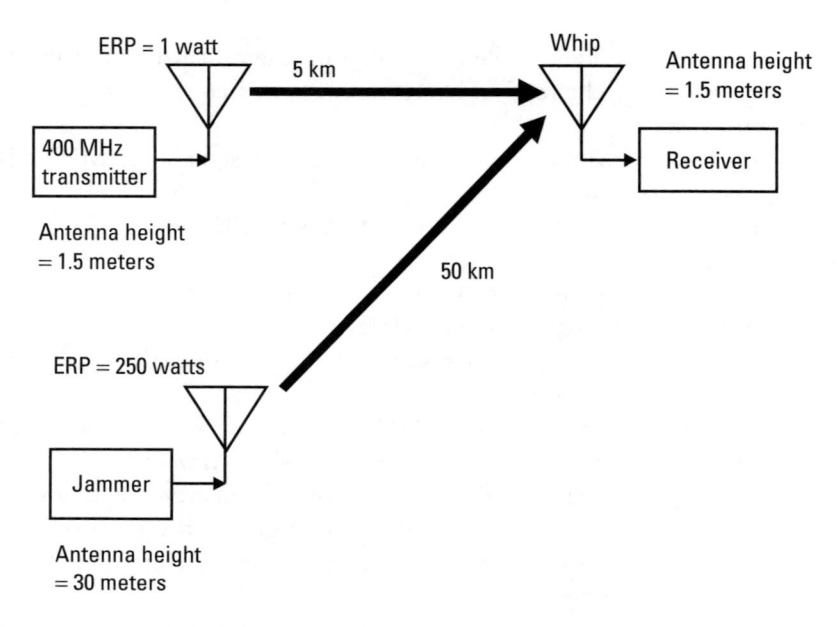

```
           ERP = 1 watt              Whip
                ▽         5 km        ▽      Antenna height
                                             = 1.5 meters
    ┌──────────┐ │
    │ 400 MHz  │─┘
    │transmitter│                           ┌────────┐
    └──────────┘                            │Receiver│
                                            └────────┘
    Antenna height
    = 1.5 meters                  50 km


    ERP = 250 watts
          ▽
    ┌────────┐ │
    │ Jammer │─┘
    └────────┘

    Antenna height
    = 30 meters
```

Figure A.25 Link diagram for Problem 17.

The jammer to receiver FZ is $(30 \times 1.5 \times 400)/24{,}000 = 750$m, so propagation is two-ray.

The transmitter ERP is 30 dBm.
The jammer ERP is 54 dBm.
The transmitter-to-receiver loss is:

$$120 + 40\log(5) - 20\log(1.5) - 20\log(1.5)$$
$$= 120 + 28 - 3.5 - 3.5 = 141\,\text{dB}$$

The jammer-to-receiver loss is:

$$120 + 40\log(50) - 20\log(30) - 20\log(1.5)$$
$$= 120 + 68 - 29.5 - 3.5 = 155\,\text{dB}$$

$$J/S = ERP_J - ERP_S - Loss_J + Loss_S$$
$$= 54 - 30 - 155 + 141 = 10\,\text{dB}$$

This would be adequate for analog target signals. 0 dB would be enough if they use digital modulation.

Problem 18: Communication J/S

From Section 9.1
 Consider the jamming geometry shown in Figure A.26.
 This is "stand-in jamming" using a low power jammer (one-quarter-watt) deployed very near the receiver (250m). What is J/S?

Solution

The transmitter to receiver FZ is $(1.5 \times 2.5 \times 1{,}600)/24{,}000 = 250$m, so propagation is two-ray.
 The jammer to receiver FZ is $(1 \times 1.5 \times 1{,}600)/24{,}000 = 100$m, so propagation is two-ray.
 The transmitter ERP is 30 dBm.
 The jammer ERP is $10 \log (250) = 24$ dBm.
 The transmitter-to-receiver loss is:

$$120 + 40 \log(5) - 20 \log(1.5) - 20 \log(2.5)$$
$$= 120 + 28 - 3.5 - 8 = 136.5 \text{ dB}$$

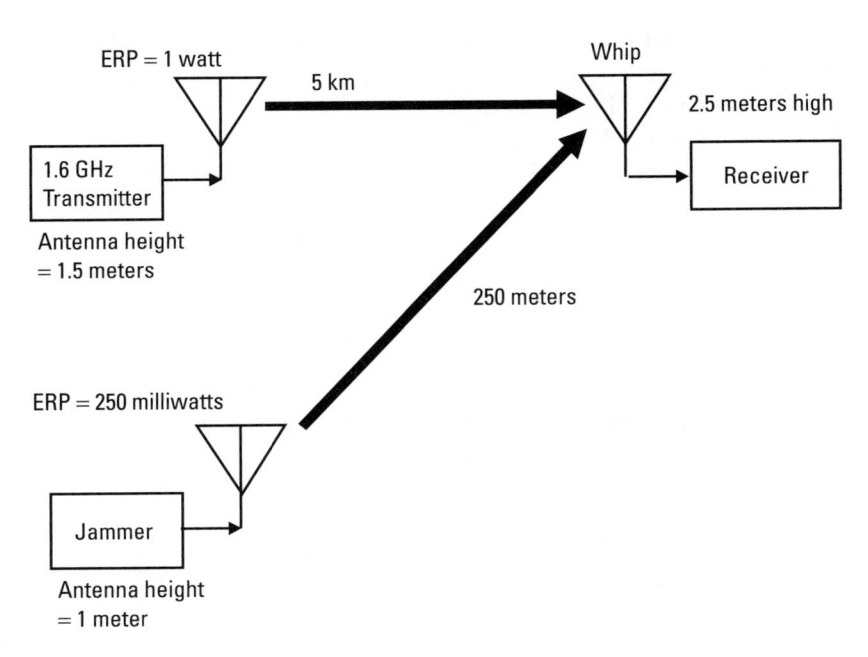

Figure A.26 Link diagram for Problem 18.

$$J/S = ERP_J - ERP_S - Loss_J + Loss_S$$
$$= 24 - 30 - 88 + 136.5 = 42.5\,\text{dB}$$

The jammer-to-receiver loss is:

$$120 + 40\log(0.25) - 20\log(1) - 20\log(2.5)$$
$$= 120 - 24 - 0 - 8 = 88\,\text{dB}$$

Problem 19: Communication J/S

From Section 9.1

Consider the jamming geometry shown in Figure A.27.

This problem uses a jammer in a helicopter which is hovering at 1,000m, 50 km from an enemy push-to-talk net. What is J/S?

Solution

The transmitter to receiver FZ is $(1.5 \times 1.5 \times 100)/24,000 = 9.3$m, so propagation is two-ray.

The jammer-to-receiver FZ is $(1,000 \times 1.5 \times 100)/24,000 = 6.25$ km, so propagation is two-ray.

The transmitter ERP is 30 dBm.
The jammer ERP is 50 dBm.
The transmitter-to-receiver loss is:

$$120 + 40\log(5) - 20\log(1.5) - 20\log(1.5)$$
$$= 120 + 28 - 3.5 - 3.5 = 141\,\text{dB}$$

The jammer-to-receiver loss is:

$$120 + 40\log(50) - 20\log(1000) - 20\log(2.5)$$
$$= 120 + 68 - 60 - 3.5 = 124.5\,\text{dB}$$

$$J/S = ERP_J - ERP_S - Loss_J + Loss_S$$
$$= 50 - 30 - 124.5 + 141 = 36.5\,\text{dB}$$

Appendix A 299

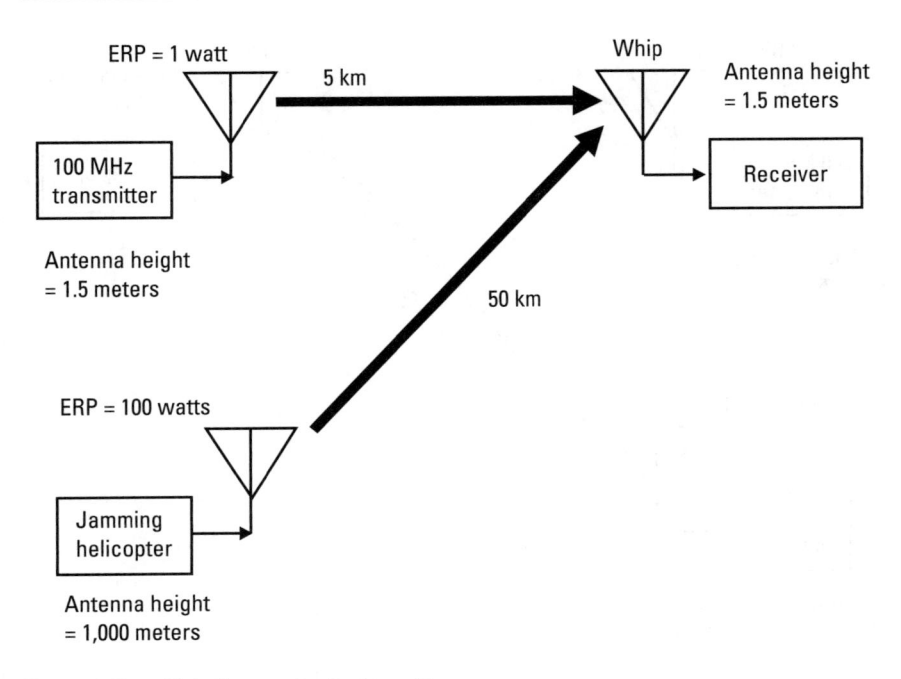

```
        ERP = 1 watt                    Whip
           ▽           5 km              ▽         Antenna height
                    ───────▶                       = 1.5 meters
   ┌──────────┐                        ┌──────────┐
   │ 100 MHz  │                        │ Receiver │
   │transmitter│                       └──────────┘
   └──────────┘
   Antenna height
   = 1.5 meters                    50 km


        ERP = 100 watts
           ▽
   ┌──────────┐
   │ Jamming  │
   │helicopter│
   └──────────┘
   Antenna height
   = 1,000 meters
```

Figure A.27   Link diagram for Problem 19.

Problem 20: Communication J/S

From Section 9.1

Consider the jamming geometry shown in Figure A.28.

This is the same as Problem 19, but the target net is now operating at 900 MHz. What is J/S?

Solution

The transmitter to receiver FZ is $(1.5 \times 1.5 \times 900)/24{,}000 = 84$m, so propagation is two-ray.

The jammer to receiver FZ is $(1{,}000 \times 1.5 \times 900)/24{,}000 = 56$ km, so propagation is line-of-sight.

The transmitter ERP is 30 dBm.
The jammer ERP is 50 dBm.
The transmitter-to-receiver loss is:

$$120 + 40 \log(5) - 20 \log(1.5) - 20 \log(1.5) = 120 + 28 - 3.5 - 3.5 = 141 \, \text{dB}$$

The jammer-to-receiver loss is:

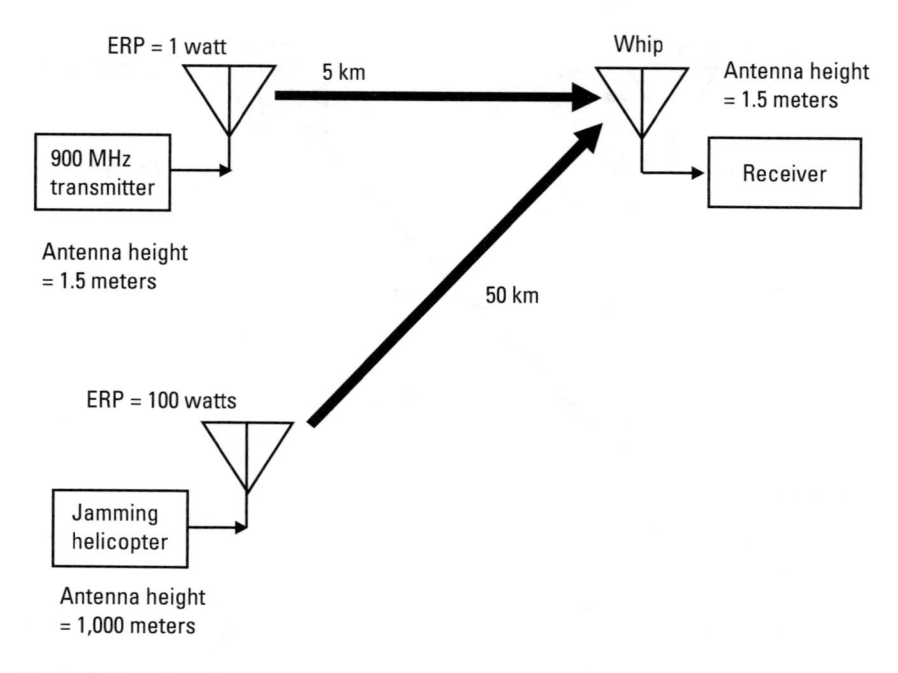

Figure A.28 Link diagram for Problem 20.

$$32.4 + 20\log(900) + 20\log(50) = 32.4 + 59 + 34 = 125.4$$

$$J/S = ERP_J - ERP_S - LOSS_J + LOSS_S = 50 - 30 - 120.4 + 141 = 35.6 \text{ dB}$$

Problem 21: Communication J/S

From Section 9.1

Consider the jamming geometry shown in Figure A.29.

This is the same situation covered in Problem 19, but now the link is using direct sequence spread spectrum (DSSS) signals which have a spreading code at 10 Mbps and an information bit rate of 10 kbps. This will create a 30-dB processing gain in the receiver.

Solution

From Problem 19, the J/S without DSSS would be 36.5 dB. However, the 30-dB processing gain reduces the actual J/S to 6.5 dB. The target signal is digital and the J/S is greater than 0 dB, so the J/S is adequate. Also, since the duty cycle is 33%, the digital target signal is effectively jammed.

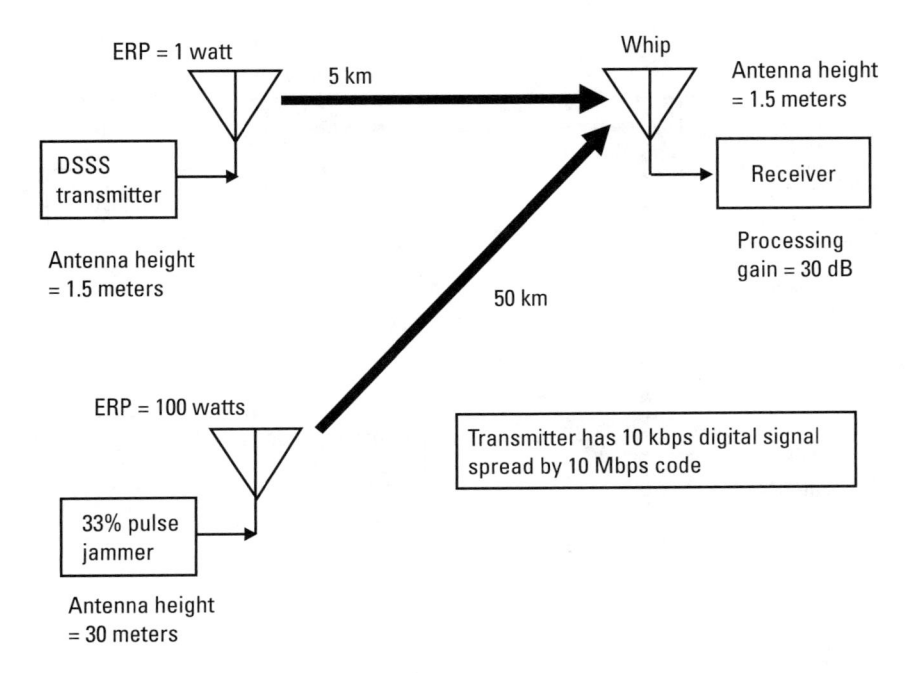

Figure A.29 — Link diagram for Problem 21

- DSSS transmitter: ERP = 1 watt, Antenna height = 1.5 meters
- Distance to receiver: 5 km
- Whip antenna (Receiver): Antenna height = 1.5 meters, Processing gain = 30 dB
- 33% pulse jammer: ERP = 100 watts, Antenna height = 30 meters
- Jammer to receiver distance: 50 km
- Transmitter has 10 kbps digital signal spread by 10 Mbps code

Figure A.29 Link diagram for Problem 21.

Problem 22: Partial band jamming

From Section 9.3.1

Consider the jamming geometry shown in Figure A.30.

This is a frequency hopping net with 25-kHz channel bandwidth that hops over 30 to 58 MHz. The transmitter ERP is unknown because emission control is used. The transceivers in the net are 5 km apart.

The jammer is 40 km from the net and has a receiving capability. It has a 10-dB log periodic antenna on a 30-m mast. The transmitted target signal is received at the jammer receiver at –110 dBm.

What percentage of the jamming range of the target signal can be jammed with 0 dB J/S?

Solution

The average transmitted frequency is 59 MHz.

The transmitter to receiver FZ is $(1.5 \times 1.5 \times 59)/24{,}000 = 5.5$m, so propagation is two-ray.

The transmitter to jammer FZ is $(1.5 \times 30 \times 59)/24{,}000 = 110$m, so propagation is two-ray.

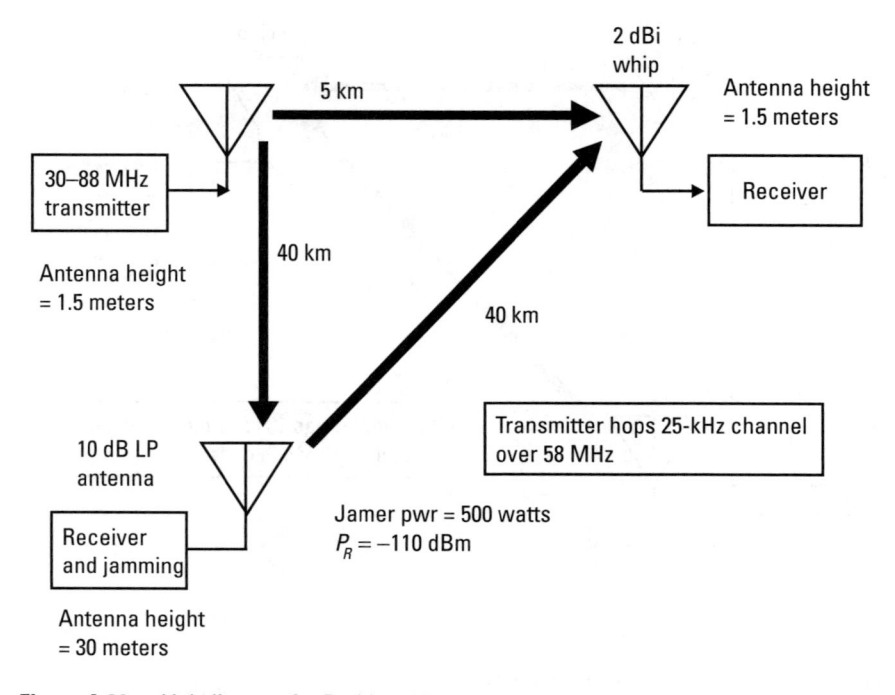

Figure A.30 Link diagram for Problem 22.

The jammer to receiver FZ is the same, so two-ray propagation applies.
The jammer ERP is 500 watts + antenna gain = 57 + 10 = 67 dBm.
The total number of transmitter hop frequencies is 58 MHz/25 kHz = 2,320.
The transmitter ERP can be determined from the received power at the jammer location.

$$P_R = ERP - [120 + 40\log(40) - 20\log(1.5) - 20\log(30)] + G_R$$

Solving for ERP

$$ERP = P_R + [120 + 40\log(40) - 20\log(1.5) - 20\log(30)] - G_R$$
$$= -110 + 120 + 64 - 3.5 - 29.5 - 10 = 31\,\text{dBm}\ (1.3\ \text{watts})$$

The received desired signal power at the target receiver is:

$$P_R \text{(called } S) = 31 - [120 + 40 \log(5) - 20 \log(1.5) - 20 \log(1.5)] + 2$$
$$= 31 - 120 - 28 + 3.5 + 3.5 + 2 = -108 \text{ dBm}$$

The received jammer signal power at the target receiver is:

$$P_R \text{(called } J) = 67 - [120 + 40 \log(40) - 20 \log(30) - 20 \log(1.5)] + 2$$
$$= 67 - 120 - 64 + 29.5 + 3.5 + 2 = -82 \text{ dBm}$$

J/S dB = J dBm − S dBm = −82 − (−108) = 26 dB. This is the single channel J/S, ignoring the hopping.

If the jamming signal is spread over enough channels to provide 0 dB per channel, the number of channels covered would be found from:

$$\text{Antilog } [26/10] = 400 \text{ channels}$$

This means that the jammer would be spread over 10 MHz.

Note that only 400 of 2,320 channels or 17% are jammed. This is not enough to be considered effective jamming, but would provide the best jamming performance possible with this jammer in this jamming geometry.

Problem 23: Jamming cell phone uplink

From Section 9.5

Consider the jamming geometry shown in Figure A.31.

You are jamming the uplink of a cell phone, which means that you must jam the receiver in the cell tower. The cell phone has 1-watt ERP at 1.8 GHz and is 1 km from the tower, 1m above the ground. The tower is 50m high. The jammer has 250 watts ERP from an antenna which is 3m above the ground 1 km from the tower. What is the J/S?

Solution

The transmitter to receiver FZ is (1 × 50 × 1,800)/24,000 = 3.75 km, so propagation is line-of-sight.

The jammer to receiver FZ is (3 × 50 × 1,800)/24,000 = 11.3 km, so propagation is line-of-sight.

The transmitter ERP is 30 dBm.

The jammer ERP is 54 dBm.

The transmitter-to-receiver loss is:

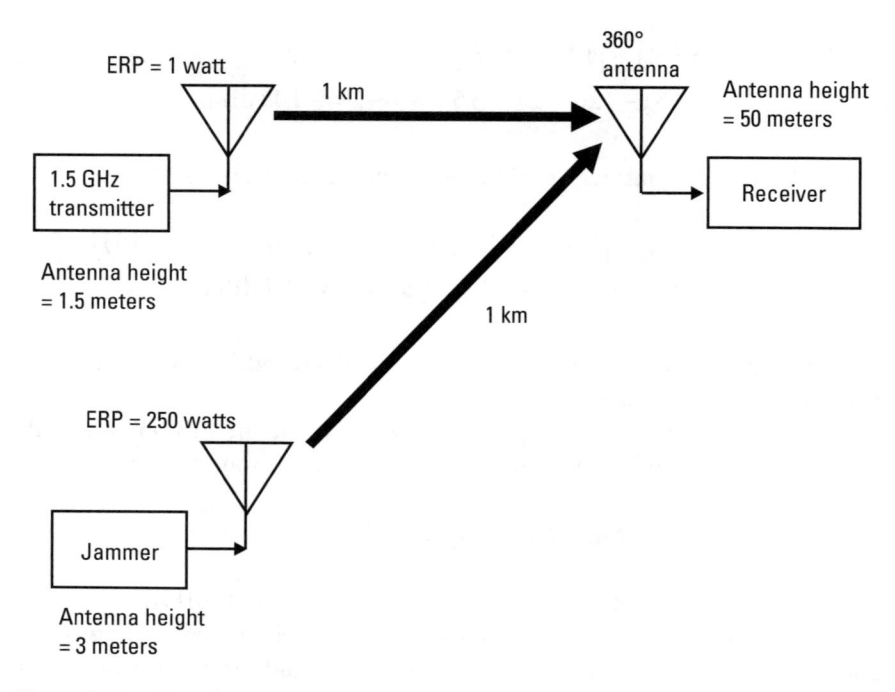

Figure A.31 Link diagram for Problem 23.

$$32.4 + 20\log(1{,}800) + 20\log(1) = 32.4 + 65 + 0 = 97.4 \text{ dB}$$

The jammer-to-receiver loss is the same as the signal loss = 97.4 dB

$$J/S = ERP_J - ERP_S - LOSS_J + LOSS_S = 54 - 30 - 97.4 + 97.4 = 24 \text{ dB}$$

If the cell phone is a GSM type, it has digital modulation (requiring only 0-dB *J/S*), so this is plenty of *J/S* to spread over several channels.

Problem 24: Jamming cell phone downlink

From Section 9.5

Consider the jamming geometry shown in Figure A.32.

You are jamming the downlink of a cell phone, which means that you must jam the receiver in the cell phone. The cell tower has 50-watt ERP at 1.8 GHz and is 1 km from the cell phone which is 1m above the ground.

Appendix A

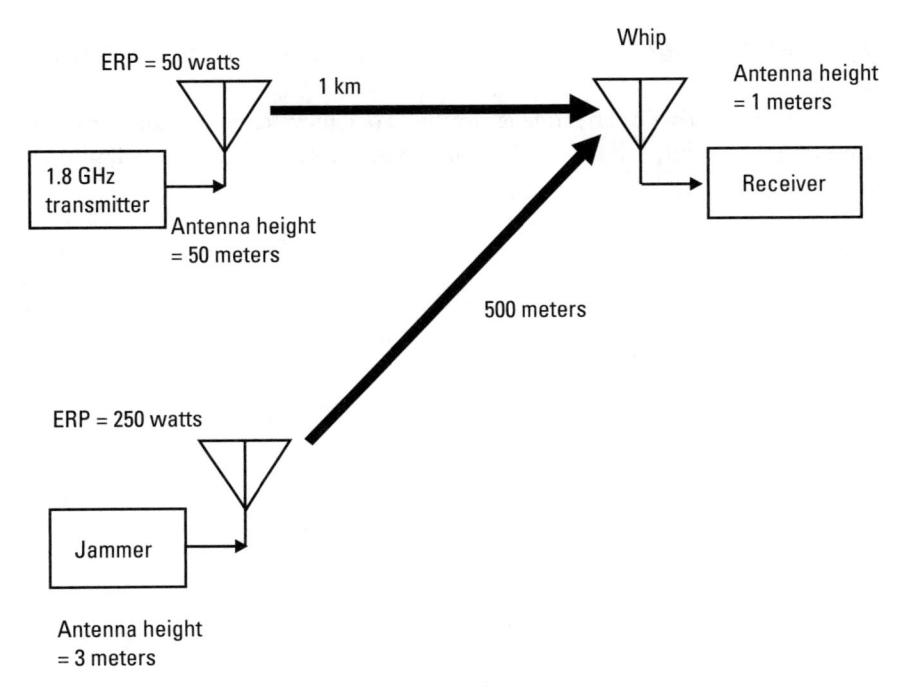

ERP = 50 watts, 1 km, **Whip**, Antenna height = 1 meters

1.8 GHz transmitter, Antenna height = 50 meters

Receiver

500 meters

ERP = 250 watts

Jammer

Antenna height = 3 meters

Figure A.32 Link diagram for Problem 24.

The tower is 50m high. The jammer has 250-watts ERP from an antenna which is 3m above the ground 500m from the cell phone. What is the J/S?

Solution

The transmitter-to-receiver FZ is $(50 \times 1 \times 1{,}800)/24{,}000 = 3.75$ km, so propagation is line-of-sight.

The jammer-to-receiver FZ is $(3 \times 1 \times 1{,}800)/24{,}000 = 225$m, so propagation is two-ray.

The transmitter ERP is 47 dBm.
The jammer ERP is 54 dBm.
The transmitter-to-receiver loss is:

$$32.4 + 20 \log(1{,}800) + 20 \log(1) = 32.4 + 65 + 0 = 97.4 \text{ dB}$$

The jammer-to-receiver loss is:

$$120 + 40 \log(0.5) - 20 \log(3) - 20 \log(1) = 120 - 6 + 0 = 114 \text{ dB}$$

$$J/S = ERP_J - ERP_S - LOSS_J + LOSS_S = 54 - 47 - 114 + 97.4 = -9.6 \text{ dB}$$

This is clearly not effective jamming. To achieve effective jamming, the jammer needs to have significantly more power or be closer to the cell phone.

Appendix B
Bibliography

The following books are suggested references related to communication electronic warfare. Most of the books cover a wider subject area, but have information which is felt to be helpful when dealing with communication EW issues. These books were used as references in the process of writing this book. New editions of books are often published, usually by the same publisher. There are certainly additional books which you may find helpful; this list should be considered only a starting point.

Usually, the books referenced here explain concepts and processes with more mathematical content than is used in this book. It is hoped that the explanations in this book will help you get started on the deeper coverage in these referenced books.

> *EW 101* by David Adamy
> ISBN 1-58053-169-5, Artech House, 2001.
> Covers the RF aspects of the electronic warfare (EW) field using little math. Based on the EW 101 columns in the *Journal of Electronic Defense* since October 1994.

> *EW 102* by David L. Adamy
> ISBN 1-58053-686-7, Artech House, 2004.
> A companion book to EW 101, covers threats, radar principles, IR and EO, communications EW, emitter location accuracy, and comm sat

links. Also based on EW 101 columns in the *Journal of Electronic Defense*.

Electronic Warfare in the Information Age by D. Curtis Schleher
ISBN 0-89006-526-8, Artech House, 1999.
Covers electronic warfare field using both physical and mathematical characterizations. Includes many examples worked in MATLAB 5.1. Although the book covers most of the EW field, it has significant information (scattered through the book) helpful in communication EW.

Applied ECM by Leroy Van Brunt
ISBN 0-931728-00-2 (Vol. 1, 1978); ISBN 0-931728-01-0 (Vol 2, 1982) ISBN 0-931728-04-5 (Vol. 1, 1995) ISBN 0-931728-05-3 (3 volume set)
A complete and rigorous coverage of electronic countermeasures in three volumes. Available only from the publisher (EW Engineering Inc., P.O. Box 28, Dunn Loring, VA 22027).

Introduction to Electronic Defense Systems by Filippo Neri
ISBN 0-89006-553-5, Artech House, 1991.
Nonmathematical coverage of whole EW field. Almost entirely radar EW, but has some material on communication EW in Chapters 4, 5, and 6.

Spread Spectrum Systems with Commercial Applications by Robert Dixon
ISBN 0 471-59342-7, John Wiley & Sons, 1994.
Overviews and mathematical characterizations of spread spectrum signals.

Detectability of Spread Spectrum Signals by Robin and George Dillard
ISBN 0-89006-299-4, Artech House, 1989.
Thorough coverage of energy detection approaches to the detection of spread spectrum signals. Includes excellent coverage of energy detection techniques.

Spread Spectrum Communications Handbook by Marvin K. Simon et al.
ISBN 0-07-057629-7, McGraw-Hill, 1994.
Compilation of authoritative papers on spread spectrum communication by experts in the field.

Advanced Techniques for Digital Receivers by Phillip Pace
ISBN 1-58053-053-2, Artech House, 2000.
Graduate-level coverage of digital signals and receivers-design and performance analysis.

Introduction to Communication Electronic Warfare Systems by Richard Poisel
ISBN 1-58053-344-2, Artech House, 2002.
Comprehensive coverage of communication signals and propagation as well as the principles and practice of EW against those signals.

Practical Communication Theory by David Adamy
ISBN 1-8885897-04-9, Lynx Publishing, 1994.
Describes the one-way communication link and gives simple dB formulas for working practical intercept problems.

Tactical Battlefield Communications Electronic Warfare by Dave Adamy
ISBN 1-885897-17-0, Lynx Publishing, 2005.
Includes the same antenna and propagation slide rule included with this book. Covers communication band propagation, communication jamming, and antenna parameter trade-offs. Includes instructions on using the slide rule. This 28-page booklet is sometimes used in briefing military EW personnel preparing for deployment.

Electronic Warfare for the Digitized Battlefield by Michael R. Frater and Michael Ryan
ISBN 1-58053-271-3, Artech House, 2001.
Operational focus on the modern electronic battlefield and the appropriate EW techniques. Operational level descriptions of important new communications EP.

Introduction to Electronic Warfare Modeling and Simulation by David L. Adamy
ISBN 1-58053-495-3, Artech House, 2003.
Broad introduction to EW modeling and simulation. Covers terms, concepts, and applications. Includes an introduction to EW sufficient to support the primary material. Mostly focused on radar EW simulation and modeling but with a little material relevant to communications EW.

The Communications Handbook, edited by Jerry D. Gibson
ISBN 0-8493-8349-8, CRC Press (in cooperation with IEEE), 1977.

Papers on a wide range of communication subjects, including a thorough coverage of propagation models.

Information Warfare: Principles and Operations by Edward Waltz
ISBN: 0-89006-511-X, Artech House, 1998.
This is the best book I have found on information warfare. It covers the official definitions, vocabulary, and concepts. It includes fine scale subdivisions of the field and the supporting technologies. Talks about strategies for all stages of conflict.

The Comprehensive Guide to Wireless Technologies by Laurence Harte et al.
ISBN: 0-965-06584-7, APDG Publishing, 2000.
Chapter 4 gives an overview of cell phone systems and operation.

ECM and ECCM Techniques for Digital Communication Systems by Ray Pettit
ISBN 0-534-97932-7, Wadsworth, Inc., 1982.
Covers modulations (including spread spectrum), codes, error-correcting codes, and other related subjects of importance in communications EW.

Telecommunications Primer by E. Bryan Carne
ISBN 0-13-206129-5, Prentice-Hall, 1995.
In-depth discussions of communication signals and systems using physical descriptions with little math.

Appendix C
Using the Included CD

The CD provided with this book will allow you to work problems by plugging in parametric values and reading out the answers to mathematical procedures. The CD requires only that you have Microsoft Excel software on your computer, XP, or compatible release.

The procedure is to upload the desired Excel file to your computer hard drive, plug in the required values in the locations, and read out the answers in the indicated locations. The procedures on the CD are:

- Line-of-sight one-way link equation;
- Two-ray one-way link equation;
- Knife-edge diffraction;
- Fresnel zone distance;
- Receiver sensitivity;
- Receiver system noise figure;
- Analog receiver dynamic range;
- Digital receiver dynamic range;

- CEP from RMS error;
- CEP from EEP;
- Communication J/S;
- Frequency spread for partial-band jamming;
- Received power in one-way link;
- Effective range in one-way link.

File LOS.xls: Line-of-sight one-way link equation

Input link distance (in km) at A4
Input frequency (in MHz) at A5
Read line-of-sight loss (in dB) at A10

File 2ray.xls: Two-ray one-way link equation

Input link distance (in km) at A4
Input transmitting antenna height (in meters) at A5
Input receiving antenna height (in meters) at A6
Read two-ray loss (in dB) at A10

File KED.xls: Knife-edge diffraction

Input distance from transmitter to ridge line (in km) at A4
Input distance from ridge line to receiver (in km) at A5
Input height of ridge line (above or below line of sight—in meters) at A6
State whether ridge line is above or below line of sight (A or B) at A7
Read knife-edge diffraction loss (in dB) at A10

File FZ.xls: Fresnel zone distance

Input transmitting antenna height (in meters) at A4
Input receiving antenna height (in meters) at A5

Input frequency (in MHz) at A6
Read Fresnel zone distance (in km) at A10

File Sens.xls: Receiver sensitivity

Input effective receiver system bandwidth (in MHz) at A4
Input system noise figure (in dB) at A5
Input required predetection signal-to-noise ratio at A6
Read system sensitivity at A10

File NF.xls: Receiver system noise figure

Input loss before preamplifier (in dB) at A4
Input preamplifier gain (in dB) at A5
Input preamplifier noise figure (in dB) at A6
Input loss between preamplifier and receiver (in dB) at A7
Input receiver noise figure (in dB) at A8
Read system noise figure (in dB) at A10

File ADR.xls: Analog receiver dynamic range

Input receiver system sensitivity (in dBm) at A4
Insert net gain between receiver input and output of preamplifier (in dB) at A5
Input second-order preamplifier second-order intercept point (in dBm) at A6
Input preamplifier third-order intercept point (in dBm) at A7
Read receiver system second-order spurious free-dynamic range at A10
Read receiver system third-order spurious free-dynamic range at A11

File DDR.xls: Digital receiver dynamic range

Input number of quantizing bits at A4
Read receiver dynamic range (in dB) at A10

File CEP_RMS.xls: CEP from RMS error

Input RMS error (in degrees) at A4
Input range from both DF sites (in km) at A5
Read CEP (in km) at A10

File CEP_EEP.xls: CEP from EEP

Input long axis of EEP (in km) at A4
Input short axis of EEP (in km) at A5
Read CEP (in km) at A10

File JtoS.xls: Communication J/S

Input ERP of desired signal transmitter (in dBm) at A4
Input ERP of jammer (in dBm) at A5
Input loss from desired signal transmitter to target receiver (in dB) at A6
Input loss from jammer to target receiver (in dB) at A7
Input target receiving antenna gain toward desired signal transmitter (in dB) at A8
Input target receiving antenna gain toward jammer (in dB) at A9
Read jamming-to-signal ratio (J/S) (in dB) at A10

File PBJ.xls: Frequency spread for partial band jamming

Input target signal information bandwidth (in kHz) at A4
Input target signal hopping range (in MHz) at A5
Input single channel J/S (in dB) at A6
Read optimum jamming bandwidth (in MHz) at A10
Read jamming duty cycle (in %) at A11

File RcvPwr.xls: Received power in one-way link

Input link distance (in km) at A4
Input transmitting antenna height (in meters) at A5
Input receiving antenna height (in meters) at A6
Input frequency (in MHz) at A7
Read received signal power (in dBm) at A10

File EffRng.xls: Effective range in one-way link

Note that this file determines the propagation mode (line of sight or two-ray) and calculates the effective range. This assumes that wide beam-width antennas are used and that the terrain is clear. If antennas are directional or propagation is down a valley, the effective range is the "LOS distance." Atmospheric and rain attenuation are not included.

Input transmitter power (in dBm) at A5
Input transmit antenna gain toward receiver (in dB) at A6
Input receiving antenna gain toward transmitter (in dB) at A7
Input transmitter antenna height (in meters) at A8
Input receiving antenna height (in meters) at A9
Input frequency (in MHz) at A10
Input receiver system sensitivity (in dBm) at A11
Read effective range (in km) at A18

With narrow antennas or propagation down a valley, read effective range (LOS) at B18.

About the Author

David L. Adamy is an internationally recognized expert in electronic warfare. He has been writing the EW 101 columns for many years. However, in addition to writing the columns, he has been an EW professional (proudly calling himself a "Crow") in and out of uniform for 46 years. As a systems engineer, project leader, program technical director, program manager, and line manager, Mr. Adamy has directly participated in EW programs from just above DC to just above light. Those programs have produced systems which were deployed on platforms from submarines to space and met requirements from "quick and dirty" to high reliability.

Mr. Adamy holds B.S.E.E. and M.S.E.E. degrees, both with communication theory majors. In addition to the EW 101 columns, he has published many technical articles in EW, reconnaissance, and related fields and has 11 books in print, including this one. Mr. Adamy teaches EW-related courses all over the world, and consults for military agencies and EW companies. He is a longtime member of the National Board of Directors and a past president of the Association of Old Crows.

He has been married to the same long-suffering wife for 47 years (she deserves a medal for putting up with a classical nerd that long) and has four daughters and eight grandchildren. He claims to be an OK engineer, but one of the world's truly great fly fishermen.

Index

A
Accuracy
    1-burst radius, 213
    angular, uncertainty versus, 195
    calibration, 198, 224
    circular error probable (CEP), 195–97
    elliptical error probable (EEP), 197
    high techniques, 207–12
    moderate techniques, 203–6
    RMS error, 193–95
    See also Emitter location
Airborne intercept systems, 241–42
    illustrated, 241
    maximum intercept range, 242
Airborne receiving systems, 168
Ambiguity resolution, 213
Amplitude modulated (AM) signals, 13
    digital, 20
    in frequency domain, 14
    in time domain, 14
Analog modulations, 13–16
Analog receiver dynamic range, 107–10
    determination, 107, 108
    digital dynamic range versus, 106–7
    intercept points, 107, 108
    problem, 280–81
    spurious-free, 109
    spurious signal isolation, 109

Analog-to-digital (ADCs) converters, 16, 17, 88
Angle-of-arrival (AOA) systems
    directional reference, 198
    direction-finding array, 199
    error, impact of reflections on, 223
    multipath errors, 223
Angular coverage, 168–69
Antennas, 55–74
    amplitude pattern, 60
    bandwidth, 56
    beam, 59–62
    beamwidth, 56, 60–61
    boresight, 60
    characteristics, 57–59
    defined, 55
    dipole, 59
    directional, 57
    Earth-coverage satellite, 149
    efficiency, 56, 61–62
    frequency coverage, 56
    gain, 56, 61, 62
    log periodic, 59
    main lobe, 60
    minimum height, 132–33
    nonsymmetrical, gain, 69–70, 277
    parabolic dish, 59, 66–74
    parameters, 55–59
    performance parameters, 56

Antennas (continued)
  phased arrays, 63–66
  polarization, 56, 62–63
  power handling, 55
  scales on slide rule, 70–72
  selection guide, 57
  side lobes, 61
  spreading loss, 179
  types of, 56–57
  very low, 133–34
  whip, 59
Artillery-delivered jammers, 255
Atmospheric loss, 140–43
  problem, 286–87
  satellite links, 150
Atomic clock, 203
Automatic gain control (AGC), 106, 244
Azimuth and elevation emitter location, 191, 192

B
Bandwidth, 23–26
  antennas, 56
  clock rate versus modulation, 25
  in narrowband search, 184
  noise and, 28
  null-to-null, 23
  SHR, 161
Base station (BS), 49
Beamwidth, 56, 60–61
  3-dB, 60–61
  defined, 56, 60
  gain versus, 67
  "n" dB, 61
  phased array, 65–66
  problems, 276–79
Bibliography, 307–10
Binary moving window, 167
Binary phase shift keying (BPSK) signals, 21, 177
  oscilloscope display, 249
  phase diagrams, 22, 178
  in time domain, 21
Bit error rate, 256
  impact of jamming on, 256
  J/S ratio versus, 257
Bit errors, 27

Boresight, 60
Bragg cell receivers, 84–85

C
Calibration, 198
  accuracy, 198, 224
  errors, 224
  stored points, 198
CD, this book, 311–14
Cell phone signals, 49–51
  analog systems, 50
  digital CDMA systems, 50–51
  digital TDMA signals, 50
  downlink, jamming, 272–73
  downlink, jamming problem, 304–6
  intercept, 249–50
  jamming, 270–71
  operation, 51
  system illustration, 49
  uplink, jamming, 271–72
  uplink, jamming problem, 303–4
Channelized receivers, 82–83
  applications, 83
  defined, 82
  frequency hopper, 230–31
  illustrated, 82
  as receiver front end, 83
  for search, 161
  See also Receivers
Channel occupancy, 169
Chip detection, 92–93
Chip detectors, 166–67, 232
  location with, 232
  spread spectrum signal detection, 167
Chips, 92
  oscilloscope, 92
  rate, 165
  transitions, 166
Chirp emitters, 231–32
Chirp signals, 41–43
  detection, 43
  frequency range, 267
  frequency versus time characteristic, 41
  illustrated, 42
  intercept, 246–47
  jamming, 267
  as LPI signal, 42–43, 177

partial band jamming, 267
transmitter, 41–42
Circular error probable (CEP), 195–97
90%, 195, 197
    approximation, 196
    defined, 195
    for FDOA system, 221
    for TDOA systems, 218
Code division multiple access (CDMA), 50–51, 270
Coherent jamming cancellers, 279
Communication antennas. *See* Antennas
Communication Handbook, 123
Communications
    nature of, 1–2
    purpose, 1–2
Communications emitters
    angle-of-arrival versus frequency, 227
    chirp, 231–32
    frequency search, 153–85
    location of, 187–233
    LPI, 232
    spread spectrum, 225–33
Communications intelligence (COMINT), 235
Communications receivers. *See* Receivers
Communications signals, 2, 13–54
    analog modulations, 13–16
    chirp, 41–43
    digital modulations, 16–27
    DSSS, 43–47
    error-correction codes, 51–54
    frequency hopping (FH), 36–41
    intercept, 235–50
    LPI, 30–51
    noise, 27–30
Compressive receivers, 85–86
    block diagram, 85
    defined, 85
    delay versus frequency and tuning rate, 86
    frequency hopper, 231
    output bandwidth, 86
    for search, 161
    See also Receivers
Conversion factor, 11
Correlative interferometers, 212, 214

Correlative radiometers, 165–66
Crystal Video receiver (CVR), 76–78
    advantages, 77–78
    defined, 76–77
    illustrated, 77
    See also Receivers

D
dB
    common definitions, 10
    converting to, with slide rule, 7–8
    converting voltage ratios to, 7
    equations, 9–11
    equivalents, 6
    logarithmic numbers, 5
    math, 4–11
    power level conversion to, 9
    signal strength in, 9, 10
    values, 5, 6
dB form
    absolute values in, 8–9
    conversion to/from, 5–7
    numbers, 5
Degradation factor, 99
Delta modulation, 89–90
Digitally tuned receivers, 161–63
Digital modulations, 16–17
    RF, 19–23
    states, changing, 24
Digital receivers, 86–88
    applications, 87–88
    defined, 86
    direction finder with, 229
    dynamic range, 110–11
    dynamic range problem, 282
    illustrated, 87
    processing, 91–92
    for search, 161, 163, 185
    See also Receivers
Digital RF memories (DRFM), 89, 267
Digital signals
    amplitude modulated, 20
    bandwidth, 23–26
    bit structure, 26
    frequency spectrum characteristics, 24
    OOK modulated, 20
    required RFSNR for, 104–5

Digital signals (continued)
   spectrum analyzer display, 24
   structure, 26–27
Digital-to-analog converter (DAC), 19
Digital waveforms, 89–90
Digitization, 17–19
   of analog waveform, 18
   in-phase and quadrature (I & Q), 91, 229
   PCM, 88
   receivers, 88–91
   techniques, 90
Digitizers, 18, 88
Dipole antennas, 59
Directional antennas, 57
Direction-finding (DF) systems, 160, 188
   airborne, 189, 200
   antenna array, 198
   Doppler, 204–6
   magnetometer, 201
   shipboard, 199
   sites, 188, 198
   Watson-Watt, 204
Direction of arrival (DOA), 185, 188
Direct search, 154–55
Direct sequence spread spectrum (DSSS) emitters, 232
Direct sequence spread spectrum (DSSS) signals, 43–47
   applications, 43
   bit rate ratio, 268
   cell phone transmission intercept, 249–50
   chip rate, 165
   defined, 43
   despreading nonspreading signals, 46–47
   frequency versus bit rate, 46
   frequency versus time characteristic, 44
   illustrated, 44
   intercept, 247–50
   jamming, 267–69
   as LPI signal, 46, 177–78
   oscilloscope chips, 92
   receiver, 45–46
   secure, long spreading codes, 247–49
   spectral lines, 248
   spectrum, 248, 249, 268
   spread spectrum, 164
   transmitter, 44–45
Direct synthesizers, 40–41
Doppler DF system, 204–6
   concept illustration, 205
   defined, 204–6
   illustrated, 206
   use, 206
Double-conversion receivers, 80–81
Dynamic range, receiver, 105–11
   analog, 107–10
   analog versus digital, 106–7
   defined, 105
   digital, 110–11
   digitizing bits versus, 112
   spurious-free, 110

E

Effective radiated power (ERP), 120
   jamming and, 271
   in partial band jamming, 261
Effective range
   LOS link problem, 289–92
   two-ray link, 292–94
Electronically steered arrays, 65
Electronic order of battle (EOB), 235
Electronic support (ES) receivers, 75
   advantages/disadvantages, 75
   multiple, 112–13
Elliptical error probable (EEP), 197
   for FDOA system, 221
   for TDOA systems, 218
Emitter location, 187–233
   accuracy definitions, 193–98
   azimuth and elevation, 191, 192
   on calculated locus, 191–92
   error budget, 222–24
   high accuracy techniques, 207–12
   moderate accuracy techniques, 203–6
   precision, 212–22
   single-site, 189–91
   site location and north reference, 198–203
   spread spectrum, 225–33
   triangulation, 188–89
   See also Communications emitters

Energy detection receivers, 164–67, 233
    binary moving window, 167
    chip detector, 166–67
    correlative detector, 165–66
    integrate and dump, 164–65
Environmental noise, 29–30
Error budget, 222–24
    calibration errors, 224
    combination of error elements, 223–24
    error related to SNR, 224
    impact of reflections on AOA error, 223
    See also Emitter location
Error detection and correction (EDC)
        codes, 51–54
    classes, 51–52
    example, 52–53
    Hamming, 52–53
    Reed Solomon, 53

F
Fast Fourier transforms (FFT), 155, 227–30
Fast hoppers, 177
FFT timing, 94–95
Fixed-tuned receivers, 82
FM improvement factor, 103–4
Follower jamming, 265–67
    fast Fourier transform (FFT) processor, 265
    fratricide and, 267
    signal environment, 266
    timing, 266
    See also Jamming
Fratricide, 181, 267
Frequency converters, 81
Frequency difference of arrival (FDOA), 218–21
    against moving transmitters, 221
    defined, 218
    EEP/CEP of, 221
    measurement, 221
    method, 218–21
    TDOA combined with, 221–22
Frequency division multiple access (FDMA), 50
Frequency hoppers
    approaches, 226

channelized receiver, 230–31
compressive receiver, 231
fast, 177
frequency range, 246
friendly, 230
intercept, 245–46
locating, 225–31
slow, 176
special receiver approaches, 230
spectrum, 246
sweeping receiver approach, 226–27
Frequency hopping (FH) signals, 36–41
    advantages, 37
    defined, 36–37
    follower jamming, 265–67
    frequency versus time characteristic, 37–38
    illustrated, 37, 225
    jamming, 262–67
    jamming effectiveness, 265
    as LPI signal, 41
    partial band jamming, 263–64
    receivers, 93–95
    settling, 40
    single channel J/S calculation, 264
    slow versus fast hop, 38
Frequency hopping transmitter, 38–41
    block diagram, 38
    fast hopping, 40
Frequency measuring receivers, 163–64
Frequency modulation (FM) signals
    advantages, 16
    defined, 15
    in frequency signal, 16
    required RFSNR for, 102–5
    in time domain, 15
Frequency ranges, 2–3
Frequency search, 153–85
    with digital receiver, 163, 185
    directed, 154–55
    fratricide, 181
    general, 154
    hand-off from wideband receiver, 185
    look through, 178–80
    LPI signals, 174–78
    narrowband, 181–85
    planning tool, 155–56

Frequency search (continued)
  practical considerations effecting, 156–58
  radio horizon, 170–74
  receivers, 160–61
  from reconnaissance aircraft, 169
  sequentially qualified, 155
  signal environment, 167–70
  strategies, 154–58
  strategy examples, 181–85
  system configurations, 158–67
  technology issues, 157–58
  with wideband frequency measurement receiver, 163–64
Frequency shift keying (FSK) signals, 20
  in frequency domain, 21
  in time domain, 21
Fresnel zone, 134–36, 239
  checking to determine propagation mode, 292
  close-up calculation scales, 136
  complex reflection environment, 136
  distance determination, 134
  formula, 134–35
  jammer to receiver, 302, 305
  problem, 286
  slide rule, 135–36
  transmitter to jammer, 301
  transmitter to receiver, 301, 303, 305
Front-back ambiguity, 212

G
Gain
  adjustment for efficiency, 70
  antenna, 56, 61, 62
  beamwidth versus, 67
  as function of diameter and frequency, 68–69
  of nonsymmetrical antenna, 69–70
  phased array, 65–66
  polarization matches and, 62
  problems, 276–79
  side-lobe, 61
General search, 154
Global positioning system (GPS)
  additional value, 203
  inertial navigation system, 202
  receiver/processor, 203
Global System for Mobile Communication (GSM), 50

H
Hamming code, 52–53
  decoder, 53
  generator, 52
Hand-off, from wideband receivers, 185
HF propagation, 143–47
  complexity, 143
  ionosphere, 144–45
  paths, 146–47
  types of, 143
  See also Propagation
High accuracy techniques, 207–12
  correlative interferometer, 212
  multiple baseline precision interferometer, 211–12
  single baseline interferometer, 207–11
  See also Emitter location
Hopped direct sequence transmitters, 48
Hopping sequence generators, 36

I
Inertial measurement units (IMUs), 203
Inertial navigation system (INS), 199, 200
  GPS-enabled, 202
  mechanically stabilized inertial platform, 201
In-phase and quadrature (I & Q) digitization, 91, 229
Instantaneous frequency measurement (IFM) receivers, 78–79
Integrate-and-dump receivers, 165
Intercept, 235–50
  airborne system, 241–42
  cell phone transmission, 249–50
  chirped signals, 246–47
  of directional transmission, 237–38
  DSSS signals, 247–50
  frequency hoppers, 245–46
  link, 236–44
  LPI signal, 245–50
  maximum range, 237, 240, 241, 242
  of nondirectional transmission, 238–41
  nonline-of-sight, 242–44
  probability, in narrow-band search, 294

received power formula, 236
of weak signal (strong signal
   environment), 244–45
Intercept receivers
   function, 235
   noise figure, 240
   received power, 236, 240, 244
   sensitivity, 238, 240
   See also Receivers
Interferometers
   antenna use, 209
   correlative, 212, 214
   multiple baseline precision, 211–12,
      214
   single baseline, 207–11
Interferometric antenna arrays
   baselines, 211
   on flat surface, 209
   ground-based, 210
   phase versus angle-of-arrival, 210
Interferometric triangle, 208
Interleaving scheme, 54
Ionosphere, 144–45
   characterization, 144
   layers, 144–45
   reflection, 145–46

J
Jaguar frequency hopping pattern, 93–94
Jammer-to-signal ratio (J/S), 252–55
   bit error rate versus, 257
   cell phone downlink, 273
   defined, 253
   formula, 253
   problems, 295–301
   single channel, 262
Jamming, 251–73
   cancellation, 180
   cell phones, 270–71
   chirp signals, 267
   coherent canceller, 179
   of combined mode spread spectrum
      signals, 269
   deceptive, 178
   digital versus analog signals, 255–58
   downlink, 272–73
   DSSS signals, 267–69

   duty cycle, 257
   error-correction coding impact on,
      270–73
   frequency hop signals, 262–67
   geometry, 252
   level, increase, 256
   LPI signals, 181
   narrowband, through despreader, 269
   partial band, 259–62
   pulse, 258
   purpose, 251
   spread spectrum signals, 258–69
   stand-in, 254–55
   UAV link geometry, 252
   uplink, 271

K
Knife-edge diffraction (KED), 137–40
   attenuation, 137, 139, 140
   calculation monograph, 138
   geometry, 138
   loss, 139
   nonline-of-sight calculation, 243
   See also Propagation

L
Line-of-sight (LOS) propagation, 124–29
   formula, 124–27
   free-space attenuation scales, 128–29
   loss calculation, 124
   loss calculation geometry, 125
   nomographs, 127
   problem, 282–84
   satellite links, 150
   slide rule, 127–29
   See also Propagation
Link margin, 122–23
Log periodic antennas, 59
Look through, 178–80
   defined, 178
   periods, 180
   time sharing, 180
Lower sideband (LSB) signals, 14
Low probability of intercept (LPI) signals,
   30–51
   cell phone, 49–51
   chirp, 41–43, 177
   combined technique signals, 47–48

LPI signals (continued)
    DSSS, 43–47, 177–78
    emission control feature, 174
    frequency hoppers, 36–41, 176–77
    frequency spectrum, 33
    generation of, 32
    jamming, 181, 258–69
    modulation spread, 174–75
    pseudo-random codes, 33–36
    search for, 174–78
    search strategies, 175
    spread spectrum strength, 175
    techniques, 31
    types, 30
    See also Communications signals
LPI signal intercept, 245–50
    chirped signals, 246–47
    DSSS signals, 247–50
    frequency hoppers, 245–46
    See also Intercept

M
Magnetometers, 201
Maximum usable frequency (MUF), 145
Minimum shift keying (MSK), 25, 26
Mobile switching center (MSC), 49, 50, 51
Moderate accuracy techniques, 203–6
    Doppler DF system, 204–6
    Watson-Watt DF system, 204
Multipath errors, 223
Multipath reflectors, 223, 224
Multiple baseline precision interferometer, 211–12, 214
Multiple receiver systems, 112–16
    electronic support, 112–13
    front-end, 115–16
    local oscillator (LO) radiation, 114–15
    receiver performance, 115
    reconnaissance, 112–13
    remote, 116

N
Narrowband search, 181–85
    add DF requirement, 185
    defined, 181
    plan diagram, 182, 183
    plan diagram with increased bandwidth, 184
    require longer dwell, 183–84
Noise, 27–30
    background, 29, 30
    bandwidth and, 28
    environmental, 29–30
    in frequency domain, 28
    thermal, 28
    in time domain, 28
Noise figure, 98–99
    defined, 98
    degradation, 100
    degradation chart, 280
    formula, 98–99
Nomographs
    LOS loss, 127
    radio horizon, 171
    two-ray loss, 130
Nondirectional transmission intercept, 238–41
    Fresnel zone, 239
    illustrated, 239
    maximum intercept range, 240, 241
    received power, 240
    receiver system noise figure, 240
    See also Intercept
Nonline-of-sight intercept, 242–44
    illustrated, 242
    KED loss determination, 243
    received power, 244
    See also Intercept
Nordic Mobile Telephone (NMT), 50
Nyquist sampling criteria, 94

O
One-way links, 119–23
    equation, 119–23
    illustrated, 120
    link margin, 122–23
    problem, 282
On-off keying (OOK), 20
Organization, this book, 3–4

P
Parabolic dish antennas, 66–74
    defined, 59
    effective antenna area, 67–68
    gain as function of diameter and frequency, 68–69

gain versus beamwidth, 67
reflector size, 66
scales on slide rule, 70–72
slide rule assumptions, 72–74
See also Antennas
Parallel integrate-and-dump receivers, 165
Partial band jamming, 259–62
 calculation, 260
 chirp signals, 267
 defined, 259
 frequency hop signal, 263–64
 geometry, 261
 ideal frequency spread calculation, 264
 illustrated, 260
 maximum bit errors per channel, 260
 problem, 301–3
 See also Jamming
Phased arrays, 63–66
 beamwidth, 65–66
 defined, 63
 electronically steered, 65
 gain, 65–66
 illustrated, 64
 linear, 64
 planar, 64
 See also Antennas
Phase-lock-loop synthesizers, 162
Phase shifters, 63, 64
Polarization, 62–63
 circular, 57, 62
 defined, 56
 EW trick, 62
 jamming versus receiving antennas, 180
 linear, 57
 losses, 62, 63
Precision emitter location, 212–22
 accuracy, 212
 against LPI emitters, 232
 combined FDOA/TDOA, 221–22
 FDOA, 218–21
 TDOA, 214–18
 See also Emitter location
Probability of intercept (POI), 154, 294–95
Problems
 analog receiver dynamic range, 280–81
 antenna gain and beamwidth, 276–77

antenna gain and beamwidth versus frequency, 277–79
atmospheric loss, 286–87
communication J/S, 295–301
digital receiver dynamic range, 282
effective range (LOS link), 289–92
effective range (two-ray link), 292–94
Fresnel zone, 286
gain of nonsymmetrical antennas, 277
jamming cell phone downlink, 304–6
jamming cell phone uplink, 303–4
line-of-sight propagation loss, 282–84
one-way link equation, 282
partial band jamming, 301–3
probability of intercept, 294–95
rain loss, 287–88
receiver system sensitivity, 279–80
satellite link loss, 288–89
two-ray propagation loss, 284–85
Propagation, 119–52
 atmospheric loss, 140–43
 Fresnel zone and, 134–36
 HF, 143–47
 indoor, 123
 knife-edge diffraction, 137–40
 line-of-sight (LOS), 124–29
 losses, 123–24
 losses, determining, 293
 models, 123
 outdoor, 123
 rain and fog loss, 140–43
 satellite links, 147–52
 two-ray, 129–34
Pseudo-random codes, 33–36
 nonlinear, 35
 shift register for generation, 34
Pulse amplitude modulation (PAM), 20
Pulse code modulation (PCM), 88, 89
Pulse jamming, 258
Pulse receivers, 76–79
 CVR, 76–78
 IFM, 78–79

Q
Quadrature amplitude modulated (QAM) signals, 22

Quadrature phase shift keying (QPSK)
  signals, 21
  phase conditions, 22
  phase diagrams, 22
  in time domain, 22

R
Radio horizon, 170–74
  4/3 Earth, chart, 171
  calculation nomograph, 171
  calculations, 172, 173, 174
  defined, 170
  geometry, 170
  lower frequencies, 173–74
Rain loss, 140–43
  problem, 287–88
  satellite links, 150–51
Read Solomon EDC code, 53
Received frequencies
  in moving receiver, 218
  in two moving receivers, 219
Receivers, 75–117
  Bragg cell, 84–85
  channelized, 82–83, 161, 230–31
  chip detection, 92–93
  compressive, 85–86, 161, 231
  Crystal Video (CVR), 76–78
  digital, 86–88, 161
  digitally tuned, 161–63
  digital waveforms, 89–90
  digitization, 88–91
  digitizing techniques, 90
  direction-finding, 160
  energy detection, 164–67
  fixed-tuned, 82
  frequency hopping signal, 93–95
  frequency measuring, 163–64
  instantaneous frequency measurement
    (IFM), 78–79
  intercept, 236, 237, 238
  multiple systems, 112–16
  performance, 115
  processing, 91–92
  pulse, 76–79
  remote systems, 116–17
  sampling rates, 89
  search, 160–61

  search and monitor, 158–59
  signal quality issues, 91–95
  single, 158
  special, 159–60
  superheterodyne, 79–81, 160–61
  system configurations, 111–17
  system dynamic range, 105–11
  system sensitivity, 95–105
  system sensitivity problem, 279–80
  tuned radio frequency (TRF), 81–82
  types list, 76
  types of, 75–88
  uses, 76
Reconnaissance systems
  receivers, 75
  receivers, multiple, 112–13
Reed Solomon block code, 270
Reference Data for Radio Engineers (RDRE),
  143
Reflectors, multipath, 223, 224
Remote receiving systems, 116–17
  with direction finding, 117
  multiple-station, 118
RF discriminators, 102–3
RFSNR (radio frequency signal-to-noise
  ratio), 101–5
  defined, 101
  for digital signals, 104–5
  for FM signals, 102–5
  SNR versus, 101–2, 104
Root mean square (RMS) error, 195
  components, 194
  defined, 193
  determining, 193
  location probability from, 194
  total, 223

S
Sampling rates, 89
Satellites
  atmospheric loss, 150
  Earth-coverage antenna, 149
  link loss example, 151–52
  link loss problem, 288–89
  links, 147–52
  LOS loss, 150
  low-Earth, range to, 147

rain loss, 150–51
synchronous, range to, 148
Search. *See* Frequency search
Search and monitor receivers system, 158–59
Sensitivity, receiver system, 95–105
   components, 97
   defined, 169
   definition locations, 96
   kTB, 97–98
   noise figure, 98–99
   in search, 169–70
Sequentially qualified search, 155
Shift register timing diagrams, 35
Signal environment
   angular coverage, 168–69
   channel occupancy, 169
   defined, 167
   sensitivity, 169–70
Signal-to-noise ratio (SNR)
   defined, 27
   low, signal, 29
   radio frequency (RFSNR), 101–5
   in receiver system, 30
   reduction, 32
   required predetection, 99–105
Single receiver systems, 158
Single sideband (SSB) signals, 14, 15
Single-site location (SSL), 189–91
Sinusoidal frequency shift keying (SFSK), 25
Slide rule
   antenna, 8
   antenna scales on, 70–72
   assumptions, 72–74
   converting to dB using, 7–8
   Fresnel zone, 135–36
   LOS propagation, 127–29
   propagation, 8
   two-ray propagation, 130–32
Slow hoppers, 176
Special receivers system, 159–60
Spreading loss, 179
Spreading modulators, 32
Spread spectrum emitters
   chirp, 231–32
   DSSS, 232
   frequency hoppers, 225–31
   locating, 225–33
   LPI, 232
   See also Communications emitters
Spread spectrum signals, 31, 33
Spread Spectrum Systems (Dixon), 25
Stand-in jamming, 254–55
Superheterodyne receivers (SHR), 79–81
   bandwidth, 161
   defined, 79, 160
   digitally tuned, 161
   double-conversion, 80–81
   illustrated, 79
   intermediate frequency (IF), 80
   local oscillator (LO), 79–80
   for search, 160–61
   See also Receivers
Sweeping receiver strategy, 156, 157, 226–27
Synchronization bits, 26
System configurations, 158–67
   digitally tuned receivers, 161–63
   digital receivers, 163
   energy detection receivers, 164–67
   frequency measuring receivers, 163–64
   search and monitor receivers, 158–59
   search receivers, 160–61
   single receiver, 158
   special receivers, 159–60
   See also Frequency search

T

Time difference of arrival (TDOA), 214–18
   EEP/CEP calculation, 218
   FDOA combined with, 218–21
   location, 217–18
   process, 216
Time division multiple access (TDMA), 50, 270
Time hopping, 48
Traveling wave tube (TWT) amplifiers, 81–82
Triangulation, 188–89
   detail illustration, 189
   geometry, 188
   from moving DF system, 190

Triangulation (continued)
   See also Emitter location
Tuned radio frequency (TRF) receivers, 81–82
Two-ray propagation, 129–34
   defined, 129
   formula, 130
   minimum antenna height, 132–33
   nomograph, 130
   problem, 284–85
   slide rule, 130–32
   very low antennas, 133–34
   See also Propagation

U
UHF transmissions, 3
Unmanned aerial vehicles (UAVs), 3
Upper sideband (USB) signals, 14

V
VHF transmissions, 3
Voltage-controlled oscillators (VCOs), 102

W
Watson-Watt DF system, 204
Whip antennas, 59

Recent Titles in the Artech House Radar Series

David K. Barton, Series Editor

Adaptive Antennas and Phased Arrays for Radar and Communications, Alan J. Fenn

Advanced Techniques for Digital Receivers, Phillip E. Pace

Advances in Direction-of-Arrival Estimation, Sathish Chandran, editor

Airborne Pulsed Doppler Radar, Second Edition, Guy V. Morris and Linda Harkness, editors

Bayesian Multiple Target Tracking, Lawrence D. Stone, Carl A. Barlow, and Thomas L. Corwin

Beyond the Kalman Filter: Particle Filters for Tracking Applications, Branko Ristic, Sanjeev Arulampalam, and Neil Gordon

Computer Simulation of Aerial Target Radar Scattering, Recognition, Detection, and Tracking, Yakov D. Shirman, editor

Design and Analysis of Modern Tracking Systems, Samuel Blackman and Robert Popoli

Detecting and Classifying Low Probability of Intercept Radar, Second Edition, Phillip E. Pace

Digital Techniques for Wideband Receivers, Second Edition, James Tsui

Electronic Intelligence: The Analysis of Radar Signals, Second Edition, Richard G. Wiley

Electronic Warfare in the Information Age, D. Curtis Schleher

ELINT: The Interception and Analysis of Radar Signals, Richard G. Wiley

EW 101: A First Course in Electronic Warfare, David Adamy

EW 102: A Second Course in Electronic Warfare, David L. Adamy

EW 103: Tactical Battlefield Communications Electronic Warfare,
David Adamy

Fourier Transforms in Radar and Signal Processing,
David Brandwood

Fundamentals of Electronic Warfare, Sergei A. Vakin, Lev N. Shustov, and Robert H. Dunwell

Fundamentals of Short-Range FM Radar, Igor V. Komarov and Sergey M. Smolskiy

Handbook of Computer Simulation in Radio Engineering, Communications, and Radar, Sergey A. Leonov and Alexander I. Leonov

High-Resolution Radar, Second Edition, Donald R. Wehner

Introduction to Electronic Defense Systems, Second Edition, Filippo Neri

Introduction to Electronic Warfare, D. Curtis Schleher

Introduction to Electronic Warfare Modeling and Simulation, David L. Adamy

Introduction to RF Equipment and System Design, Pekka Eskelinen

Microwave Radar: Imaging and Advanced Concepts, Roger J. Sullivan

Millimeter-Wave Radar Targets and Clutter, Gennadiy P. Kulemin

Modern Radar Systems, Second Edition, Hamish Meikle

Modern Radar System Analysis, David K. Barton

Modern Radar System Analysis Software and User's Manual, Version 3.0, David K. Barton

Multitarget-Multisensor Tracking: Applications and Advances Volume III, Yaakov Bar-Shalom and William Dale Blair, editors

Principles of High-Resolution Radar, August W. Rihaczek

Principles of Radar and Sonar Signal Processing, François Le Chevalier

Radar Cross Section, Second Edition, Eugene F. Knott et al.

Radar Evaluation Handbook, David K. Barton et al.

Radar Meteorology, Henri Sauvageot

Radar Reflectivity of Land and Sea, Third Edition, Maurice W. Long

Radar Resolution and Complex-Image Analysis, August W. Rihaczek and Stephen J. Hershkowitz

Radar Signal Processing and Adaptive Systems, Ramon Nitzberg

Radar System Analysis, Design, and Simulation, Eyung W. Kang

Radar System Analysis and Modeling, David K. Barton

Radar System Performance Modeling, Second Edition, G. Richard Curry

Radar Technology Encyclopedia, David K. Barton and Sergey A. Leonov, editors

Range-Doppler Radar Imaging and Motion Compensation, Jae Sok Son et al.

Signal Detection and Estimation, Second Edition, Mourad Barkat

Space-Time Adaptive Processing for Radar, J. R. Guerci

Theory and Practice of Radar Target Identification, August W. Rihaczek and Stephen J. Hershkowitz

Time-Frequency Transforms for Radar Imaging and Signal Analysis, Victor C. Chen and Hao Ling

For further information on these and other Artech House titles, including previously considered out-of-print books now available through our In-Print-Forever® (IPF®) program, contact:

Artech House
685 Canton Street
Norwood, MA 02062
Phone: 781-769-9750
Fax: 781-769-6334
e-mail: artech@artechhouse.com

Artech House
46 Gillingham Street
London SW1V 1AH UK
Phone: +44 (0)20 7596-8750
Fax: +44 (0)20 7630-0166
e-mail: artech-uk@artechhouse.com

Find us on the World Wide Web at: www.artechhouse.com